Lecture Notes in Mathematics

continuation on page 359

Lecture Notes in Mathematics

Edited by A. Dold and B. Eckmann

406

Graphs and Combinatorics

Proceedings of the Capital Conference
on Graph Theory and Combinatorics
at the George Washington University
June 18–22, 1973

Sponsored by the Graduate School of Arts and Sciences
of the George Washington University

Edited by Ruth A. Bari and Frank Harary

Springer-Verlag
Berlin · Heidelberg · New York 1974

Prof. Ruth A. Bari
George Washington University
Washington, DC/USA

Prof. Frank Harary
University of Michigan
Ann Arbor, MI/USA
and Oxford University/England

Library of Congress Cataloging in Publication Data

Capital Conference on Graph Theory and Combinatorics,
    George Washington University, 1973.
    Graphs and combinations; proceedings.

    (Lecture notes in mathematics, 406)
    Bibliography: p.
    1. Combinatorial analysis--Congresses. 2.   Graph
theory--Congresses.   I.   Bari, Ruth A., 1917-   ed.
II. Harary, Frank, ed. III. Title. IV. Series: Lecture notes in mathematics (Berlin) 406.
QA3.L28   no. 406 [QA164]   510'.8s [511'.6]   74-13955

AMS Subject Classifications (1970): 05-00

ISBN 3-540-06854-6 Springer-Verlag Berlin · Heidelberg · New York
ISBN 0-387-06854-6 Springer-Verlag New York · Heidelberg · Berlin

Offsetdruck: Julius Beltz, Hemsbach/Bergstr.

- PREFACE -

Wherefore is this conference book on graphs and combinatorics different from other such books?

One way is that it is, frankly, a progress report on recent results in the field, and does not claim to be a definitive work. Another is that all contributions have been refereed. A third way is that it contains both expository review articles and research contributions.

We wish to thank A. J. Schwenk for his invaluable editorial assistance, V. Chvátal and P. Hell for several constructive suggestions, and Mary Breen for her masterly coordination and organization of the meeting. Certainly, not least, we thank the accurate, fast, reliable typist, Debrah A. Johnson, whose neat work can be seen on every page of this book.

We also thank the Graduate School of Arts and Sciences of the George Washington University for providing the financial support which made the conference possible.

The contributions to this book are divided into three parts. In Part I, there are six invited articles of an expository nature, four by Frank Harary, the principal speaker at the conference, one by Harary and R. C. Read, and one by Harary and A. J. Schwenk. In Part II, there are four contributed survey papers, and Part III contains contributed papers which give new results on graphs and combinatorics, mostly the former.

It has been a pleasure for both of us to edit these proceedings together, especially as we have been friends for the past thirty-five years, and still are, even after actively working on all the detailed tasks involved in the production of this book.

Ruth A. Bari
Washington, D. C.

Frank Harary
Ann Arbor and Oxford

- CONTENTS -

RECENT RESULTS ON TREES

Frank Harary
University of Michigan

ABSTRACT

Our object is to present six theorems on trees discovered since the appearance of the book [4] . These deal with (1) trees with hamiltonian square, (2) path numbers, (3) the tree graph of a graph, (4) the intersection graph of subtrees of a tree, (5) cospectral trees, and (6) the probability of an endpoint.

RECENT RESULTS ON TREES

1. TREES WITH HAMILTONIAN SQUARE

The <u>square</u> of graph G is the graph G^2 obtained by adding to G all lines
joining two points u and v whose distance is 2. In an already classical but not
yet published result, Fleischner [3] proved that the square of every nonseparable
graph is hamiltonian. This immediately suggested the question of characterizing
those trees with hamiltonian square. To describe this result, we require some definitions. The <u>subdivision graph</u> of G , written S(G) , is obtained from G by
inserting just one new vertex at the "midpoint" of each edge; see Figure 1.

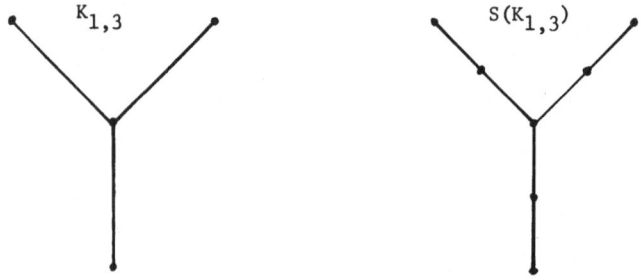

Figure 1. A Graph and Its Subdivision Graph

For any tree T with at least 3 points, the <u>derived tree</u> T' is obtained from T
by deleting all its endpoints. A <u>caterpillar</u> is a tree T for which T' is a path;
see Figure 2.

Figure 2. A Caterpillar

A <u>new crossing number</u>, denoted $\nu_2(T)$, was defined in [6] for a tree T as the minimum number of crossings which can result when T is drawn as a bipartite graph with the points in one set located on a horizontal line, the points in the other set located on a parallel line just one inch below the first one, and all the edges drawn as straight line segments. Figure 2 shows that for a caterpilliar T , $\nu_2(T) = 0$. Of course $\nu_2(G)$ is meaningful for any bipartite graph G .

We discovered the characterization [6] by verifying that the only tree T with $p \leq 7$ points with T^2 not hamiltonian is $S(K_{1,3})$ shown in Figure 1. The square is shown in Figure 3 following a drawing by V. Chvátal in which the lines of T are drawn solid, and the added lines to form T^2 are drawn dashed.

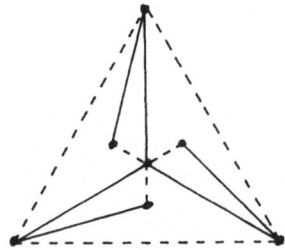

Figure 3.   The Nonhamiltonian Square of a Tree

Theorem 1.   For any tree T with at least 3 points, the following statements are equivalent:

    (1) T^2 is hamiltonian.

    (2)   T is a caterpillar.

    (3)   T does not contain $S(K_{1,3})$ as a subtree.

    (4) $\nu_2(T) = 0$.

2.   PATH NUMBERS

In [7] the <u>path number</u> of a graph $\pi(G)$ is defined as the minimum number of line-disjoint paths covering all the lines of G . A second path-covering invariant of a graph is the <u>unrestricted path number</u> $\pi^*(G)$ taken as the minimum number of paths which are not necessarily line-disjoint, needed to cover the lines of G . We immediately see that if G has p_0 points of odd degree, then since at least one path must end at each odd point, $\pi(G) \geq p_0 / 2$. For the unrestricted path number, this need not be the case, as seen by the example in Figure 4(c). It is possible to cover the tree T with two unrestricted paths not ending at either of the two points of degree 3. But of course, we must still have a path terminating at each endpoint, and so if G has e endpoints, then $\pi*(G) \geq \{e/2\}$.

T:

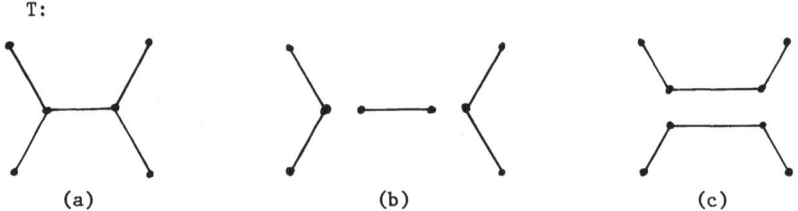

(a)   (b)   (c)

Figure 4. A Tree and Its Covering Paths

For graphs in general, these two lower bounds may or may not be attained, but we found in [7] that trees are nicely behaved, and the bounds are the exact values!

Theorem 2. For any tree T ,

$$\pi(T) = \frac{p_0}{2}$$

and

$$\pi*(T) = \{ \frac{e}{2} \} .$$

3.   THE TREE GRAPH OF A GRAPH

The tree graph T(G) of the connected graph G has one point for each spanning tree of G , and two of these points are adjacent if the corresponding trees differ by just one line. For example, the random graph depicted in Figure 5 has eight spanning trees as shown. These give rise to the tree graph given in Figure 6. It is easy to check that for each line y of $T(K_4-x)$, there exists a hamiltonian cycle containing y and a second hamiltonian cycle not containing y . This has been found to be true for all tree graphs [5].

$K_4 - x$:

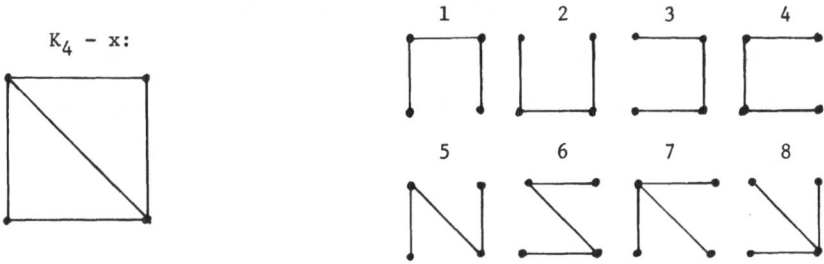

Figure 5. A Graph and Its Spanning Tree

$T(K_4 - x)$:

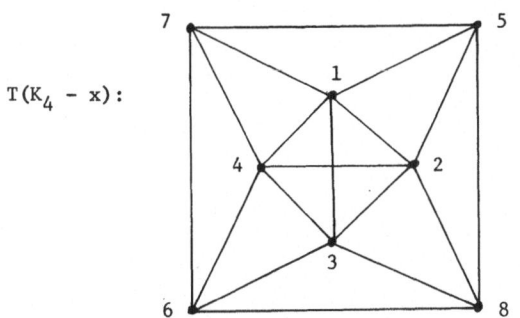

Figure 6. The Tree Graph of a Graph

Theorem 3. Let G be a connected graph containing a cycle. For every line y in the tree graph T(G) , there exists a hamiltonian cycle containing y and a second hamiltonian cycle not containing y .

Analogously, one can define the "base graph" of a matroid. Base graphs are also shown [5] to have the property described in the theorem.

4. THE INTERSECTION GRAPH OF SUBTREES OF A TREE

A chordal graph is a graph in which every cycle of length greater than three has a chord, that is, a line joining two nonconsecutive points on the cycle. In particular, every maximal planar graph is chordal. In the past, such graphs have been called "rigid circuit graphs". Given a family of distinct sets $S = \{S_1 , S_2 , \ldots, S_n\}$, the intersection graph I(S) has point set S and adjacency is defined by S_i adj S_j if $S_i \cap S_j \neq \phi$. Figure 7 depicts the intersection graph obtained from certain subtrees of a tree T . Notice that I(S) is chordal, and in fact, maximal outerplanar. Buneman [1] characterized chordal graphs in this way.

Theorem 4. A graph G is chordal if and only if G is the intersection graph of subtrees of a tree.

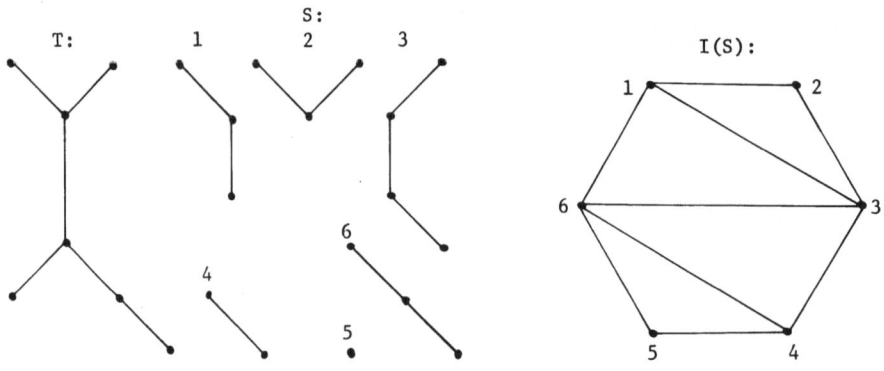

Figure 7.   The Intersection Graph of Subtrees of a Tree

5.   COSPECTRAL TREES

The spectrum S(G) of a graph G of order p is defined as the non-increasing sequence of the p real eigenvalues of the adjacency matrix of G . Two graphs are cospectral if they have the same spectrum. For example, it is no trouble to calculate that the two graphs of Figure 8 have the spectrum $S(C_4 \cup K_1) = S(K_{1,4}) = (2,0,0,0,-2)$.

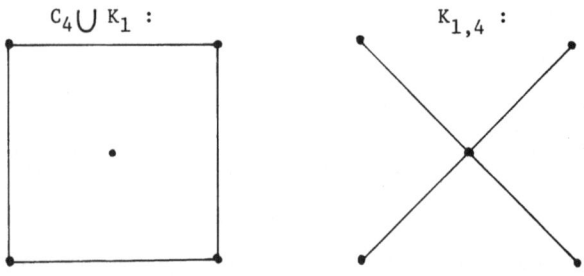

Figure 8.   The Smallest Pair of Cospectral Graphs

This example demonstrates the fallacy of the casual but ambitious conjecture that the spectrum determines the graph. Still, one might hope that the conjecture holds for some nontrivial subclass of graphs. Thus we are naturally led to search for cospectral trees. Collatz and Sinogowitz [2] found the smallest cospectral pair, shown in Figure 9. Yet, among small trees, cospectral pairs are rare, and we might

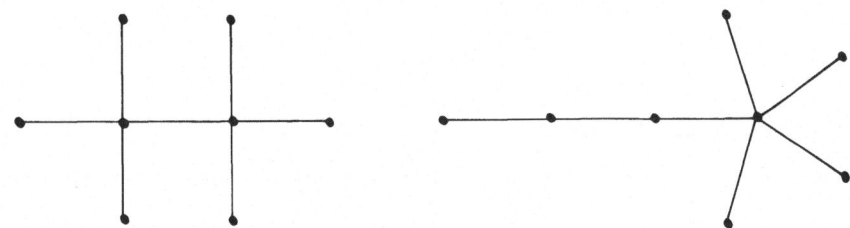

Figure 9. The Smallest Cospectral Pair of Trees

expect that most trees do not have a cospectral mate, that is, another tree with the same spectrum. Schwenk [11] has shown that quite the opposite is true.

Theorem 5. Almost every tree has a cospectral mate.

The proof of this result is a combinatorial tour de force. It relies on a construction that produces some of the possible cospectral pairs by demonstrating that a certain pair of 8-line subtrees can be substituted for one another without altering the spectrum. Then a counting argument, interesting in its own right, is given to show that almost every tree has one of these two particular subtrees. A branch at v of a tree is a maximal subtree having v as an endpoint. A limb at v is the union of one or more branches at v . The weight of a limb is the number of lines in it. Schwenk found that the number of trees having L as a limb depends solely upon the weight of L , and not upon its structure. Furthermore, as the number of points increases, the fraction of trees having a fixed L as a limb tends to one.

6. THE PROBABILITY OF AN ENDPOINT

Rényi [9] found that if a point is chosen at random from among all the labeled n point trees, the probability of an endpoint approaches 1/e as n increases.

To verify this, note that there are n ways to label a prospective endpoint, there are $(n-1)^{n-3}$ possible labeled trees of order n-1 to which this endpoint can be added and $(n-1)$ ways to join it to each such tree. Thus, there are $n(n-1)^{n-2}$ endpoints among all the n-point labeled trees. Of course, the total number of points among all labeled trees of order n is just $n \cdot n^{n-2}$, and so the probability of an endpoint is precisely

$$\left(\frac{n-1}{n}\right)^{n-2} = \left(1 - \frac{1}{n}\right)^{n-2}$$

which clearly approaches 1/e .

The corresponding question for unlabeled trees remained unsolved until very recently when B. Mayhew, a sociologist, stimulated our interest by conjecturing, on the basis of data for $n \leq 12$, that the limit in this case is 1/2 . Our approach [10] is to enumerate e_n , the total number of endpoints among n-point trees, and then determine the asymptotic behavior of the probability $p_n = \dfrac{e_n}{n \cdot t_n}$.

An illustration is helpful, to distinguish between trees and labeled trees. The two trees of order 4 are the path P_4 and the star $K_{1,3}$ shown in Figure 10.

Figure 10.   The Trees of Order 4

There are 8 points in these two trees, of which 5 are endpoints so that $p_4 = 5/8$. However, there are 12 ways to label P_4 and 4 ways to label $K_{1,3}$ so that for labeled trees of order 4, the probability of an endpoint is 9/16 .

To calculate p_n , we use Pólya's asymptotic estimate [8] for rooted trees T_n and develop analogous results for E_n, the number of endpoints in n-point planted trees, in which the plant is counted neither as an endpoint nor among the n points. Let ρ denote the common radius of convergence of the series T(x), t(x), E(x), and e(x). As found by Pólya [8], $\rho = 0.338....$

<u>Theorem 6</u>. The asymptotic probability that a point chosen at random from a random large tree is given by

$$\lim_{n\to\infty} \frac{e_n}{nt_n} = \lim_{n\to\infty} \frac{E_n}{nT_n} = \frac{\rho + \sum_{n=1}^{\infty} E_n \rho^{2n}/(1-\rho^n)}{1 + \sum_{n=1}^{\infty} nT_n \rho^{2n}/(1-\rho^n)}$$

Notice that the limits are evaluated in terms of a weighted average with weighting factor $\rho^{2n}/(1-\rho^n)$. These sums converge rapidly because E_n and nT_n each grow at a rate of roughly ρ^{-n} .

An appropriate computer program enables the precise calculation of this limiting value to 12 places, 0.438 156 235 664 ... .

REFERENCES

1. Buneman, P., A Characterization of Rigid Circuit Graphs, <u>Discrete Math.</u>, to appear.

2. Collatz, L., and Sinogowitz, U., Spektren endlicher Graphen, <u>Abh. Math. Sem. Univ. Hamburg</u> 211 (1957) 64-77.

3. Fleishner, H. J., The Square of Every Nonseparable Graph is Hamiltonian, <u>J. Combinatorial Theory</u>, to appear.

4. Harary, F., <u>Graph Theory</u>, Addison-Wesley, Reading, 1969.

5. Harary, F., and Holzmann, C., On the Tree Graph of a Matroid, <u>SIAM J. Appl. Math.</u> 22 (1972) 187-193.

6. Harary, F., and Schwenk, A. J., Trees with Hamiltonian Square, <u>Mathematika</u> 18 (1971) 138-140.

7. Harary, F., and Schwenk, A. J., Evolution of the Path Number of a Graph. <u>Graph Theory and Computing</u> (R.C. Read, ed.) Academic Press, New York, 1972, 39-45.

8. Pólya, G., Kombinatorisone Anzahlbestimmungen fur Gruppen, Graphen und chemische Verbindungen, <u>Acta Math.</u> 68 (1937) 145-254.

9. Rényi, A., Some Remarks on the Theory of Trees, <u>Publ. Math. Inst. Hungar. Acad. Sci.</u> 4 (1959) 73-85.

10. Robinson, R. W., and Schwenk, A. J., The Probability of an Endpoint in a Large Random Tree, <u>Proc. Camb. Phil. Soc.</u>, submitted.

11. Schwenk, A. J., Almost All Trees are Cospectral, <u>New Directions in the Theory of Graphs</u> (F. Harary, ed.) Academic Press, New York, 1973, 275-307.

A SURVEY OF GENERALIZED RAMSEY THEORY

Frank Harary
University of Michigan

ABSTRACT

This is a progress report on a very dynamic branch of graph theory. We begin with a historical review of the origins of generalized ramsey theory and then indicate the small graphs for which the diagonal ramsey numbers are now known. The ramsey multiplicity of a graph is taken up and applied to ramsey games. We conclude with a listing of those families of graphs for which the ramsey numbers have been determined. There still does not exist any general powerful method for computing ramsey numbers.

A SURVEY OF GENERALIZED RAMSEY THEORY

1. HISTORICAL OBSERVATIONS

An account of the generalized ramsey theory of graphs must begin with Ramsey [24]. We state two of his results strictly in terms of pairs of complete graphs.

Theorem 1. (Ramsey) If G has a countably infinite point set, then either G or $\overline{G}$ contains the complete graph of order $\aleph_0$ as a subgraph.

This result is conveniently restated in terms of coloring the lines of a graph. A 2-coloring of the complete graph K_p results when each of its lines is colored red or green. Thus Theorem 1 asserts that every 2-coloring of the complete graph K of order $\aleph_0$ must contain a monochromatic K .

Theorem 2. (Ramsey) For any two positive integers m,n there exists a minimum integer p such that every 2-coloring of K_p contains a green K_m or a red K_n .

The author's personal discovery of generalized ramsey theory for graphs took place during a survey lecture by Erdös [12] on conventional ramsey numbers given at the seminar on graph theory at University College London in October 1962. It seemed natural to me to study ramsey numbers not only for complete graphs but for arbitrary graphs. Fully eight years later, I had the pleasure of attending the scintillating doctoral dissertation defense of V. Chvátal on ramsey numbers for hypergraphs. Chvátal and I immediately launched our joint research effort which led to papers [6, 7, 8, 9]. Only later did we learn that other authors were independently discovering generalized ramsey theory for graphs, see Burr [1], Cockayne [10], Gerencsér and Gyárfás [16], and Parsons [23].

2. RAMSEY NUMBERS FOR SMALL GRAPHS

It is useful to obtain the ramsey numbers for small graphs in order to have data for making conjectures. Also, as Burr [1] observed, these small cases often provide the starting point for inductive proofs, but must be proved independently. All graphs considered have no isolated points. Chvátal and I [8,9] obtained the ramsey numbers for all pairs of graphs with at most four points. Chvátal and Clancy [5] are extending this effort to the diagonal numbers for five point graphs. A complete solution cannot be expected quickly since that would include as just one case the classical ramsey number $r(K_5)$, a most difficult unsolved combinatorial problem. Burr [2] recognized this roadblock, and so he chose to extend our results to all graphs with at most six lines. Thus, in the extreme case of $6K_2$ he permitted as many as 12 points, but avoided the complications caused by having too many lines.

Figure 1 provides the diagonal ramsey numbers for the graphs with at most four points, following the notation for graphs in the book [18].

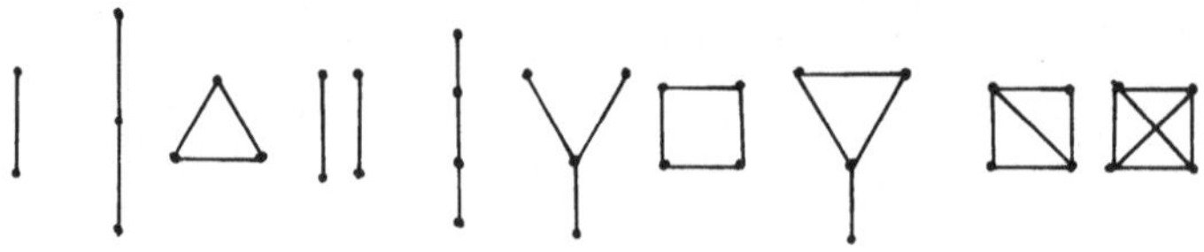

F =	K_2	P_3	K_3	$2K_2$	P_4	$K_{1,3}$	C_4	$K_{1,3}+x$	K_4-x	K_4
r(F) =	2	3	6	5	5	6	6	7	10	18
R(F) =	1	1	2	3	10	3	2	12	15	?

Figure 1. Small Diagonal Ramsey Numbers and Multiplicities

3. RAMSEY MULTIPLICITY

Once one knows the diagonal ramsey number $r(F) = p$, one may then ask just how many monochromatic F must occur when K_p is 2-colored. This invariant R(F) is called the ramsey multiplicity. The off-diagonal multiplicity $R(F_1,F_2)$ is defined similarly. Since it provides more detailed information than the usual ramsey number, it is a more difficult value to establish. Stars constitute the only family for which the ramsey multiplicities are known:

$$R(K_1,n) = \begin{cases} 1 & \text{if } n \text{ is even} \\ 2n & \text{if } n \text{ is odd} \end{cases}$$

Note that the multiplicity of $2n$ for stars with n odd was misprinted as n in [20]. The multiplicities given for the first eight graphs in Figure 1 were determined in [22]. There, a 2-coloring of K_{10} was given to demonstrate that $R(K_4-x) \le 15$.

A. J. Schwenk has just demonstrated that $R(K_4-x) \geq 15$, and we present his proof.

Consider a 2-coloring of K_{10} attaining the minimum possible number of monochromatic $K_4 - x$. Let t_{ij} be the number of monochromatic triangles containing line $v_i v_j$. It is clear that each choice of two monochromatic triangles containing $v_i v_j$ produces a unique monochromatic $K_4 - x$ having $v_i v_j$ as its diagonal. Thus, the ramsey multiplicity is given by

$$(1) \qquad R(K_4-x) = \sum \binom{t_{ij}}{2} = \sum \frac{t_{ij}^2}{2} - \sum \frac{t_{ij}}{2}$$

since we are considering an <u>optimal</u> 2-coloring of K_{10}. Now a result of Goodman [17] reformulated by Schwenk [26] demonstrates that the minimum number of monochromatic triangles is bounded by

$$(2) \qquad \binom{p}{3} - \left[\frac{p}{2} \left[\left(\frac{p-1}{2} \right)^2 \right] \right] .$$

For p = 10, this yields

$$(3) \qquad \frac{1}{3} \sum t_{ij} \geq 20 .$$

Or equivalently, multiplying by 3 we obtain

$$(4) \qquad \sum t_{ij} \geq 60 .$$

The minimum possible value for (1) occurs when each line $v_i v_j$ of K_{10} has the smallest possible value of t_{ij} subject to the provision (4). This clearly occurs when equality is attained in (4) and each individual value of t_{ij} is as close to the average value as possible. Since K_{10} has 45 lines, this requires 30 lines with $t_{ij} = 1$ and 15 with $t_{ij} = 2$. Using this information as a bound for the right side of equation (1), we obtain the desired lower bound:

$$R(K_4-x) \geq 15 .$$

Together with the 2-coloring in [22] containing only 15 monochromatic $K_4 - x$, this establishes the ramsey multiplicity. The multiplicity of the complete graph K_4 has not even been conjectured as yet.

The only graphs F with R(F) = 1 known to date are K_2 and the stars $K_{1,n}$ with n even. This led us to conjecture that these are the only graphs with ramsey multiplicity one! We offerred during a lecture in Oberwolfach, June 1972, U.S. \$100 for the first solution to this conjecture. One of the graph theorists present im- mediately protested, "No, no, please make it 300 D.M. and pay us in real money, not play money".

4. RAMSEY GAMES

In our preceding review [20] of generalized ramsey theory, we described four ramsey games called (1) avoidance (of the first monochromatic triangle), (2) achieve- ment, (3) minimization and (4) maximization. There are new developments concerning games (1) and (2). Game (1) was popularized by Gardner [15] who pointed out that Simmons [27] named it _similarly_ to himself by calling it 'SIM'. It is my considered opinion that 'RAM' would be more apt, not only after Ramsey, but because the winner has _rammed_ a triangle down the loser's throat. Simmons devised a computer program which demonstrated exhaustively that the game is a second player win and which de- feated all human opponents who made the first move. Very recently A. Rosa (not yet published) proved that the RAM player with the second move can force a victory. The RAM game for a quadrilateral C_4 is much more subtle since it often happens that a monochromatic C_4 is formed on the six given vertices without either player realizing it. Of course RAM C_5 would be even more delicate, both for recognition and because it is played on nine vertices.

Game (2) is quite trivial since the first player can construct a monochromatic triangle of his own color within four moves, and in fact without even using the sixth vertex. This achievement game for C_4 is again not as obvious.

For Games (3) and (4), an exhaustive listing of the 2-colorings of K_6 which contain exactly $R(K_3) = 2$ monochromatic triangles T_1 and T_2 is given in [19]. It is proved there that in such a 2-coloring of K_6 , the triangles T_1 and T_2 have just one common point if and only if they have different colors. Hence, there can only be a minimum total score draw when the two players form one triangle each, with one point in common.

5. FAMILIES OF GRAPHS

Although Burr [1] presents a comprehensive list of those ramsey numbers which have been found for families of graphs, we include here, for the sake of completeness, a smaller list.

Whenever two members of the same family are compared we take the convention $m \leq n$. But when different families are confronted, such as paths and cycles, m and n are arbitrary.

Stars [20] $R(K_{1,m}, K_{1,n}) = \begin{cases} m+n & \text{if } m \text{ or } n \text{ is odd} \\ m+n-1 & \text{if } m \text{ and } n \text{ are both even} \end{cases}$

Paths [3,16] $r(P_m, P_n) = n-1 + [m/2]$

Stripes [11] $r(mK_2, nK_2) = 2n + m - 1$

Stars and stripes [11] $r(K_{1,m}, nK_2) = \max\{2n, n+m\}$

Trees and complete graphs [4] $r(T, K_n) = (m-1)(n-1) + 1$

Cycles and paths [13] $r(C_m, P_n) = \max\{n-1+m/2, m-1 + [n/2], (2n-1)(m-2[m/2])\}$

Cycles [14,25] $r(C_m, C_n) = \max\{6, n-1+m/2, (2m-1)(n-2[n/2]), (2n-1)(m-2[m/2])\}$

The last two formulas in this list are not usually written in this way, but were so recast by the fiendish efforts of A. J. Schwenk. Compare the formulas in Burr [1].

Ramsey numbers for digraphs were introduced in [21] where it was observed that $r(D_1, D_2)$ exists if and only at least one of D_1, D_2 is acyclic. The ramsey numbers of all acyclic digraphs with $p \leq 4$ points were determined. Also the ramsey numbers for transitive tournaments, $R(T_m, T_n)$ were demonstrated to be equal to those for complete graphs, $r(K_m, K_n)$.

The only family of digraphs for which the ramsey numbers have been determined are the directed paths, [28]:

$$r(\vec{P}_m, \vec{P}_n) = \begin{cases} n & \text{for } n = 2 \\ n + m - 3 & \text{for } 3 \leq m \leq n. \end{cases}$$

REFERENCES

1. Burr, S. A., Generalized Ramsey Theory for Graphs - a Survey, this volume p. 52

2. Burr, S. A., Diagonal Ramsey Numbers for Small Graphs, to appear.

3. Burr, S. A., and Roberts, J. A., On Ramsey Numbers for Linear Forests, Discrete Math, to appear.

4. Chvátal, V., Tree - Complete Graph Ramsey Numbers, _J. Combinatorial Theory_, to appear.

5. Chvátal, V., and Clancy, M., Diagonal Ramsey Numbers for Most 5-point Graphs, to appear.

6. Chvátal, V., and Harary, F., Generalized Ramsey Theory for Graphs, _Bull. Amer. Math. Soc_. 78 (1972) 423-426.

7. _____ , Generalized Ramsey Theory for Graphs I, Diagonal Numbers, _Periodica Math. Hungar_. 3(1973) 113-122.

8. _____ , Generalized Ramsey Theory for Graphs, II, Small Diagonal Numbers, _Proc. Amer. Math. Soc_. 32 (1972) 389-394.

9. _____ , Generalized Ramsey Theory for Graphs, III, Small Off-Diagonal Numbers, _Pacific J. Math_. 41 (1972) 335-345.

10. Cockayne, E. J., An Application of Ramsey's Theorem, _Canad. Math. Bull_. 13 (1970) 145-146.

11. Cockayne, E. J., and Lorimer, P. J., On Ramsey Numbers for Stars and Stripes, _Canad. Math. Bull_., to appear.

12. Erdös, P., Applications of Probabilistic Methods to Graph Theory, _A Seminar on Graph Theory_ (F. Harary, ed.) Holt, 1967, 60-64.

13. Faudree, R. J., Lawrence, S. L., Parsons, T. D., and Schelp, R. H., Path-cycle Ramsey Numbers, to appear.

14. Faudree, R. J., and Schelp, R. H., All Ramsey Numbers for Cycles in Graphs, to appear.

15. Gardner, M., Mathematical Games, _Scientific American_ 228 (January, 1973) 108-111.

16. Gerencśer, L., and Gyárfás, A., On Ramsey-type Problems,_Ann. Univ. Sci. Budapest Eötvös_ 10 (1967) 167-170.

17. Goodman, A. W., On Sets of Acquaintances and Strangers at any Party, _Amer. Math. Monthly_ 66 (1959) 778-783.

18. Harary, F., _Graph Theory_, Addison-Wesley, Reading, 1969.

19. Harary, F., The Two-Triangle Case of the Acquaintance Graph, _Math. Mag_. 45 (1972) 130-135.

20. Harary, F., Recent Results on Generalized Ramsey Theory for Graphs, _Graph Theory and Applications_ (Y. Alavi, et al. eds.), Springer, Berlin, 1972, 125-138.

21. Harary, F., and Hell, P., Generalized Ramsey Theory for Graphs, V, Ramsey Numbers for Digraphs, <u>Bull. London Math. Soc.</u>, to appear.

22. Harary, F., and Prins, G., Generalized Ramsey Theory for Graphs IV, Ramsey Multiplicities, <u>Networks</u>, to appear.

23. Parsons, T. D., Path-star Ramsey Numbers, <u>Proc. Amer. Math. Soc.</u>, to appear.

24. Ramsey, F. P., On a Problem of Formal Logic, <u>Proc. London Math. Soc.</u> 30 (1930) 264-286.

25. Rosta, V., On a Ramsey Type Problem of J. A. Bondy and P. Erdös, I, II, <u>J. Combinatorial Theory</u>, 15B (1973), 94-104, 105-120.

26. Schwenk, A. J., Acquaintance Graph Party Problem, <u>Amer. Math. Monthly</u>, 79 (1972) 1113-1117.

27. Simmons, G. J., The Game of Sim, <u>J. Recreational Math</u>. 2 (1969) 66.

28. Williamson, J., A Ramsey-type Problem for Paths in Digraphs, <u>Math. Ann</u>. 203 (1973) 117-118.

A SURVEY OF THE RECONSTRUCTION CONJECTURE

Frank Harary
University of Michigan
and Oxford University

ABSTRACT

We begin by tracing the history of the Reconstruction Conjecture (RC) for graphs. After describing the RC as the problem of reconstructing a graph G from a given deck of cards, each containing just one point-deleted subgraph of G, we proceed to derive information about G which is deducible from this deck.

Various theorems proving the RC for trees are then taken up. The status of the RC for digraphs is reported. Several variations of the RC are stated and partial results obtained to date are indicated. We conclude with a summary of structures other than trees which have been reconstructed, and some remarks on the reconstruction of countably infinite graphs.

A SURVEY OF THE RECONSTRUCTION CONJECTURE

1. HISTORICAL NOTES

The author first heard of this fascinating problem when Kelly [30] proved the theorem for trees in 1957. This result was obtained in Kelly's doctoral dissertation which was written under Ulam, who published [45] a statement of the problem in 1960 (although it was already known to him in 1929, when he assiduously collected mathematical problems posed by his fellow graduate students and professors in Lwów, Poland.) This has led to some confusion concerning whose name should be attached to this conjecture. The solution which I recommend heartily is to refer to this problem henceforth as the Reconstruction Conjecture.

In Ulam [45, p. 29], the following statement appears: "Algebraic Problem 1. Suppose that in two sets A and B, each of n elements, there is defined a distance function ρ for every pair of distinct points, with values either 1 or 2, and $\rho(p,p) = 0$. Assume that for every subset of n - 1 points of A , there exists an isometric system of n - 1 points of B , and that the number of distinct subsets isometric to any given subset of n - 1 points is the same in A as in B . Are A and B isometric?"

In terms of graphs, this becomes the following:

Problem 1'. If G and H are graphs with points sets $\{v_1,\ldots,v_p\}$ and $\{u_1,\ldots u_p\}$, $p \geq 3$, and if for all i , the subgraphs $G - v_i$ and $H - u_i$ are isomorphic, then G and H are themselves isomorphic.

This is the form of the problem which Kelly [30] solved for <u>two</u> trees T and S. It seemed to me that more information than necessary was entailed in this statement of Problem 1' and I proposed in [21] the following alternate version:

Reconstruction Conjecture (RC). If G is a graph with $p \geq 3$ points and if the p subgraphs $G - v_i$ are given, then the entire graph G can be reconstructed, uniquely up to isomorphism, from these point-deleted subgraphs.

Speaking very technically, the RC involves the determination of the structure

of G, whereas Problem 1' only asks for a proof of the existence of an isomorphism.
Nevertheless, these two statements are logically equivalent.

It is convenient and intuitively helpful to associate with a graph G of
order p, a deck of p numbered cards with the i'th card containing a drawing of
the point-deleted subgraph $G_i = G - v_i$. We assume throughout that each given
deck of cards is legitimate, that is, it actually comes from a graph in this way.
Incidentally, the problem of characterizing legitimate decks appears to be as
difficult as the RC itself.

Why is $p \geq 3$ taken as part of the hypothesis of the RC? Because the two
graphs of order 2, K_2 and $\overline{K}_2$, have identical decks of cards. Trivially the
only smaller graph K_1 has a deck consisting of one blank card. Henceforth, graph
G always has at least 3 points.

Since these beginnings, the literature on the RC has grown explosively; see
[1] – [45]. This bibliography of the RC was carefully compiled by B. Manvel, as
an extension of the list of references in his elegant doctoral dissertation, [34].

2. INFORMATION ABOUT G INFERRABLE FROM THE G_i .

One approach to the reconstruction problem (other than trying at one fell
swoop to prove or disprove it) is to glean as much information about G as possible
from the deck of subgraphs. Each additional invariant of G which is found from
the G_i serves a double purpose: supporting our faith in the conjecture and providing the remote possibility that this item of information might just be the
crucial step needed to prove that G is reconstructible.

We find at once the number p of points in G either by simply counting the
cards or by counting the points on one card and adding 1. Determining the number
q of lines in G is slightly more complicated. Let q_i be the number of lines in G_i ,
which is known by inspecting card i , and let d_i be the degree of v_i , which is not
yet known. Then clearly $q = q_i + d_i$, and summing these equalities over i yields

$$pq = \Sigma q_i + \Sigma d_i .$$

But the first theorem of graph theory [20, p. 14] asserts that $\Sigma d_i = 2q$, which enables us to solve the above equality for q:

$$q = \frac{\Sigma q_i}{p - 2}$$

This may be viewed as finding the number of subgraphs K_2 in G. A similar argument [19, 21, 24] counts the number of occurrences of any subgraph H with fewer points than G. Thus, for example, we can find the number of cycles C_n and complete subgraphs K_n provided n < p (but not the number of hamiltonian cycles or paths).

A little theorem [21] notes that G is connected if and only if at least two of the G_i are. For if G is connected, two points at maximum distance cannot be cutpoints, and so those two G_i are connected. Conversely, if G_i is connected, either G is connected or else v_i is isolated in G. But if two G_i are connected, both points cannot be isolated, and so G must be connected.

Consider a given legitimate deck. We begin by examining the deck to determine if G is connected. If not, we can reconstruct G, as noted by Kelly [32], Harary [21], and Chartrand and Kronk[9]. On the other hand, if G is connected, the question, of course, remains open. Therefore, we can assume henceforth that G is connected.

Also note that since the complement of $G - v_i$ is identical to $\overline{G} - v_i$, we may replace any legitimate deck by its complementary deck and attempt to reconstruct $\overline{G}$, from which we can obtain G by complementation. Thus, Kelly [30] also concluded that a graph with disconnected complement is reconstructible.

Other deducible information about G includes the cycle rank m = q - p + 1, at once, since p and q are known. The connectivity κ of G is by definition given by

$$\kappa = 1 + \min_i \kappa_i .$$

Obviously the number c of cutpoints of G is the number of disconnected G_i. To obtain the number b of blocks, let k_i be the known number of components in G_i. It was shown in [18] that b satisfies the formula:

$$b = 1 + \Sigma (k_i - 1) .$$

3. TREES.

Like so many other problems in graph theory, the Reconstruction Conjecture when stated for trees becomes much easier. As mentioned earlier, Kelly proved the RC for trees. One should note that he need not be told in advance that the graph G he is attempting to reconstruct is a tree, for he can quickly determine that G is connected and $q = p - 1$. However, Harary and Palmer [27] observed that as soon as we know that T is a tree, we no longer need many of the cards in the deck. In fact, we can throw away all the cards not resulting from the removal of an endpoint of T (all the disconnected T_i) and still reconstruct T! Bondy carried this a step further. A point of tree T is called <u>peripheral</u> if it is an endpoint of some longest path. Clearly a peripheral point must be an endpoint, but not all endpoints are peripheral. For example, the tree in Figure 1 has 9 points, of which 5 are endpoints, of which 3 are peripheral points. Bondy [3] found that trees can be reconstructed from their peripheral point-deleted subgraphs.

<u>Figure 1.</u> A tree.

Another variation of the RC occurs if we throw out multiple occurrences of the same card, so that each remaining card is unique. This variation is called "set-reconstruction," since we are given the <u>set</u> of point-deleted subgraphs, but we are not told the multiplicity of each subgraph. Manvel [36] showed that trees are endpoint set-reconstructible except for the two pairs shown in Figure 2.

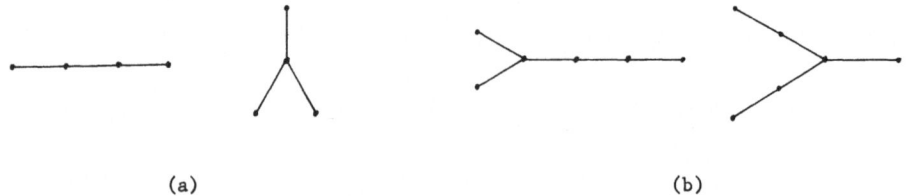

(a)   (b)

<u>Figure 2.</u> The endpoint set-nonreconstructible pairs of trees.

It is rumored that it has been proved that the cards obtained from endpoints can be thrown away, and then T can be reconstructed from the remaining cutpoint-deleted subgraphs.

In general, we seem to have much more information than necessary to reconstruct trees, as observed by A.J. Schwenk. For example, we could discard a card at random from the deck. A count of the lines in the remaining deck would still suffice to decide whether the discarded card was obtained from an endpoint or a cutpoint. In either case, the opposite subdeck of cards, i.e., the cutpoint cards or the endpoint cards respectively, remains intact and can be used to recontruct T (if this rumor, which I started, is in fact true). This suggests the following question: How many cards can be discarded at random from a full deck obtained from a tree T and still guarantee that T is recontructible? For 2 discards, there are nonreconstructible pairs on 4 and 5 points (see Figure 3), but all larger trees seem to be reconstructible.

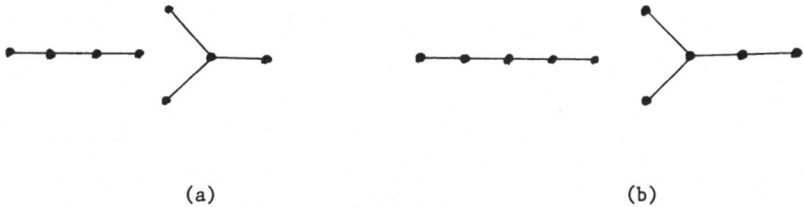

(a)   (b)

Figure 3. Two pairs of trees whose deck may be identical after 2 discards.

4. DIGRAPH RECONSTRUCTION.

Since disconnected graphs are reconstructible, we formulated the analogous problem for digraphs and in [24], we demonstrated that every digraph D with $p \geq 5$ which is not strongly connected is reconstructible from its point-deleted subdigraphs D_i. The reason for the restriction $p \geq 5$ is that the two tournaments T_1 and T_2 of Figure 4 serve as a counterexample for $p = 4$. In each of T_1 and T_2 the subdigraphs consist of one cyclic triple and three transitive triples.

T_1: 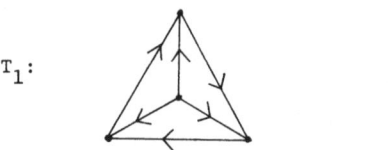 T_2: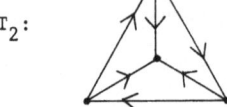

Figure 4. A pair of nonreconstructible tournaments

Beineke and Parker [2] found two additional pairs of nonreconstructible strong tournaments, of order 5 and 6. Hence we now believe that the following statement is true.

RC for strong Digraphs. Every strong digraph of order $p \geq 7$ is reconstructible.

5. VARIATIONS

We have just examined several variations on the standard reconstruction problem for trees. Similarly, Table 1 suggests two dozen reconstruction problems: to reconstruct each of the six types of structures listed from one of the four types of decks given. Of course, there is practically no limit to the additional variations one might propose.

Reconstruct:

graphs

multigraphs

digraphs

tournaments

relations

networks

From:

point-deleted subgraphs

line- (or arc-) deleted subgraphs

set of point-deleted subgraphs

set of line- (or arc-) deleted subgraphs

Table 1. Twenty four reconstruction problems

Manvel and Stockmeyer [43] reconstructed n x n matrices from the n principal minors. It might seem that this should imply the RC by reconstructing the adjacency matrix A(G), but a second glance reveals that the minors retain the relative order of the n - 1 remaining rows and columns whereas the deck of point-deleted subgraphs loses that information. Thus, it is not surprising that square matrices are much easier to reconstruct than graphs.

Lovász [33] recently verified the Line-Reconstruction Conjecture when G has at least half the possible lines, that is, when

$$q \geq \frac{1}{2} \binom{p}{2} \ .$$

The unrestricted L-RC remains unsolved.

There are structures for which the RC has been settled.[*] Manvel extended the RC to structures other than trees by proving it for unicyclic graphs in [35]. This was soon superseded by a proof for casti by Geller and Manvel [42].

Manvel [39] and Giles [14] proved the RC for maximal outplanar graphs. And now Giles [15] has a proof for any outerplanar graph. A proof for planar graphs continues to be elusive, but everyone believes the RC to be true for planar graphs and in fact to be true for graphs.[**]

Square-celled animals were reconstructed by Harary and Manvel [23].

Finally, a result for matroids by Brylawski [8] specializes to graphs as follows. Given the q chromatic polynomials of the line-deleted subgraphs $G - x_j$, the chromatic polynomial of G can be determined.

6. INFINITE GRAPHS

There is nothing to keep us from stating the RC for countably infinite graphs, but the methods and partial results are so different from (finite) graphs that it is proper to consider the problems separately. The first result on the infinite RC was a nonreconstructible pair due to Fisher [12]. Later, at the author's farewell party from his 1970 Visiting Professorship at the University of Waterloo,

---

[*]The definitions of these structures can be found in the book [20].

[**]The reconstruction of labeled graphs was also studied by Harary and Manvel [22].

we found [13] a much simpler nonreconstructible pair: Let G be the infinite

tree which is regular of countable degree, and let H be the forest consisting of

two copies of G. All the point-deleted subgraphs of G and H are identical, namely,

a countable number of copies of G. Thus G and H are a nonreconstructible pair.

Notice that we cannot even distinguish between a connected and a disconnected

infinite graph. In fact, the information we obtain so easily for (finite) graphs

always seems to be much more difficult for infinite graphs.

An infinite graph whose points have finite degree is called <u>locally finite</u>.

If only finitely many points have infinite degree, we say it is <u>almost locally</u>

<u>finite</u>. Similarly, G is <u>almost r-regular</u> if only finitely many points have degree

$\neq$ r. Using this condition to forbid the offending pair described above, we recon-

structed [28] almost r-regular infinite forests. Bondy and Hemminger [5,6] also

succeeded in reconstructing infinite forests with certain severe restrictions on the

number of infinite paths present. Both of these approaches effectively limit the

permitted structural variations to such an extent as to make the reconstruction

possible. In general, it also appears true that all infinite <u>trees</u> are reconstruc-

tible.

REFERENCES

1. Arjomandi, E. and Corneil, D. G., "Unicyclic graphs satisfy Harary's conjecture," <u>Canad. Math. Bull</u>., to appear.

2. Beineke, L. W., and Parker, E. M., "On nonreconstructable tournaments,"
   <u>J. Combinatorial Theory</u> 9 (1970) 324-326.

3. Bondy, J. A., "On Kelly's congruence theorem for trees," <u>Proc. Cambridge
   Philos. Soc.</u> 65 (1969), 1-11.

4. Bondy, J. A., "On Ulam's Conjecture for separable graphs," <u>Pacific J. Math</u>. 31
   (1969), 281-288.

5. Bondy, J. A., and Hemminger, R. L., "Almost reconstructing infinite graphs,"
   <u>Pacific J. Math</u>., to appear.

6. Bondy, J. A., and Hemminger, R. L., "Reconstructing infinite graphs," Unpublished.

7. Bryant, R. M., "On a conjecture concerning the reconstruction of graphs,"
   <u>J. Combinatorial Theory</u> 11 (1971), 139-141.

8. Brylawski, T. A., "Reconstructing combinatorial geometries," this volume p. 226.

9. Chartrand, G., and Kronk, H. V., "On reconstructing disconnected graphs," Ann. N. Y. Acad. Sci. 175 (1970), 85-86.

10. Chinn, P. Z., "A graph with p points and enough distinct (p-2) order subgraphs is reconstructable," Recent Trends in Graph Theory, (M. Capobianco et al., eds.), Springer, Berlin, 1971, 71-73.

11. Faber, V., "Reconstruction of graphs from indexed p-2 point subgraphs," Notices Amer. Math. Soc. 18 (1971), 807.

12. Fisher, J.,"A counterexample to the countable version of a conjecture of Ulam," J. Combinatorial Theory, 7 (1969), 364-365.

13. Fisher, J., Graham, R. L., and Harary, F., "A simpler counterexample to the reconstruction conjecture for denumerable graphs," J. Combinatorial Theory 12B (1972), 203-204.

14. Giles, W. B., "On reconstructing maximal outerplanar graphs," Discrete Math., to appear.

15. Giles, W.B., "The reconstruction of outerplanar graphs," Michigan Math. J., to appear.

16. Greenwell, D. L., "Reconstructing graphs," Proc. Amer. Math. Soc. 30 (1971), 431-433.

17. Greenwell, D. L., and Hemminger, R. L.,"Reconstructing graphs," The Many Facets of Graph Theory (G. T. Chartrand and S. F. Kapoor, eds.) Springer, Berlin, 1969, 91-114.

18. Harary, F., "An elementary theorem on graphs," Amer. Math. Monthly 66 (1959) 405-407.

19. Harary, F., "Graphical reconstruction," A Seminar on Graph Theory (F. Harary, ed.), Holt, New York, 1967, 18-20.

20. Harary, F., Graph Theory, Addison-Wesley, Reading, 1969.

21. Harary, F., "On the reconstruction of a graph from a collection of subgraphs," Theory of Graphs and its Applications (M. Fiedler, ed.) Academic Press, New York, 1964, 47-52.

22. Harary, F., and Manvel, B., "The reconstruction conjecture for labeled graphs," Combinatorial Structures and Their Applications (R. K. Guy, ed.) Gordon and Breach, New York, 1969, 131-146.

23. Harary, F., and Manvel, G., "Reconstruction of square-celled animals," Bull. Soc. Math. Belg. 25 (1973) 80-84.

24. Harary, F., and Palmer, E. M., "On the problem of reconstructing a tournament from subtournaments," Monatsh. Math. 71 (1967), 14-23.

25. Harary, F., and Palmer, E. M., "A note on similar lines of a graph," Rev. Roum. Math. Pures et Appl. 10 (1965), 1489-1492.

26. Harary, F., and Palmer, E.M., "On similar points of a graph," J. Math. Mech. 15 (1966), 623-630.

27. Harary, F., and Palmer, E.M., "The reconstruction of a tree from its maximal proper subtrees," Canad. J. Math. 18 (1966), 803-810.

28. Harary, F., Schwenk, A. J., and Scott, R. L., "On the reconstruction of countable forests," Publ. Math. Inst. (Beograd) 13 (1972), 39-42.

29. Hemminger, R. L., "On reconstructing a graph," Proc. Amer. Math. Soc. 20 (1969), 185-187.

30. Kelly, P. J., "A congruence theorem for trees," Pacific J. Math. 7 (1957), 961-968.

31. Kelly, P. J., "On some mappings related to graphs," Pacific J. Math. 14 (1964), 191-194.

32. Lovász, L., "Operations with structures," Acta Math. 18 (1967), 321-328.

33. Lovász, L., "A note on the line reconstruction problem," J. Combinatorial Theory, 13B (1972), 309-310.

34. Manvel, B., "On reconstruction of graphs," Doctoral dissertation, Univ. of Michigan, 1970.

35. Manvel, B., "Reconstruction of unicyclic graphs," Proof Techniques in Graph Theory (F. Harary, ed.) Academic Press, New York, 1969, 103-107.

36. Manvel, B., "Reconstruction of trees," Canad. J. Math 22 (1970), 55-60.

37. Manvel, B., "On reconstruction of graphs," The Many Facets of Graph Theory (G. T. Chartrand and S. F. Kapoor, eds.) Springer, Berlin, 1969, 207-214.

38. Manvel, B., "Reconstructing the degree pair sequence of·a digraph," J. Combinatorial Theory, 15B, (1973) 18-31.

39. Manvel, B., "Reconstruction of maximal outerplanar graphs," Discrete Math. 2 (1972), 269-278.

40. Manvel, B. Determining Connectedness from Subdigraphs," J. Combinatorial Theory, to appear.

41. Manvel, B., "Some basic observations on Kelly's Conjecture for graphs," Discrete Math., to appear.

42. Manvel, B., and Geller, D. P., "Reconstruction of cacti," Canad. J. Math. 21 (1969), 1354-1360.

43. Manvel, B., and Stockmeyer, P., "Reconstruction of matrices," Math. Mag., 44 (1971), 218-221.

44. O'Neil, P. V., "Ulam's conjecture and graph reconstructions," Amer. Math. Monthly 77 (1970), 35-43.

45. Ulam, S. M., A Collection of Mathematical Problems, Wiley, New York, 1960, p. 29.

<u>RECENT RESULTS ON GRAPHICAL ENUMERATION</u>

<u>Frank Harary</u>
University of Michigan

<u>ABSTRACT</u>

We present four of our counting results which are to appear soon in various journals. These comprise: (1) the number of caterpillars, (2) the number of self-complementary configurations, (3) the number of achiral trees, and (4) the probability of an endpoint in a large random tree. We conclude with a brief mention of four miscellaneous results.

RECENT RESULTS ON GRAPHICAL ENUMERATION

1. CATERPILLARS

For a connected graph G with at least 3 points, the underlined derived graph G' is formed by deleting the endpoints of G. A underlined caterpillar is a tree T whose derived graph T' is a path. Theorem 1 of [5] was presented as a characterization of trees with hamiltonian square, but it can equally well be viewed as characterizing caterpillars. We restate it here from this point of view.

Theorem 1. For any tree T with at least 3 points, the following statements are equivalent:

    (1) T is a caterpillar.

    (2) T^2 is hamiltonian.

    (3) T does not contain the subdivision graph $S(K_{1,3})$ as a subtree.

Of course, many caterpillars may have the same path P_n as their derived tree. However, it is possible to mark each point of P_n with the number of endpoints that were joined to it in T. Having done this, there is clearly a one-to-one correspondence between such weighted paths and caterpillars. In fact, it is convenient to add 1 to each weight to account for that point itself in addition to the endpoints joined to it, so that each weight is a positive integer. Certainly, this does not disturb the correspondence, and it has the advantage of making a weighted path P_n with total weight p correspond to a caterpillar T with p points.

Since the series $x/(1 - x) = x + x^2 + x^3 + \ldots$ provides the possible weights of $1, 2, 3, \ldots$, we may apply Pólya's theorem [6, Chapter 2] with $x/(1 - x)$ as the figure counting series and $\Gamma(P_n)$ as the configuration group to enumerate caterpillars of weight p whose derived tree is P_n. Let c_p be the number of p-point caterpillars, $p \geq 3$, and define the generating function $c(x) = \sum c_p x^p$. Then we have just seen that

(1) $$c(x) = x^2 \sum_{n=1}^{\infty} Z(\Gamma(P_n)); \ \frac{x}{1 - x}) \ .$$

But the path P_n has just two automorphisms, and this formula simplifies to

(2)
$$c(x) = \frac{x^3(1 - 3x^2)}{(1 - 2x)(1 - 2x^2)} \quad .$$

Writing $p = n + 4$, (2) yields

(3)
$$c_{n+4} = 2^n + 2^{[n/2]}$$

as the explicit number of caterpillars. The formula (3) can also be obtained by a direct argument, given in [9] which is much less elegant.

2. SELF-COMPLEMENTARY CONFIGURATIONS

The result in [3] is illustrated nicely by the graphs of order 4. The number of such graphs is $g_4 = 11$ and we write g_{pq} for the number of graphs with p points and q lines. Let s_p be the number of self-complementary graphs of order p . Then we have the two equations

(4)
$$g_p = \sum_q g_{pq} \quad ,$$

(5)
$$s_p = \sum_q (-1)^q g_{pq} \quad ,$$

the first by definition and the second as proved in [3] by applying a result of Read [11] . For $p = 4$, equation (5) gives

$$s_4 = 1 - 1 + 2 - 3 + 2 - 1 + 1 = 1 \quad ,$$

and the only self-complementary graph of order 4 is the path P_4 . A similar formula holds for digraphs. Let the number of self-complementary digraphs of order p be $\vec{s}_p$. Then with d_{pq} denoting the number of digraphs with p points and q arcs, we find from [3] that

(6)
$$\vec{s}_p = \sum_q (-1)^q d_{pq} \quad ,$$

which gives for $p = 3$

$$\vec{s}_3 = 1 - 1 + 4 - 4 + 4 - 1 + 1 = 4 \quad .$$

Thus all 4 digraphs with 3 points and 3 arcs are self-complementary. Both (5) and (6) are special cases of a general formula for the number of self-complementary configurations with equivalence determined by a specified permutation group.

3. ACHIRAL TREES

A plane tree is an embedding of a tree in the plane. A chiral plane tree is one whose mirror image is a different plane tree; otherwise it is achiral. These concepts are illustrated in Figure 1 which shows three different plane embeddings of the same tree.

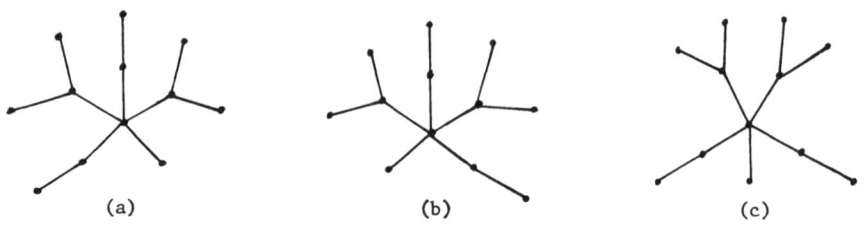

Figure 1. Three Different Plane Trees

The plane trees (a) and (b) are both chiral and are reflections of each other, while (c) is an achiral plane tree.

My interest in this topic stems from the November 1970 conference in Paris on "Chirality from Biology to Mathematics" which was attended predominantly by theoretical chemists, some of whom presented ideas such as the chiral and achiral letters of the Roman alphabet. During a drive shortly thereafter from Ann Arbor to Waterloo with R. W. Robinson, we explored the possibility of applying chirality to plane trees, but did not have the opportunity to write up our results untill two years later, when he was visiting professor at the University of Michigan.

Surprisingly, we found [8] that the number of planted (rooted at an endpoint) achiral plane trees of order n equals the number of unrooted ones, and that the number of rooted ones is 2^{n-2}. However, in spite of much effort, we could not find any straightforward method for proving that this power of 2 is the counting formula.

We define a tree T to be <u>partially achiral</u> if some plane embedding is achiral; it is (totally) <u>achiral</u> if every plane embedding is achiral. In [8], we also counted both partially achiral trees and achiral trees, but the formulas are not simple.

4. THE PROBABILITY OF AN ENDPOINT

We determined in [12] the probability that a point chosen arbitrarily from a large random tree is an endpoint and found that this value approaches .438156235664.... We gave a brief account of this result in [5], describing the asymptotic aspect of the analysis. We describe the enumerative aspects of this result.

Let $t(x)$ and $T(x)$ denote the generating functions for trees and rooted trees. The usual method for counting trees [6, Chap. 2] relies on the equations

$$(7) \qquad t(x) = T(x) - \frac{1}{2} T^2(x) + \frac{1}{2} T(x^2) \quad ,$$

$$(8) \qquad T(x) = x \exp \sum_{1}^{\infty} T(x^i)/i \quad .$$

Obviously the number of planted trees on $n + 1$ points is identical to the number of rooted trees on n points. Thus if we define $P(x)$ to be the generating function for planted trees in which the plant is <u>not</u> counted as a point, we have the identity $P(x) = T(x)$. Consequently, it is possible to replace T by P in (7) and (8) to obtain an equivalent pair of equations.

In order to compute the probability of an endpoint, we must find e_n , the total number of endpoints among all n-point trees, and divide this by nt_n , the total number of points present. To find e_n , we define two new generating functions on two variables. Let $t(x,y) = \Sigma \Sigma t_{nm} x^n y^m$, where t_{nm} is the number of trees with n points and m endpoints, and let $P(x,y) = \Sigma \Sigma P_{nm} x^n y^m$ where P_{nm} is the number of planted trees with n points and m endpoints, in which again the plant has not been counted either as a point or as an endpoint. This exclusion of the plant serves to simplify

the resulting formulas. There should be no confusion between the generating functions

$t(x)$ and $t(x,y)$ because the number of variables indicates the context; similarly

for P . Notice that $t(x) = t(x,1)$ and $P(x) = P(x,1)$.

Arguments analogous to the traditional ones used to obtain (7) and (8) yield

$$(9) \quad t(x,y) = P(x,y) - \frac{1}{2} P^2(x,y) + \frac{1}{2} P(x^2,y^2) + x(1-y) (1-P(x,y)) \quad ,$$

$$(10) \quad P(x,y) = xy - x + x \exp \sum_{i=1}^{\infty} P(x^i , y^i) / i \quad .$$

It is easy to see that these reduce to (7) and (8) on substituting $y = 1$.

We now display a simple but useful enumerative device. Take the partial derivative of $t(x,y)$ with respect to y and then evaluate it at $y = 1$ to obtain

$$(11) \qquad\qquad t_y (x,1) = \Sigma \Sigma \, mt_{nm} \, x^n \quad .$$

That is, each tree with m endpoints is counted with weight m toward the coefficient

of x^n . In other words, e_n , the total number of endpoints among all n-point trees,

is the coefficient of x^n in $t_y(x,1)$. This is expressed concisely by defining

$e(x) = \Sigma \, e_n x^n$, so that $e(x) = t_y(x,1)$.

Equations (9) and (10) are used to evaluate the partial derivative and arrive at

an asymptotic estimate for e_n which leads to the estimation of the limiting value

of the probability given above.

5. MISCELLANEOUS RESULTS

A tapeworm is defined as a simplified variation on square-celled animals [6,p. 234]

in which the growth pattern is "linear" but need not be planar. We found in [7] that the

number of tapeworms with an even number $2n + 2$ of cells is $(3^{2n} + 2.3^n + 1)/4$ while

the number with $2n + 3$ cells is $(3^{2n+1} + 4.3^n + 1)/4$.

In order to develop an application of the counting of achiral trees described in Section 3 above to hydrocarbon compounds as they occur in euclidean 3-space, we worked out [1] analogous formulas for "enantiomeric steric trees". These are chiral trees in 3 dimensions in which the endpoints representing hydrogen atoms are distributed symmetrically about their adjacent carbon atom. The resulting formulas are similar in appearance to equations (7) and (8) above, with differences which reflect the presence of points of degrees 1 and 4 only and the chiral character of the configuration .

Palmer and Robinson [10] succeeded in deriving a formula for the cycle index of the exponentiation group (see [6, p. 98]) of two given permutation groups. I had first introduced this binary operation on permutation groups for the purpose of counting bicolored graphs, but the general formula for this cycle index eluded me. Just as the other binary operations on permutation groups (product, cartesian product, composition) help to solve relevant problems in graphical enumeration by applying Pólya's Theorem using the respective cycle index, so does the cycle index of the exponentiation group settle some previously open questions.

Finally, we mention a very recent result due to Chao [2]. In a symmetric graph G (see [4, p. 171]) all points are similar with regard to the group of G and also all lines are similar. Chao showed that the number of labeled symmetric graphs of prime order p is:

$$2 + (p - 2)! \sum d ,$$

where the sum is taken over all divisors d of $(p - 1)/2$.

We conclude with the observation that the results presented here are indeed recent since only 1/4 of the references are now in print.

REFERENCES

1. Balaban, A. T., Harary, F., and Robinson, R. W., Enantiomeric Steric Trees, Tetrahedron, submitted.

2. Chao, C-Y., The Number of Labeled Symmetric Graphs. Discrete Math. to appear.

3. Frucht, R., and Harary, F., Self-complementary Generalized Orbits of a Permutation Group. Canad. Math. Bull., to appear, 1974.

4. Harary, F., Graph Theory. Addison-Wesley, Reading, 1969.

5. Harary, F., Recent Results on Trees, this volume p. 1.

6. Harary, F., and Palmer, E. M., Graphical Enumeration. Academic Press, New York, 1973.

7. Harary, F., and Robinson, R. W., The Number of Tapeworms, to appear.

8. Harary, F., and Robinson, R. W., The Number of Achiral Trees, J. Reine Agnew. Math., to appear.

9. Harary, F., and Schwenk, A. J., The Number of Caterpillars. Discrete Math., to appear, 1974.

10. Palmer, E. M., and Robinson, R. W., Enumeration Under Two Representations of the Wreath Product, Acta Math., to appear.

11. Read, R. C., On the Number of Self-complementary Graphs and Digraphs. J. London Math. Soc., 38 (1963) 99-104.

12. Robinson, R. W., and Schwenk, A. J., The Distribution of Degrees in a Large Random Tree. Proc. Cambridge Phil. Soc., submitted.

IS THE NULL-GRAPH A POINTLESS CONCEPT?

Frank Harary
University of Michigan
and Oxford University

Ronald C. Read
University of Waterloo

ABSTRACT

The graph with no points and no lines is discussed critically. Arguments for and against its official admittance as a graph are presented. This is accompanied by an extensive survey of the literature. Paradoxical properties of the null-graph are noted. No conclusion is reached.

IS THE NULL-GRAPH A POINTLESS CONCEPT?

1. INTRODUCTION

It is well-known in set theory that the introduction of the seemingly paradoxical null-set, having no elements, results in a considerable simplification of the statement of theorems, relieving an author of having to make special provision for the possibility of there being no objects possessing the property which defines the elements of a certain set. Some writers in graph theory, presumably by imitation, have introduced the concept of a (or the) "null-graph" - having no points and no lines; see Figure 1. The advantages of such a concept are less obvious than those of the null-set, and we can reasonably question whether such advantages as it may have are not outweighed by extra complications that it introduces. We discuss here the pros and cons of the use of the null-graph. Note that it is not a question of whether the null-graph "really exists"; it is simply a question of whether there is any point in it.

Figure 1. The Null Graph

2. CURRENT USAGE

Let us start with a brief survey of how some well-known writers on graph theory have handled the null-graph, noting in particular that if a writer does not explicitly state that the set of points of a graph is non-empty, he has, by default, admitted the null-graph, even though he may subsequently pay no attention to it.

Ore [7] makes no mention of the null-graph in the sense that we are using the term, but instead uses the adjective "null" to describe a graph consisting only of isolated points. One presumes, therefore, that since he regards as trivial a graph whose set of lines is empty, he would, a fortiori, discount the graph whose set of points is empty, even though he does not actually say so.

Berge [1] and Wagner [13] do not explicitly state that the set of points is non-empty, and hence, by default, admit the null-graph. However, the null-graph is never mentioned as such, and hence one may assume that they do not intend it to be regarded as a graph.

König [5] distinguishes between the null-graph and "proper" graphs (eigentlichen Graphen). He adds that the adjective "proper" will be omitted when the context makes it clear what is meant. He thus recognizes the problem, but surmounts it by what are perhaps somewhat unsatisfactory methods. Sedlaček [11], Zykov [14], and Busacker and Saaty [2] all define the set of vertices of a graph to be non-empty, thus explicitly excluding the null-graph.

In complete contrast, Tutte [12] specifically admits the null-graph and refers to it as a graph. Many of his theorems apply to the null-graph as well as other graphs, but many do not. In consequence, the statements of a large number of theorems in his book contain the proviso that the graphs to which they refer must not be the null-graph. Sachs [10] also explicitly defines the null-graph.

Perhaps the most enthusiastic proponents of the null-graph are Maxwell and Reed [6], who in their own words, give to the null-graph "full-fledged membership in the 'society' of graphs." They realize, however, that "the acceptance of the null-graph has the aspect of both simplifying and complicating the development of the theory of graphs." In Section 2.9 of their book, they present an amusing but somewhat confused discussion of "this paradoxical beast - the null-graph." In fact, they ask, "Does the null-graph possess all properties, or in fact, does it have no properties?", and infer that such questions "lead to a situation of absolute and utter chaos." Despite their apparent affection for it, they conclude by saying, "except on specific occasions, ... the null-graph is ignored."

As for the authors of this paper, one of us [4] is decisively opposed to the null-graph; the other is willing to afford it grudging recognition in certain contexts, but nevertheless, when called upon to give a definition of a graph [9] has required the point-set to be non-empty.

3. PRO

We present three arguments in favor of the null-graph involving set theoretic considerations, the chromatic polynomial, and graphical enumeration.

The most convincing argument in favor of admitting the null-graph is that this assures that the intersection of any two subgraphs of a given graph is always a subgraph. In particular, the collection of all subgraphs of G then forms a boolean algebra, in which the null-graph is the zero element.

It is convenient to introduce a notation for the null-graph. By definition, a graph is <u>complete</u> if every pair of points are adjacent. Thus the null-graph is obviously, though vacuously, complete. It is becoming standard notation to write K_p for the complete graph with p points; hence the null-graph is K_0. A good example of this set theoretical use of K_0 occurs in the theory of chromatic polynomials. In the expository paper [8], the following results are given:

(1) If a graph G has two components A and B, then

$$P_G(\lambda) = P_A(\lambda) \cdot P_B(\lambda)$$

where $P_G(\lambda)$ denotes the chromatic polynomial of G.

(2) If two graphs A and B intersect in a complete graph K_p, then the chromatic polynomial of their union is

$$P_A(\lambda) \ P_B(\lambda) \ / \ P_{K_p}(\lambda) \ .$$

If K_0 is admitted as a graph, these two results can be combined, (1) being the special case of (2) for p = 0 . All that is required is that we define $P_{K_0}(\lambda)$ to be 1. This is perfectly natural; no matter what the number λ of colors, there is only one way to color the points of K_0, namely, do nothing! The chromatic polynomial of K_p is

$$\lambda^{(p)} = \lambda(\lambda - 1) \ \cdots \ (\lambda - p + 1)$$

and we are therefore led to write $\lambda^{(0)} = 1$. This agrees with standard practice in combinatorial analysis.

Perhaps the most striking way in which the null-graph arises "naturally" in graph-theoretical research occurs in the theory of graphical enumeration. Consider, for example, the exponential generating function $T(x)$ for labeled trees, defined by

$$(3) \qquad T(x) = \sum_{r=1}^{\infty} T_n x^n / n!$$

where T_n is the number of trees on n labeled points. It is well-known that $T_n = n^{n-2}$. Another result, not quite so well-known, is that the generating function $F(x)$ for labeled forests is obtained by taking the formal exponential of $T(x)$:

$$(4) \qquad F(x) = \exp T(x)$$

Now the left-hand side of (4) contains the term "1", and if this refers to anything at all, it can only refer to the null-graph.

A relation similar to (4) holds in more general applications between the numbers of <u>labeled</u> graphs and of <u>connected labeled</u> graphs of a certain kind. Let $G(x)$ be the exponential generating function for such graphs and $C(x)$ the corresponding function for the connected graphs. Then these two functions are related by an equation which is often written as

$$(5) \qquad 1 + G(x) = \exp C(x),$$

or equivalently as

$$(6) \qquad C(x) = \log(1 + G(x)).$$

Here the terms in both G and C correspond to non-null graphs only.

In most enumeration problems, it is the counting result for all graphs of a specified type (not necessarily connected) which is obtained first, the result for connected graphs being derived from it by use of (6). The decision to take powers from 1 upwards is then a natural one; it avoids introducing the null-graph, but at the expense of having to insert a "1" into equation (5) in order to "make the answer come out right." On the other hand, when it is the <u>connected</u> result which is obtained first (as in the labeled tree example above), the null-graph can be ignored only by hustling it out of sight after it has already made its appearance.

A similar phenomenon occurs with enumeration problems [3] for unlabeled graphs; for example, with the relation between the numbers g_{pq} for graphs that are not necessarily connected. This relation is

$$\sum_{p=0}^{\infty} \sum_{q=0}^{\infty} g_{pq} \, x^p \, y^q = \prod_{\substack{m=1 \\ n=0}}^{\infty} (1 - x^m \, y^n)^{-g_{mn}}$$

Here again the null-graph $(p = q = 0)$ occurs naturally on the left-hand side, but not on the right-hand side (we cannot have $m = n = 0$, or the whole right-hand side vanishes!)

These examples show that the null-graph can arise naturally, without being specifically introduced. They also indicate that the null-graph must be regarded as not being connected! We shall consider below the paradox which this statement seems to imply.

4. CONTRA

The main objection to admitting the null-graph is that it has flagrantly contradictory properties. By one definition, a graph G is <u>connected</u> if every pair of points are joined by a path. As K_0 has no points, this definition is vacuously satisfied. Thus K_0 is connected.

Clearly K_0 is also acyclic since it contains no cycles; hence K_0 is a tree. This opinion is shared by Tutte [12] who asserts in his Theorem 3.31 that "All null-graphs, vertex-graphs, and link-graphs are trees," by which he means K_0, K_1, and K_2 are trees.

However, by another definition, a graph is a tree if it is connected, and $p = q + 1$. For K_0, $p = q = 0$; hence K_0 is <u>not</u> a tree. But the graph K_0 is acyclic and so by definition is a forest, but not a tree. One naturally concludes that K_0 is in fact <u>disconnected</u>! This has already been suggested by the enumeration argument in the preceding section. Confirmation of this conclusion is obtained if we take the definition of connected to mean that a graph has exactly one component; but K_0 has no components at all.

Perhaps surprisingly there is also an enumerational argument against the recognition of a null-graph. This is entailed in equation (6) above, where fortunately, the explicit presence of the term "1" enables one to expand the right-hand member and thus express $C(x)$ in powers of $G(x)$.

5. CONCLUDING REMARKS

It can be seen from these examples that the main reason for the paradoxical nature of the null-graph is as follows. Many concepts in graph theory can be defined in several ways. When there are several possible definitions for a concept, it is necessary that they be consistent, and for non-null graphs this is so. But since some authors admit the null-graph while others do not, it often happens that different definitions of a concept disagree when applied to the null-graph. It would presumably be possible to frame a set of definitions that would be consistent even when applied to the null-graph, and would give to the null-graph the properties that it requires for those applications when it occurs naturally. To achieve this would be of limited usefulness, and the present authors have no intention of attempting to do so.

REFERENCES

1. Berge, C., The Theory of Graphs and its Applications. Wiley, New York, 1962.

2. Busacker, R. and Saaty, T., Finite Graphs and Networks. McGraw-Hill, New York, 1965.

3. Harary, F., The Number of Linear, Directed, Rooted, and Connected Graphs. Trans. American Mathematical Society, 78 (1955), pp. 445-463.

4. Harary, F., Graph Theory. Addison-Wesley, Reading, Massachusetts, 1969.

5. König, D. Theorie der endlichen und unendlichen Graphen. Leipzig, 1936. Reprinted Chelsea, New York, 1950.

6. Maxwell, L. M., and Reed, M. B., Theory of Graphs: A Basis for Network Theory, Pergamon Press, New York, 1971.

7. Ore, O., Theory of Graphs. American Mathematical Society Colloq. Publ., Volume 38, Providence, 1962.

8. Read, R. C., An Introduction to Chromatic Polynomials. J. Combinatorial Theory 4 (1968), pp. 52-71.

9. Read, R. C., A Mathematical Background for Economists and Social Scientists, Prentice-Hall, Englewood Cliffs, 1971.

10. Sachs, H., Einführung in die Theorie der endlichen Graphen, Teil I. Teubner, Leipzig, 1970.

11. Sedlaček, J., Einführung in die Graphentheorie, Teubner, Leipzig, 1968.

12. Tutte, W. T., The Connectivity of Graphs. Toronto University Press, Toronoto, 1967.

13. Wagner, K., Graphentheorie, Bibliographisches Institut, Mannheim, 1970.

14. Zykov, A. A., Theory of Finite Graphs (Russian), Nauka, Novosibirsk, 1969.

WHICH GRAPHS HAVE INTEGRAL SPECTRA?

Frank Harary*
University of Michigan
and Oxford University

Allen J. Schwenk*
University of Michigan
and Oxford University

ABSTRACT

The spectrum $S(G)$ of a graph G of order p is defined as the non-increasing sequence of the p real eigenvalues of the adjacency matrix of G. It has been found that certain graphs have an integral spectrum, i.e., every eigenvalue is an integer. Thus, it is natural to ask just which graphs have this property. We develop a systematic approach to this question based on operations on graphs. The general problem appears intractable.

*Research supported in part by grant 73-2502 from the Air Force Office of Scientific Research.

WHICH GRAPHS HAVE INTEGRAL SPECTRA?

1. INTRODUCTION

Some graphs have a spectrum consisting entirely of integers. Our purpose is to ask which graphs have this integral spectrum property and to exhibit several families of such integral graphs. To illustrate, we begin by showing in Figure 1 all integral cycles, together with their spectra. Our notation and terminology follow the book [2].

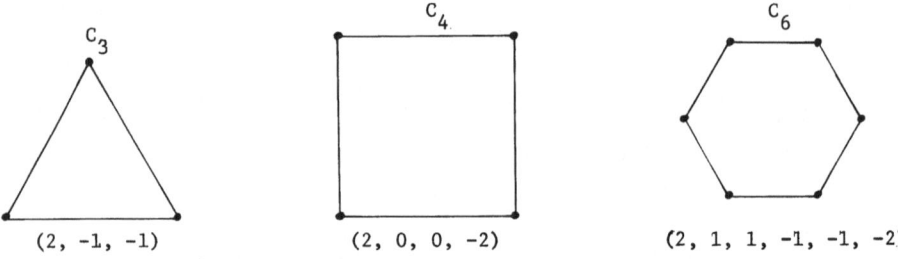

C_3

(2, -1, -1)

C_4

(2, 0, 0, -2)

C_6

(2, 1, 1, -1, -1, -2)

Figure 1. The Three Integral Cycles

That these are the only cycles C_n with integral spectrum follows at once from equation (11) of Collatz and Sinogowitz [1] which gives $2 \cos 2\pi/n$ as one of the eigenvalues.

2. OPERATIONS

This is a haphazard way of identifying integral graphs. A more systematic approach results from the use of operations on graphs. For given graphs G, G_1, G_2 , recall that $\overline{G}$ is the complement of G and that $G_1 \times G_2$, $G_1 \wedge G_2$, $G_1 * G_2$, $G_1 + G_2$ denote respectively the cartesian product, the conjunction, the strong product (which is the union of the preceding two), and the join of G_1 and G_2 . Schwenk [5] proved the following results concerning these operations.

Theorem 1. If G is regular of degree r, then the spectrum of $\overline{G}$ is

$$(1) \qquad S(\overline{G}) = \{p - 1 - r\} \bigcup \{- 1 - S(G)\} - \{- 1 - r\} .$$

We note with amusement that the minus sign is used in four different ways in (1), but we trust that the meaning will be clear by context.

Corollary 1. If a regular graph G is integral, then so is $\overline{G}$.

Example 1. Obviously $S(\overline{K}_p)$ consists entirely of zeros. By Corollary 1, every complete graph K_p is integral and it is easily verified that

(2) $$S(K_p) = (p-1, -1, -1, \ldots, -1) \ .$$

Example 2. The triangular prism, Figure 2, must be integral by Corollary 1 since its complement is the cycle C_6 (Figure 1).

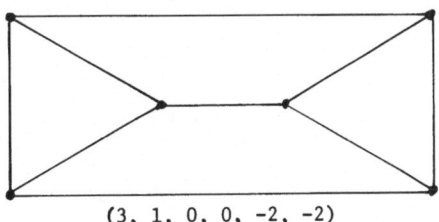

$$(3, 1, 0, 0, -2, -2)$$

Figure 2. The Triangular Prism

Theorem 2. The spectrum of the cartesian product is given by

(3) $$S(G_1 \times G_2) = S(G_1) + S(G_2) \ ,$$

which is the sum of each number in $S(G_1)$ with every number in $S(G_2)$.

Corollary 2. If G_1 and G_2 are integral, then so is $G_1 \times G_2$.

Example 3. Since the complete graphs are integral by Example 1, it follows from Corollary 2, that every cartesian product $K_m \times K_n$ is integral. In particular, the triangular prism (Figure 2) is integral since it is $K_3 \times K_2$.

Example 4. The cube Q_n is the iterated cartesian product of n copies of K_2 . Hence, it must be integral by Corollary 2 and it is easy to see that its spectrum contains the eigenvalues n, $n-2$, $n-4$, ... where $n-2k$ has multiplicity $\binom{n}{k}$.

Theorem 3. The spectrum of the conjunction is given by

(4) $$S(G_1 \wedge G_2) = S(G_1) \cdot S(G_2) \ ,$$

the product of each number in $S(G_1)$ with every number in $S(G_2)$.

Corollary 3. If G_1 and G_2 are integral, then so is $G_1 \wedge G_2$.

Example 5. Every conjunction $K_m \wedge K_n$ is integral. We have already encountered one member of this family since $C_6 = K_3 \wedge K_2$. Because $K_m \wedge K_n$ is regular and $\overline{K_m \wedge K_n} = K_m \times K_n$, we see that Examples 3 and 5 are equivalent.

<u>Theorem 4.</u> The spectrum $S(G_1 * G_2)$ of the strong product $G_1 * G_2 = G_1 \times G_2 \cup G_1 \wedge G_2$ is the sequence of values $\lambda\mu + \lambda + \mu$ with $\lambda \varepsilon S(G_1)$ and $\mu \varepsilon S(G_2)$.

<u>Corollary 4.</u> If G_1 and G_2 are integral, so is $G_1 * G_2$.

It is well known [1] that if G is regular of degree r , then r is in $S(G)$.

<u>Theorem 5.</u> If G_1 and G_2 are regular of degrees r_1 and r_2 , then the spectrum of their join is

$$(5) \quad S(G_1 + G_2) = S(G_1) \cup S(G_2) - \{r_1 , r_2\} \cup \frac{r_1 + r_2 \pm \sqrt{(r_1 - r_2)^2 + 4 \, p_1 \, p_2}}{2}$$

<u>Corollary 5.</u> The join $G_1 + G_2$ of two regular graphs is integral if and only if both G_1 and G_2 are integral and $(r_1 - r_2)^2 + 4p_1 \, p_2$ is a perfect square.

<u>Example 6.</u> A complete bipartite graph $K_{m,n}$ is integral if and only if mn is square.

<u>Example 7.</u> The join $\overline{K}_m + K_n$ is integral if and only if $(n-1)^2 + 4 \, mn$ is square, by Corollary 5. This always holds for $n = 2$ and $m = \binom{k}{2}$. In particular $\overline{K}_3 + K_2$ (Figure 3) is integral as already noted in [1].

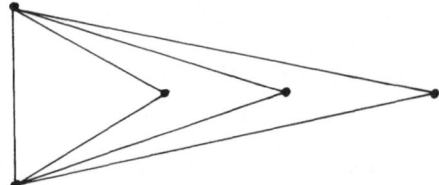

<u>Figure 3.</u> An Integral Join

<u>Theorem 6.</u> (Hoffman [3]) If G is regular of degree $r \geq 2$, then the spectrum of the line graph of G is given by

$$(6) \qquad S(L(G)) = \{(r - 2) + S(G)\} \cup (-2 , -2 , \ldots , -2) \quad ,$$

where the number of occurrences of -2 is $q - p = p(r - 2)/2$.

<u>Corollary 6.</u> If G is integral, so is its line graph.

Example 8. Every line graph $L(K_p)$ is integral. In particular, the octahedron is integral since it is $L(K_4)$. Furthermore the Pertersen graph P is seen to be integral by applying both Corollary 1 and Corollary 6 since $P = \overline{L(K_5)}$.

3. INTEGRAL TREES

There are several families of integral trees. One of these consists of trees of diameter 3. The smallest two members of this family are now displayed. For the tree T_1 of Figure 4, $S(T_1) = (2 , 1 , 0 , 0 , -1 , -2)$. Other members of this family are not easily found. The next such tree is T_2 in the figure.

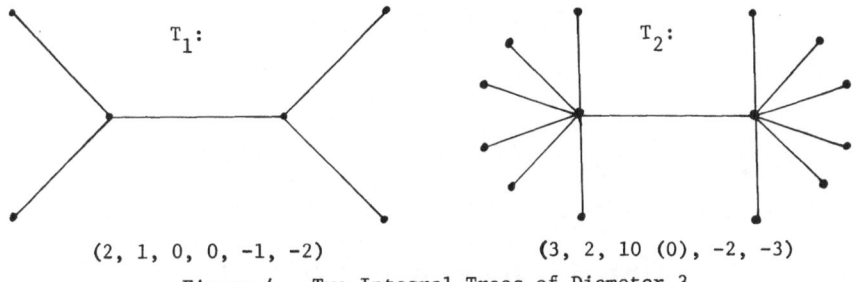

T_1: T_2:

(2, 1, 0, 0, -1, -2)   (3, 2, 10 (0), -2, -3)

Figure 4.   Two Integral Trees of Diameter 3

Another family of integral trees is provided by the subdivision graphs of certain stars $K_{1,n}$. Theorem 2 of [5] can be used to calculate the characteristic polynomial of the subdivision graph of $K_{1,n}$ and it is

$$x(x^2 - 1)^{n-1} (x^2 - 1 - n) .$$

Example 9. The subdivision graph of star $K_{1,n}$ is integral if and only if $n + 1$, the number of points of the star, is a square.

Thus the smallest member of this family is the tree of Figure 5, which by the way is also the smallest tree whose square is not hamiltonian.

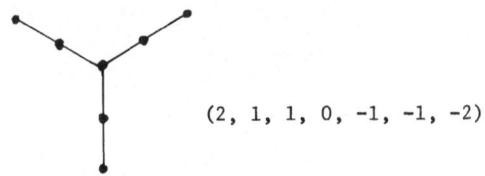

(2, 1, 1, 0, -1, -1, -2)

Figure 5.   Another Integral Tree

4. OTHER INTEGRAL GRAPHS

Finally, there are two other integral graphs [1] which we have not been able to identify as a member of a family. These are shown in Figure 6. The first graph G_1 has spectrum $S(G_1) = (3 , 1 , 1 , 0 , -1 , -2 , -2)$. The second is $L_2(K_4)$ the line graph of the subdivision graph of K_4 , as shown in [2, p. 80] .

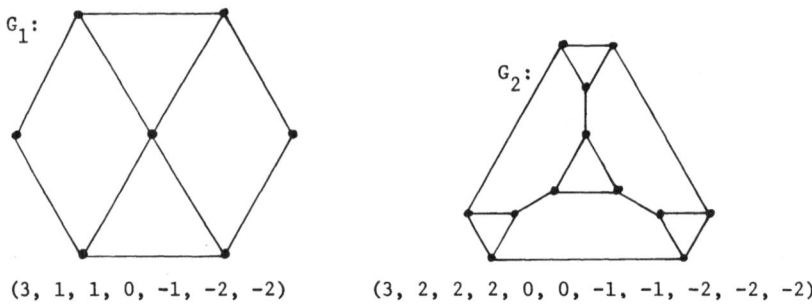

G_1:

(3, 1, 1, 0, -1, -2, -2) G_2:   (3, 2, 2, 2, 0, 0, -1, -1, -2, -2, -2)

Figure 6. Two More Integral Graphs

5. UNSOLVED PROBLEMS

1. Which graphs have zero as an eigenvalue? For a tree T , it
    is well known that $0 \in S(T)$ if and only if T has no 1-factor.
    Otherwise, the question is open.

2. Which graphs have distinct eigenvalues? The only known result
    is that the group of such a graph must be boolean, as shown by
    Mowshowitz [4] .

3. Which nonsymmetric strong digraphs D are integral? Are there
    any? It is well known that S(D) is the union of the spectra
    of its strong components.

4. Which multigraphs are integral?

REFERENCES

1. Collatz, L., and Sinogowitz, U., Spektren endlicher Graphen, <u>Abh. Math. Sem. Univ. Hamburg</u>, 21 (1957) 64-77.

2. Harary, F., <u>Graph Theory</u>, Addison-Wesley, Reading, 1969.

3. Hoffman, A. J., Some Recent Results on Spectral Properties of Graphs, <u>Beitrage zur Graphentheorie</u> (H. Sachs et al., eds.) Teubner, Leipzig (1968), 75-80.

4. Mowshowitz, A., The Group of a Graph Whose Adjacency Matrix Has All Distinct Eigenvalues, <u>Proof Techniques in Graph Theory</u>, (F. Harary, ed.), Academic Press, New York (1969), 109-111.

5. Schwenk, A. J., Computing the Characteristic Polynomial of a Graph, this volume p. 153.

GENERALIZED RAMSEY THEORY FOR GRAPHS - A SURVEY

Stefan A. Burr
Bell Laboratories
Madison, N. J.

ABSTRACT

Almost nonexistent a few years ago, the field of generalized Ramsey theory for graphs is now being pursued very actively and with remarkable success. This survey paper will emphasize the following class of problems: Given graphs $G_1,\ldots,G_c$, determine or estimate the Ramsey number $r(G_1,\ldots,G_c)$, the smallest number p such that if the lines of a complete graph K_p are c-colored in any manner, then for some j there exists a subgraph in color j which is isomorphic to G_j. Ramsey numbers have now been evaluated completely in a large number of cases, particularly when c = 2. The most strikingly general result is due to Chvátal: If T is a tree on m points, then $r(T,K_n) = mn-m-n+2$. Also of interest is the study of asymptotic questions about $r(G_1,\ldots,G_c)$. For instance, Burr, Erdős, and Spencer have shown that if G has m points, none of them isolated, and if i is the maximal number of independent points in G, then $(2m-i)n-1 \leq r(nG,nG) \leq (2m-i)n+C$, where C is a constant depending only on G.

GENERALIZED RAMSEY THEORY FOR GRAPHS - A SURVEY

1. INTRODUCTION

Almost nonexistent a few years ago, the field of generalized Ramsey theory for graphs is now being pursued very actively and with remarkable success. This work has gone in various directions. In this survey paper we will discuss results in the direction which most resembles that of classical Ramsey theory. The following definition is an obvious generalization (when $c > 2$) of that of Chvátal and Harary [16].

Definition. If $G_1,\ldots,G_c$ are graphs, define the Ramsey number $r(G_1,\ldots,G_c)$ to be the least number p such that if the lines of the complete graph K_p are colored in any fashion by c colors, than for some i the ith colored subgraph contains a G_i. Of course, K_p denotes the complete graph on p points; here, and at most other places in this paper, the notation of [38] is followed. These numbers generalize the classical Ramsey numbers for graphs, in which all the G_i are complete graphs. In the classical case, only seven non-trivial Ramsey numbers are known, six with $c = 2$ and one with $c = 3$; see [36,37]. As will be seen, very much more is already known about generalized Ramsey numbers, in spite of the relative newness of the field.

The term Ramsey number of course comes from Ramsey's theorem [49]. That generalized Ramsey numbers are finite follows immediately from the fact that the classical Ramsey numbers are finite.

Most of the work done in this field has been in the case of two colors. In this case another definition is useful.

Definition. The diagonal Ramsey numbers are given by

$$r(G) = r(G,G)$$

Except as noted, all graphs in this paper will always be assumed to have no isolated points. One consequence of this is that when a graph on n points is mentioned (as, say P_n, the path on n points), $n \geq 2$ will be assumed automatically. In addition, cycles C_n carry the automatic assumption $n \geq 3$.

The following is the plan of this paper. The second section makes some observations about the nature of the critical colorings that give the lower bounds for Ramsey numbers. The third section gives a list of what the author believes to be almost all known exact general values of Ramsey numbers. The fourth section comments on the results of the third and discusses possible directions for further work. The final section considers some asymtotic questions. A list of references to other work is also given in that section.

2. CRITICAL COLORINGS

Let us call a critical coloring for $G_1, \ldots, G_c$ one using c colors on $r(G_1, \ldots, G_c) - 1$ points and not containing a G_i in color i for any i. In other words, such a coloring sets the lower bound that determines the Ramsey number. (Of course, critical colorings are not necessarily unique.) A striking fact about the evaluations of generalized Ramsey numbers to date is that almost all involve very simple critical colorings, in contrast to the situation that appears to pertain in classical Ramsey theory.

As usual, the situation for c = 2 is clearest. Two classes of critical coloring are especially predominant when c = 2, and will therefore be called <u>canonical colorings</u>. (See Figure 1.)

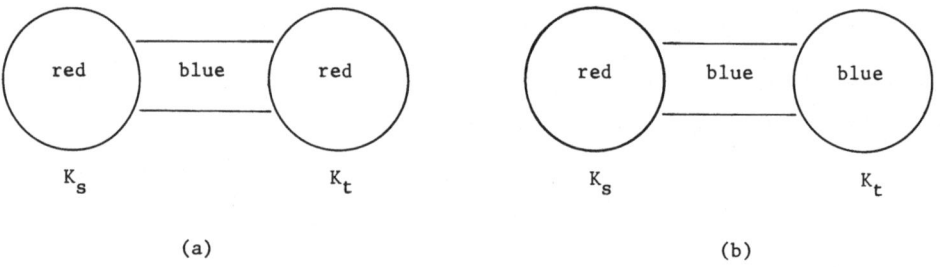

(a)   (b)

Figure 1. Canonical Colorings

In each case, the graph consists of two monochromatic complete graphs on s and t points with all lines between them blue (say). In the first case, both complete graphs are red; in the second case, one is red and the other blue. These two cases will be called Types I and II respectively. Type II is of somewhat secondary importance; in particular, it is rarely necessary when G_1 and G_2 are both connected, as will be seen. However, it often comes into its own for graphs which are not connected. Note, incidentally, that there exist trivial canonical colorings in which all lines are the same color; these may be considered to be of either type.

For any specific graph or class of graphs, it is generally very easy to determine the best lower bound the canonical colorings can provide, although it would be very tedious to state a fully general result. However, it is possible to give a reasonably economical statement of the situation in which the graphs are connected. (The following result also yields some information when only one of the graphs is connected.)

Lemma 1. Let G be a connected graph on m points, and let H be a graph on n points. Then if H is not bipartite and has point independence number $\beta_0(H) = i$,

$$(1) \qquad r(G,H) \geq 2m - 1 ,$$

and also

$$(2) \qquad r(G,H) \geq m + n - i - 1 .$$

Otherwise, let H be connected and bipartite, the two parts having k and ℓ points, with $k \geq \ell$. Then

$$(3) \qquad r(G,H) \geq 2 \min(m,k) - 1 ,$$

and also

$$(4) \qquad r(G,H) \geq m + \ell - 1 .$$

Moreover, if G and H are both connected, these results, together with the fact that $r(G,H) = r(H,G)$, yield the best lower bounds that canonical colorings can provide.

Proof. To prove (1), consider a canonical coloring of Type I, with $s = t = m - 1$. It is clear the red graph cannot contain G and the blue graph cannot contain H. To prove (2), use a coloring of Type II, with $s = m - 1$, $t = n - i - 1$. Clearly, there is no red G; on the other hand a blue H would have to use at least $i + 1$ points out of the K_s, which is impossible. Note that this argument applies even when H is bipartite; but in this case it will be shown later that relations (3) and (4) yield better results. Of more significance is the fact that (2) holds even when G is not connected.

To prove (3), use a coloring of Type I with $s = t = \min(m,k) - 1$. As before, the red graph cannot contain G; the blue graph cannot contain H, since neither of the two parts is as large as k. To prove (4), first consider a coloring of Type I with $s = m - 1$, $t = \min(m,\ell) - 1$. Again, there is no red G; there can be no blue H because $t < \ell$. Thus,

$$(5) \qquad\qquad r(G,H) \geq m + \min(m,\ell) - 1 \ ,$$

and (4) is established unless $m < \ell$. In this case, reverse the roles of G and H in (5), yielding $r(G,H) = r(H,G) > k + \ell - 1 \geq \ell + \ell - 1 > m + \ell - 1$. Therefore, (4) holds.

We now show that no better results than these can be obtained using canonical colorings, together with $r(G,H) = r(H,G)$. Because of the symmetry condition, we may assume the colors are assigned as they are in Figure 1. First consider colorings of Type I. It is clear that the formula (1) is best possible, since we must have $s < m$, $t < m$. Consequently, let the conditions pertaining to (3) and (4) be satisfied. If both s and t are at least ℓ, neither can be as large as k, for otherwise there will be a blue H; hence, in this case, (3) is the best one can do. On the other hand, if (say) $t < \ell$ we must still have $s < m$, for otherwise there will be a red G. Hence in this case (4) is the best one can do.

It remains to consider colorings of Type II. We must certainly have $s < m$. Likewise, we must have $n - t > \min(m,i)$. If $i \leq m$, we see immediately that (2) is the best we can do. If $i \geq m$, then $n - t > m > s$, so $s + t < n$, leading to a trivial canonical coloring. One final step is now necessary, namely to justify the earlier assertion that when H is bipartite, (2) although true, does not lead to a better result than (3) and (4). But we see immediately that $i = \beta_0(H) = k$, so $m + n - i - 1 = m + \ell - 1$, which is the lower bound of (4). This completes the proof.

It should be noted that relation (2) will rarely, if ever, be useful, since one can usually find some other simple coloring which does better.

Although they predominate, the canonical colorings do not exhaust the set of simple colorings useful in generalized Ramsey theory. In particular, the canonical colorings can be generalized. The simplest such generalization is like a Type I cannonical coloring except that there are more than two red complete graphs, all of the connecting lines being blue. (Note that some of the "red" graphs might consist of a single point.)

The following result, which is Lemma 4 of [19], is based on such a generalized canonical coloring. It is one of very few general lower bounds that have been given, and often yields the exact result.

Lemma 2. Let G and H be graphs (without isolates). Let χ be the chromatic number of G and let n be the number of points in a largest connected component of H. Then

$$r(G,H) \geq (\chi-1)(n-1) + 1 .$$

The generalized canonical colorings yielding this result has for its red graph $(\chi-1)K_{n-1}$, with all other lines blue. Note that when $\chi = 3$, Lemma 2 leads to relation (1) of Lemma 1.

In the case of more than two colors, generalizations of the canonical colorings have also been found useful.

3. EXACT RESULTS

This section is devoted to a list of known general exact evaluations of $r(G_1,\ldots,G_c)$. Only a few relatively trivial exact results, mainly the author's, have been intentionally omitted. The author apologizes for any inadvertent omissions. In order to make this section more useful for purposes of reference, comments will be held to a minimum. In particular, since in the majority of cases the critical colorings are canonical, mention will only be made when they are not. Some expanded comments will appear in the next section. The results of this section are divided into two parts. The first part is devoted to results that are relatively complete as they stand, the second to partial results. Each part breaks down naturally into various subparts.

We mention here, but do not dignify by stating as a theorem, the fact that if G has n points, then $r(G,K_2) = n$. It is perhaps also worth stating for the record the relation between Ramsey numbers for graphs with and without isolated points: Let H_i, $1 \leq i \leq c$, be formed from G_i by adding some (possibly no) isolates. Then $r(H_1,\ldots,H_c) = \max(r(G_1,\ldots,G_c),m)$, where m is the maximum number of points in any of the H_i.

Theorem 1. (Gerencsér - Gyárfás [34]): If $m \geq n$, then

$$r(P_m,P_n) = m + [\tfrac{n}{2}] - 1 .$$

(Recall that P_n is the path on n points, not lines.)

Theorem 2. (Rosta, Faudree - Schelp [50, 51, 29]): If $m \geq n$, n odd, and $(m,n) = \neq (3,3)$, then

$$r(C_m,C_n) = 2m - 1 .$$

If $m \geq n$, m, n even, and $(m,n) \neq (4,4)$, then

$$r(C_m,C_n) = m + \tfrac{n}{2} - 1 .$$

If $m > n$, m odd, n even, then

$$r(C_m,C_n) = \max(m + \tfrac{n}{2} - 1, 2n - 1) .$$

Finally,

$$r(C_3,C_3) = r(C_4,C_4) = 6 .$$

All these values are the same as the lower bounds given by Lemma 1, except the two special values at the end. Rosta's work and that of Faudree and Schelp are independent of each other; furthermore, each depends on partial results obtained by Chartrand and Schuster [11; see also 10] and Bondy and Erdős [3].

Theorem 3. (Faudree-Lawrence-Parsons-Schelp [28]): If $m \leq n$, m odd, then

$$r(C_m,P_n) = 2n - 1 .$$

If $m \leq n$, m even, then

$$r(C_m,P_n) = n + \tfrac{m}{2} - 1 .$$

If $m \geq n$, m odd, then

$$r(C_m,P_n) = \max(m + \lceil\tfrac{n}{2}\rceil - 1, 2n - 1) .$$

If $m \geq n$, m even, then

$$r(C_m,P_n) = m + [\tfrac{n}{2}] - 1 .$$

Theorem 4. (Burr - Roberts [8]): If $n = n_1 + \ldots, + n_k$, and j is the number of n_i that are odd, then

$$r(P_{n_1} \cup \cdots \cup P_{n_k}) = \frac{3n-j}{2} - 1 .$$

Theorem 5. (Chvátal - Harary [19]):

$$r(K_n, 2K_2) = n + 2 ;$$

if G has n points and is not complete, then

$$r(G, 2K_2) = n + 1 .$$

Theorem 6. (Chvátal - Harary [19]): Let G have n points. If $\overline{G}$ has a 1-factor, then

$$r(G, P_3) = n .$$

If $\overline{G}$ does not have a 1-factor, then

$$r(G, P_3) = 2n - 2\beta_1(\overline{G}) - 1.$$

The second half of this result does not come from a canonical coloring.

The Ramsey numbers for stars are not hard to evaluate for any number of colors, and this result will be given later in Theorem 15. An important problem that is much more difficult is that of unions of stars. V. Rosta (personal communication) has found the diagonal Ramsey numbers for unions of two stars.

Theorem 7. (Rosta): If $m \geq n$, then

$$r(K_{1,m} \cup K_{1,n}) = \max(2m + 1, m + 2n) .$$

Theorem 8. (Parsons [46]):

$$r(P_2, K_{1,n}) = n + 1 .$$

If $n \leq m \leq 2n - 1$, then

$$r(P_m, K_{1,n}) = 2n - 1 .$$

If $m \geq 2n - 1$,

$$r(P_m, K_{1,n}) = m .$$

For $n \geq m \geq 3$, the following recurrence holds:

$$r(P_m, K_{1,n}) = \max(r(P_{m-1}, K_{1,n}), r(P_m, K_{1,n-m+1}) + m - 1) .$$

The complexity of this result shows that the critical colorings cannot be canonical; but they are of the generalized canonical type.

Theorem 9. (Chvátal [14]): If T is a tree on m points, then

$$r(T,K_n) = (m-1)(n-1) + 1 .$$

A proof of this, different from Chvátal's, will be given in the next section. The
value is determined by the lower bound of Lemma 2. This striking result generalizes
earlier ones by Parsons [48] and Schwenk (given in [39]). Chvátal's result has
itself recently been generalized.

Theorem 10. (Stahl [54]): Let F be a forest $T_1 \cup \ldots \cup T_k$, where each T_i is a
tree on m_i points, and where $m_1 \leq \ldots \leq m_k$. Then

$$r(F,K_n) = \max_{1 \leq i \leq k} \left((m_i-1)(n-1) + \sum_{j=i+1}^{k} m_i \right) + 1 .$$

The critical colorings pertaining to this result are of generalized canonical type.

The next four results involve multiple copies of graphs, and are all due to
Burr, Erdös, and Spencer [7]. Only the first requires a non-canonical critical
coloring, and even that coloring comes from adding a single point to a canonical
coloring.

Theorem 11. If $m \geq n$, $m \geq 2$, then

$$r(mK_3,nK_3) = 3m + 2n .$$

The author understands that this result, or part of it, was also discovered by
Seymour at Oxford.

Theorem 12. If $m \geq n$, $m \geq 2$, then

$$r(mK_{1,3}, nK_{1,3}) = 4m + n - 1 .$$

Theorem 13. If G has m points, then

$$r(nG,nK_2) = (m+1)n - 1 .$$

Theorem 14. Let $n \geq 2$. Then

$$r(nK_m,nP_3) = (m+2)n - 1 .$$

If $G \neq K_m$ and G has m points, then

$$r(nG,nP_3) = (m+1)n - 1 .$$

The next three results all involve multiple colors. Only the last uses critical colorings that generalize the canonical colorings.

Theorem 15. (Burr - Roberts [9]): Let t be the number of m_i, $1 \leq i \leq c$, that are even. Then

$$r(K_{1,m_1}, \ldots, K_{1,m_c}) = \sum_{i=1}^{c} m_i - c + \varepsilon_t ,$$

where $\varepsilon_t = 1$ if t is even and positive, $\varepsilon_t = 2$ otherwise. The two-color case of this result has been considered by Chvátal and Harary [18] and Harary [39].

Theorem 16. (Cockayne - Lorimer [24]): If $\sum_{i=1}^{c-1} m_i \geq n$, then

$$r(K_{1,m_1}, \ldots, K_{1,m_{c-1}}, nK_2) = 2n ;$$

otherwise, letting t be the number of m_i that are even,

$$r(K_{1,m_1}, \ldots, K_{1,m_{c-1}}, nK_2) = \sum_{i=1}^{c-1} m_i + n - c + \varepsilon_t ,$$

where ε_t is as defined in Theorem 15.

Theorem 17. (Cockayne - Lorimer [25]):

$$r(m_1K_2, \ldots, m_cK_2) = \sum_{i=1}^{c} m_i + \max_{1 \leq i \leq c} m_i - c + 1 .$$

The second part of this section lists some partial results.

Theorem 18. (Lawrence [43]): If $m \geq 2n$, then

$$r(C_m, K_{1,n}) = m .$$

If $m \leq 2n + 1$, m odd, then

$$r(C_m, K_{1,n}) = 2n + 1 .$$

(The case m = 2n, not mentioned in [43], comes from personal communication.) The remaining cases of $r(C_m, K_{1,n})$ are much more difficult and appear to be intimately connected to combinatorial configurations. Only a few results are known.

Theorem 19. (Parsons [47]): If $n = q^2 + 1$, where q is a prime power, then

$$r(C_4, K_{1,n}) = q^2 + q + 2 .$$

Theorem 20. (Lawrence - Parsons; see [47]): If $n = q^2$, where q is a prime power, then

$$r(C_4, K_{1,n}) = q^2 + q + 1 \; .$$

Dealing with $r(C_m, K_n)$ is even more difficult, and only the following result is known.

Theorem 21.   (Bondy - Erdös [3]):   If $m \geq n^2 - 2$, then

$$r(C_m, K_n) = (m-1)(n-1) + 1 \; .$$

The lower bound comes about in the same way as that in Theorem 7.

Theorem 22.   (Bondy - Erdös [3]):   Let $K_{s;n}$ denote the complete n - partite graph $K_{s,\ldots,s}$, with n subscripts.   Then, for s, n fixed and m sufficiently large,

$$r(C_m, K_{s;n}) = (m-1)(n-1) + s \; .$$

The critical colorings are of generalized canonical type.   It is possible to give an explicit bound on how big m has to be for the formula to be true.

Theorem 23.   If T is a tree on m points, and $(m - 1) \mid (n - 1)$, then

$$r(T, K_{1,n}) = m + n - 1 \; .$$

A proof of this based on ideas in Parsons' proof of Theorem 8, will be given in the next section.   The critical colorings are of generalized canonical type. E. J. Cockayne has recently extended Theorem 23, also using ideas in Parson's proof.   The full result will not be given, since it is rather technical.

Theorem 24.   (Cockayne [23]):   Let T be a tree on m points which has a point of degree one adjacent to a point of degree two.   Then

$$r(T, K_{1,n}) = m + n - 2 \; ,$$

provided that one of the following four conditions holds:

$$m \equiv 0, 2 \pmod{m-1},$$

$$n \not\equiv 1 \pmod{m-1} \text{ and } n \geq (m-3)^2 \; ,$$

$$n \not\equiv 1 \pmod{m-1} \text{ and } n \equiv 1 \pmod{m-2} \; ,$$

or

$$n \equiv -1 \pmod{m-1} \text{ and } n > m-2.$$

The critical colorings are of generalized canonical type.   The statement of the full theorem involves a more complicated condition on T.   However, almost all trees satisfy even the stronger condition of Theorem 24, as can be seen from work of Schwenk [53].

The next two results, announced by Faudree and and Schelp in [32] and proved in [31], involve multiple colors. The critical colorings are generalizations of canonical colorings.

__Theorem 25.__ If $n_1 \geq 2 \left(\sum_{i=2}^{c} n_i \right)^2$ and $\delta = 0$ or 1, then

$$r(P_{n_1}, P_{2n_2+\delta}, P_{2n_3}, \ldots, P_{2n_c}) = \sum_{i=1}^{c} n_i - c + 1.$$

__Theorem 26.__ If $\ell \geq \frac{7}{2} (m+n)^2$, then

$$r(P_\ell, P_m, P_n) = \ell + \left[\frac{m}{2} \right] + \left[\frac{n}{2} \right] - 2 .$$

It should be noted that Faudree and Schelp [30] have solved, and used in the proof of Theorem 27, a problem of great interest, although one outside the scope of this paper. We do not state their result here; but it can be said that they have determined precisely for what p_1, p_2 a two-colored complete bipartite graph K_{p_1, p_2} must contain a red P_m or a blue P_n.

4. SOME COMMENTS

This section discusses some matters involving exact results. We begin by mentioning the existence of three collections of special values. Such values are useful for suggesting theorems, finding counterexamples, starting inductions, etc. In [19], Chvátal and Harary evaluate $r(F,G)$ for all pairs F, G with no more than four points each. (In their preceding paper [18] they considered the diagonal case only.) In [4], Burr evaluates $r(G)$ for all graphs G with no more than six lines. In [15], Chvátal and Clancy evaluate $r(F,G)$ for most (but not all) pairs F, G with F having no more than five points and G having no more than four.

We now turn to the two proofs promised in the previous section. Both proofs spring from similar ideas; both theorems also follow easily from results in [6]. __Proof of Theorem 9.__ That $r(T,K_n) \geq (m-1)(n-1) + 1$ comes immediately from Lemma 2. We now must prove the inequality in the other direction. This is trivial when m or n is 2; the rest will be proved by induction on $m + n$. Assume the theorem to have been proved for all $m' + n' < m + n$ and consider any two-colored complete graph on $(m-1)(n-1) + 1$ points. Let T' be a tree formed from T by the removal of an endpoint. By the induction hypothesis there exists either a red T' or a blue K_n, so we may assume there is a red T'. Remove the $m - 1$ points of this T', leaving a two-colored complete graph on $(m-1)(n-2) + 1$ points. By the induction hypothesis, this remaining graph contains either a red T or a blue K_{n-1}, and we may assume the latter.

Consequently, we have a red T' and a blue K_{n-1} disjoint from it. Some point of T' has the property that if any line leading from it to the blue K_{n-1} is red, the original graph contains a red T. Hence, all lines from this point to the K_{n-1} can be assumed to be blue, giving a blue K_n. This completes the proof.

Proof of Theorem 23. We first produce a generalized canonical coloring yielding $r(T, K_{1,n}) \geq m + n - 1$. Since $(m - 1) | (n - 1), k = \frac{n-1}{m-1}$ is an integer. The coloring to be considered has for its red graph $(k+1)K_{m-1}$, which has $n - 1 + m - 1 = m + n - 2$ points, and contains no T. Every point in the complementary (blue) graph has degree $k(m-1) = n - 1$, so there is no blue $K_{1,n}$. Thus $r(T, K_{1,n}) > m + n - 2$.

The proof that $m + n - 1$ is an upper bound is by induction on m. The result is trivial for $m = 2$. Now assume the result to have been proved for some $m - 1$, $m \geq 3$, and consider any two-colored complete graph on $m + n - 1$ points. Let T' be a tree formed from T by the removal of an endpoint. By the induction hypothesis we may assume that the graph contains a red T'. Now, that T' contains some point with the property that if any line from that point to the n points not in the T' is red, the graph contains a red T. So we may assume all these lines are blue; but this gives a blue $K_{1,n}$, completing the proof.

Note that we have proved that $m + n - 1$ is an upper bound for $r(T, K_{1,n})$ for any m and n.

It is interesting to consider the proof techniques used in the theorems of Section 3. It is possible to be fairly systematic about the critical colorings. Of the twenty-one theorems involving two colors, ten use canonical colorings exclusively (except Theorem 2, which uses non-canonical colorings for two isolated cases). Of the eleven others, eight use generalized canonical colorings, often using Lemma 2. Two others (Theorems 6 and 11) use quite simple non-canonical colorings. Only Theorems 19 and 20 use complicated colorings, involving such structures as Moore graphs and block designs. Of the five theorems involving multiple colors, three use critical colorings that can be said to be of general-ized canonical type. The remaining two (Theorems 15 and 16) use rather simple colorings involving partitions into regular graphs. It should also be noted that a substantial majority of the results in the collection [19] comes from either canonical colorings or Lemma 2, and that likewise a substantial majority of the results in the collection [4] come just from canonical colorings. Many of the results in the collection [15] also come from canonical colorings.

It is harder to speak systematically about the techniques used to prove that the values in the theorems in Section 3 are upper bounds. A few general approaches can be identified. One is based on the idea that if one has a red (say) path or cycle, either that path or cycle can be lengthened or there exist enough blue lines to force an appropriate blue path or cycle. This idea figures in the proofs of Theorems 1, 2 and several others. In view of the obvious connection, it is perhaps surprising that extremal theory is not used more heavily than it is; but it is used fairly frequently. See for instance Faudree and Schelp's proof of Theorem 2 and Chvátal's proof of Theorem 9, as well as the proofs of Theorems 21 and 22. There is a fairly coherent principle behind the proofs of Theorems 11 - 14 on multiple copies, which is also seen in Theorem 4, namely that if one can find a small graph containing both a red G and a blue H, one can apply induction. Counting arguments are sometimes useful, as in Theorems 15, 19, and 20. A final method is that used here in the proofs of Theorems 9 and 23, which also has been used in the proofs of Theorems 8 and 10.

There are many attractive directions for future work. Each partial result in this section suggests such a direction. Two such are of particular interest, namely to find further values for $r(C_m, K_{1,n})$ and $r(C_m, K_n)$. Limited progress can probably be made but complete solutions seem extremely difficult, especially for the second problem. (If m = 3, the second problem becomes one in classical Ramsey theory.)

Another potentially valuable direction in which to work would be to try to understand why the canonical colorings are so powerful in determining the values of Ramsey numbers. The insight so gained might lead to the establishment of rather general conditions for the Ramsey number of a graph to be determined by a canonical coloring.

A very important problem is the determination, or even accurate estimation, of Ramsey numbers for star forests. The problem is surprisingly difficult in view of the ease of determining Ramsey numbers for stars. Theorem 7 solves the problem in the diagonal case for unions of two stars, but even that result is not easy to prove. Very little is known for more than two stars.

It should certainly be possible to determine $r(mG,nH)$ in more cases than Theorems 11 - 14. In addition, it seems likely that one could evaluate $r(G,K_{1,n})$ or $r(G,P_n)$ for general classes of G.

It would be very interesting, though probably difficult, to evaluate $r(T)$ for T a tree. Very likely, Lemma 1 determines $r(T)$ when T does not contain a large star (in which case the Ramsey number would be the Ramsey number of the star). Theorem 8 demonstrates that the off-diagonal case must be much more com-plicated.

It is likely that there remain interesting and tractable problems involving multiple colors, beyond the five results in Section 3.

Finally, it would be valuable to extend the collections of special values in [4, 15, 19].

5. ASYMPTOTIC PROBLEMS

Although not as much has been done on asymptotic questions on generalized Ramsey theory as on exact questions, a few results have been obtained. The subject is deserving of attention for itself; but furthermore it has in several instances illuminated exact problems as well. This section considers four types of problem: multiple copies, multiple colors, sequences of graphs whose Ramsey numbers grow linearly in the number of points, and what might be called extremal Ramsey theory.

We have already encountered Theorems 11 - 14, taken from [7], giving some exact results involving multiple copies. The same paper considers the corresponding asymptotic questions as well. It is possible to obtain a surprisingly sharp result.

<u>Theorem 27.</u> Let $p(G) = k$, $p(H) = \ell$, $\beta_o(G) = i$, $\beta_o(H) = j$. Then

$$km + \ell n - \min(mi,nj) - 1$$

$$\leq r(mG,nH)$$

$$\leq km + \ell n - \min(mi,nj) + C ,$$

where C depends only on G and H.

Although an explicit bound for C can be given, it is certainly much too large. It may be possible to give a better one. The lower bound in the theorem comes from canonical colorings of Type II, and indeed from (2) of Lemma 1, which, as noted, applies to arbitrary G and H. The ideas of [7] can certainly be applied to other cases than those of Section 3 to yield exact results. When m = n, it is possible to improve Theorem 27, although in a non-constructive way.

<u>Theorem 28.</u> Under the conditions of Theorem 27, there exist n_0 and C_0 such that, whenever $n \geq n_0$,

$$r(nG,nH) = (k + \ell - \min(i,j)) \, n + C_0 .$$

It has not been possible to evaluate C_0 in general, nor even to obtain a bound for n_0. Nevertheless, it does not seem altogether beyond hope to do either or both.

Another class of problems on which some progress has been made are those in which the number of colors is large. Set $r(G;c) = r(G,\ldots,G)$ with c arguments. One very general lower bound is of interest here.

<u>Theorem 29.</u> (Chvátal - Harary [17]): Let G have p points and q lines, and let s be the number of symmetries of G. Then

$$r(G;c) > (sc^{q-1})^{1/p} .$$

The proof is probabilistic, and hence non-constructive; it is of great value when q is large relative to p.

The other results on r(G;c) discussed here apply to fairly special situations, in which q is relatively small. The next five results are due to Erdös and Graham [27]. It should be noted that the results are not presented in the same form as in that paper; various theorems and remarks have been gathered into the five results stated below.

Theorem 30. If T is a tree on n points, then

$$(c-1) \ [\tfrac{n-1}{2}] < r(T;c) \le 2cn + 1 \ .$$

Actually, Erdös and Graham prove a better lower bound than that above, which for fixed n is asymptotically equal to cn. (In fact, the c-coloring they exhibit has no monochromatic connected graph on n points.) In fact, it seems likely that r(T;c) $\sim$ cn as c $\to \infty$.

The situation for forests is rather more complicated, and is less well understood.

Theorem 31. If F is forest on n lines (not points), then

$$\frac{c(\sqrt{n} - 1)}{2} \ < r(F;c) < 4cn \ .$$

Moreover, if c $\le$ n^2, then

$$r(F;c) > A \ \sqrt{cn} \ ,$$

where A is a positive universal constant.

Thus for every forest F, r(F;c) displays essentially linear growth in c for large c, as is the case with trees throughout the whole range of c. However, the second formula holds open the possibility that r(F;c) grow as $\sqrt{c}$ for small c. The next theorem shows that this is indeed the case for the forest $mK_{1,m}$, as well as giving rather sharp upper bound when c is large.

Theorem 32. There is a constant A_1 such that, if c $\le$ m, then

$$r(mK_{1,m};c) < A_1 \ \sqrt{c} \ m^2 \ .$$

Also, if c $\ge$ 3m^2, then

$$r(mK_{1,m};c) \le 3cm \ .$$

Since $mK_{1,m}$ has m^2 lines, we see that the upper bounds for this forest come within constant factors of their lower bounds, except conceivably for some middle range of c. The next theorem shows that the phenomenon of having two regimes of behavior applies also to even cycles.

Theorem 33. For any c and n,

$$r(C_{2n};c) > (c-1)(n-1) .$$

Furthermore, if $c \leq 10^n/201n$, then

$$r(C_{2n};c) \leq 201cn .$$

Finally, there exists on A > 0 and a positive function B such that for any $\varepsilon > 0$,

$$Ac^{1+\frac{1}{2n}} < r(C_{2n};c) < B(n,\varepsilon)c^{1+\frac{1+\varepsilon}{n-1}}$$

The import of this result is that $r(C_{2n};c)$ grows linearly in c out to a rather large value, and then grows roughly like some power > 1 of c.

Much more is known about $r(C_4;c)$, but first we consider the question of odd cycles. The behavior of odd cycles is very different, and the problem seems much less tractable.

Theorem 34. $2^c \cdot n < r(C_{2n+1};c) < 2(c+2)! \ n$.

Thus the growth of $r(C_{2n+1};c)$ is very rapid. In [27] it is also shown that $r(C_{m+1};c) < Ac^3 nr^2 (C_3;k)$ for some A. An old question is whether or not $r(C_3;c) < A^c$ for some A. Erdős has suggested that it might be easier to show that $r(C_{2n+1};c) < A^c$ for some 2n + 1 > 3.

We now return to the special case C_4 of even cycles. Rather sharp results are known.

Theorem 35. (Chung [12]): $r(C_4;c) \leq c^2 + c + 1$.

In [12], Chung also considers the classical Ramsey numbers $r(C_3;c)$, but this is beyond the scope of this survey.

Theorem 36. (Chung – Graham [13]): If c–1 is a power of a prime, then $r(C_4;c) \geq c^2 - c + 2$.

These two results taken together yield immediately that $r(C_4;c) \sim c^2$ as c → ∞. In fact, this statement can be greatly generalized.

Theorem 37. (Chung – Graham [12]): $r(K_{2,n};c) \sim (n-1)c$ as $c \to \infty$.

Chung and Graham also apply their method to estimating $r(K_{m,n};c)$ in general. Their work is not completed at present, so it is not certain how sharp the estimates will be.

In order to introduce the next topic we make two definitions. We first define an L-set $\{G_1,G_2,\ldots\}$ of graphs to be one for which there is a constant C such that

$$r(G_i) \leq C \cdot p(G_i)$$

for all i. We also define the edge-density $\rho(G)$ of a graph G by

$$\rho(G) = \max_{H \subseteq G} \frac{q(H)}{p(H)} .$$

Thus $\rho(G)$ is approximately the arboricity of G. We can now state the following conjecture.

Conjecture. (Erdős): Any set of graphs with bounded edge-density is an L-set.

This may be stated in alternative form: There is a function f such that $r(G) \leq p(G)f(\rho(G))$. This statement seems less appealing than the first. Although the conjecture is a long way from being settled, some rather suggestive results are known. Any sequence of trees is an L-set, as can be seen, for instance, from Theorem 30. The following two results are due to Burr and Erdős [6].

Theorem 38. If $\{G_i\}$ is an L-set of graphs (with isolates permitted) with bounded edge-density, so is the set $\{G_i + K_1\}$.

Thus, wheels form an L-set, for instance. It is useful to permit the G_i to have isolates in this one case because the set of possible $G_i + K_1$ is thereby expanded. The ideas used in the proof of Theorem 38 lead directly to the proofs in this paper of Theorems 9 and 23, which are exact results.

Theorem 39. For any k, $\left\{C_i^k\right\}$ is an L-set.

Theorem 38 and 39, together with the trivial fact that $F \subseteq G$ implies $r(F) \leq r(G)$, enable one to show that a very great many sets of graphs are L-sets. An interesting problem would be to characterize those sets that can be treated in this way.

Lest one be led to think that Erdős' conjecture tells the whole story, it is observed in [6] that $\left\{4^i K_i\right\}$ is an L-set (as is very easy to check); but the set does not have bounded edge-density. In fact, $\rho\left(4^i K_i\right) = (i-1)/2$. A somewhat similar example can be given which is connected.

Note that there is a constant C such that if $G_i = 4^i K_i$, then $\rho(G_i) \leq C \log p(G_i)$. This is a necessary condition for an L-set, since Theorem 29 shows that for any L-set, $\rho(G_i)/\log p(G_i)$ must be bounded. This fact has led Erdös to ask whether the set of cubes $\{Q_i\}$ is an L-set, since this set also (barely) satisfies the necessary condition. The necessary condition is certainly not sufficient, since Lemma 2 shows that the set $\left\{4^i K_i + K_1\right\}$ is not an L-set.

Another topic of interest, closely related to the previous one, is the following question: For the set of graphs S satisfying some property, what is the minimum of $r(G)$ for $G \in S$? One can, of course, also ask for maxima, and ask questions about off-diagonal numbers. It seems natural to call this topic extremal Ramsey theory, even at the risk of clouding the issue of the connection between extremal theory and Ramsey theory. One such question, posed by Erdös, is: Of all graphs with chromatic number χ, does K_χ have the smallest Ramsey number? One might mean "smallest" χ in the weak or strong sense. A test case currently being considered is whether or not the 5-spoked wheel $C_5 + K_1$ has a Ramsey number greater than $r(K_4) = 18$. Chvátal and Schwenk [20] have shown that $17 \leq r(C_5 + K_1) \leq 21$.

In [5], Burr and Erdös consider several extremal questions, and in particular give the next three results.

Theorem 40. If G is connected, not bipartite, and has n points, then
$$r(G) \geq 2n - 1 \ .$$
Also, if $n \geq 4$, there is such a G for which $r(G) = 2n - 1$.

This exact result is out of place in this section; however, the two other extremal Ramsey results given here are asymptotic in character, so Theorem 40 is included here. The extremal graphs consist of either odd cycles, or odd cycles to which one point and one line have been added.

Theorem 41. If G is connected and has n points, then
$$r(G) \geq 2 \left\lceil \frac{2n+2}{3} \right\rceil - 1.$$
Also, there is such a G for which $r(G) \leq \frac{4}{3} n + 4$.

The upper bound can be improved, but it has not been possible to close the gap altogether. It is conjectured that the lower bound is sharp. The upper bound comes from graphs consisting of a path of length roughly 2n/3, connected to a star of order roughly n/3.

<u>Theorem 42</u>. There exists a C_o such that if G has n points (and no isolates), then

$$r(G) \geq n + \log_2 n - C_o \log \log n .$$

Also, there is such a G and a C_1 for which $r(G) \geq n + C_1 \sqrt{n}$.

Here the gap is even larger. It is conjectured that the lower bound tells essentially the true story. It is clear that it is only necessary to consider G that are star forests. The upper bound is given essentially by graphs of the form $mK_{1,m}$. The lower bound is given roughly by

$$K_{1,[n/2]} \cup K_{1,[n/4]} \cup K_{1,[n/8]} \cup \cdots \cup K_{1,1} .$$

In [5] consideration is also given to off-diagonal versions of extremal Ramsey theory. As one simple example, note that Lemma 2, taken together with Theorem 9, constitutes such an extremal result.

This paper has considered the evaluation and estimation of the numbers $r(G_1,\ldots,G_c)$. However, there are many other questions which generalize classical Ramsey theory for graphs; these questions have been studied in numerous papers, some of them already mentioned. As a partial listing of such papers, see [1, 2, 7, 16, 17, 21, 22, 26, 30, 33, 34, 35, 39, 40, 41, 42, 44, 45, 47, 55].

REFERENCES

1. Andrásfai, B., Remarks on a Paper of Gerencsér and Gyárfás, <u>Ann. Univ. Sci</u>. <u>Budapest. Eötvos Sect. Math. 13</u> (1970), 103-107.

2. Bermond, J. C., Some Ramsey Numbers for Directed Graphs, to appear.

3. Bondy, J. A. and Erdös, P., Ramsey Numbers for Cycles in Graphs, <u>J. Combina-torial Theory</u> 14B (1973), 46-54.

4. Burr, S. A., Diagonal Ramsey Numbers for Small Graphs, to appear.

5. Burr, S. A. and Erdös, P., Extremal Ramsey Theory for Graphs, to appear.

6. Burr, S. A. and Erdös, P., On the Magnitude of Generalized Ramsey Numbers for Graphs, Proc. of the International Symposium on Infinite and Finite Sets, Keszthély, Hungary, to appear.

7. Burr, S. A., Erdös, P. and Spencer, J. H., Ramsey Theorems for Multiple Copies of Graphs, to appear.

8. Burr, S. A. and Roberts, J. A., On Ramsey Numbers for Linear Forests, <u>Discrete Math</u>., to appear.

9. Burr, S. A. and Roberts, J. A., On Ramsey Numbers for Stars, <u>Utilitas Math</u>., to appear.

10. Chartrand, G. and Schuster, S., On a Variation of the Ramsey Number, <u>Trans</u>. <u>Amer. Math. Soc</u>. 173 (1972), 353-362.

11. Chartrand, G. and Schuster, S., On the Existence of Specified Cycles in Complementary Graphs, <u>Bull. Amer. Math. Soc</u>. 77 (1971), 995-998.

12. Chung, F. R. K., On Triangular and Cyclic Ramsey Numbers with k Colors, this volume p. 236.

13. Chung, F. R. K. and Graham, R. L., Multicolor Ramsey Theorems for Complete Bipartite Graphs, to appear.

14. Chvátal, V., On the Ramsey Numbers $r(K_m, T)$, to appear.

15. Chvátal, V. and Clancy, M. C., Some Small Ramsey Numbers, to appear.

16. Chvátal, V. and Harary, F., Generalized Ramsey Theory for Graphs, <u>Bull. Amer</u>. <u>Math. Soc</u>. 78 (1972), 423-426.

17. Chvátal, V. and Harary, F., Generalized Ramsey Theory for Graphs I, Diagonal Numbers, <u>Periodica Math. Hungar</u>. 3 (1973), 113-122.

18. Chvátal, V. and Harary, F., Generalized Ramsey Theory for Graphs II, Small Diagonal Numbers, <u>Proc. Amer. Math. Soc</u>. 32 (1972), 389-394.

19. Chvátal, V. and Harary, F., Generalized Ramsey Theory for Graphs, III. Small Off-diagonal Numbers, <u>Pacific J. Math</u>. 41 (1972), 335-345.

20. Chvátal, V. and Schwenk, A. J., On the Ramsey Number of the Five-spoked Wheel, this volume p. 247.

21. Cockayne, E. J., An Application of Ramsey's Theorem, <u>Canad. Math. Bull</u>. 13 (1970), 145-146.

22. Cockayne, E. J. Colour Classes for r-graphs, Canad. Math. Bull. 15 (1972) 349-354.

23. Cockayne, E. J., Some Tree-star Ramsey Numbers, J. Combinatorial Theory, to appear.

24. Cockayne, E. J. and Lorimer, P. J., The Ramsey Numbers $r(S_k,f)$ Where f is a Forest, Canad. Math. Bull., to appear.

25. Cockayne, E. J. and Lorimer, P. J., The Ramsey Graph Number for Stripes, J. Australian Math. Soc., to appear.

26. Erdös, P., Graph Theory and Probability, Canad. J. Math. 11 (1959), 34-38.

27. Erdös, P. and Graham, R. L., On Partition Theorems for Finite Graphs, Proceeding of the International Symposium on Infinite and Finite Sets, Keszthély, Hungary, to appear.

28. Faudree, R. J., Lawrence, S. L., Parsons, T. D. and Schelp, R. H., Path-cycle Ramsey Numbers, Discrete Math. to appear.

29. Faudree, R. J. and Schelp, R. H., All Ramsey Numbers for Cycles in Graphs, Discrete Math., to appear.

30. Faudree, R. J. and Schelp, R. H., Path-path Ramsey Type Numbers for the Complete Bipartite Graph, to appear.

31. Faudree, R. J. and Schelp, R. H., Path Ramsey Numbers in Multicolorings, to appear.

32. Faudree, R. J. and Schelp, R. H., Ramsey Type Results, Proceedings of the International Symposium on Infinite and Finite Sets, Keszthély, Hungary, to appear.

33. Folkman, J., Graphs with Monochromatic Complete Subgraphs in Every Edge Coloring, SIAM J. Appl. Math. 18 (1970), 19-24.

34. Gerencsér, L. and Gyárfás, A., On Ramsey-type Problems, Ann. Univ. Sci. Budapest. Eötvös Sect. Math. 10 (1967), 167-170.

35. Graham, R. L. and Spencer, J. H., On Small Graphs with Forced Monochromatic Triangles, Recent Trends in Graph Theory, Springer-Verlag, Berlin 1971, 137-141.

36. Graver, J. E. and Yackel, J., Some Graph Theoretic Results Associated with Ramsey's Theorem, J. Combinatorial Theory 4 (1968), 125-175.

37. Greenwood, R. E. and Gleason, A. M., Combinatorial Relations and Chromatic Graphs, Canad. J. Math. 7 (1955), 1-7.

38. Harary, F., Graph Theory, Addison-Wesley, Reading, Mass., 1969.

39. Harary, F., Recent Results on Generalized Ramsey Theory for Graphs, Graph Theory and Applications (Y. Alavi, D. R. Lick, and A. T. White, eds.), Springer-Verlag, Berlin 1972, 125-138.

40. Harary, F. and Hell, P., Generalized Ramsey Theory for Graphs V. The Ramsey Number of a Digraph, Bull. London Math. Soc., to appear.

41. Harary, F. and Prins, G., Generalized Ramsey Theory for Graphs IV, the Ramsey Multiplicity of a Graph, Networks, to appear.

42. Irving, R. W., On a Bound of Graham and Spencer for a Graph-colouring Constant, to appear.

43. Lawrence, S. L., Cycle-star Ramsey Numbers, <u>Notices Amer. Math. Soc</u>. 20 (1973), A-420.

44. Lin, S., On Ramsey Numbers and K_r-coloring of Graphs, <u>J. Combinatorial Theory</u> 12B (1972), 82-92.

45. Moon, J. W., Disjoint Triangles to Chromatic Graphs, <u>Math. Mag</u>. 39 (1966), 259-261.

46. Parsons, T. D., Path-star Ramsey Numbers, <u>J. Combinatorial Theory</u> B, to appear.

47. Parsons, T. D., Ramsey Graphs and Block Designs, I, to appear.

48. Parsons, T. D., The Ramsey Numbers $r(P_m, K_n)$, <u>Discrete Math</u>., 6 (1973), 159-162.

49. Ramsey, F. P., On a Problem of Formal Logic, <u>Proc. London Math. Soc</u>. 30 (1930), 264-286.

50. Rosta, V., On a Ramsey Type Problem of J. A. Bondy and P. Erdös. I, <u>J. Combinatorial Theory</u> 15B (1973), 94-104.

51. Rosta, V., On a Ramsey Type Problem of J. A.Bondy and P. Erdös, II, <u>J. Combinatorial Theory</u> 15B (1973), 105-120.

52. Schwenk, A. J., Acquaintance Graph Party Problem, <u>Amer. Math. Monthly</u>, 79 (1972), 1113-1117.

53. Schwenk, A. J., Almost All Trees are Cospectral, <u>New Directions in the Theory of Graphs</u> (F. Harary, ed.),Academic Press, New York 1973, 275-301.

54. Stahl, S., On the Ramsey Numbers $r(F, K_n)$, Where F is a Forest, to appear.

SOME RECENT RESULTS IN TOPOLOGICAL GRAPH THEORY

Paul C. Kainen
Case Western Reserve University

ABSTRACT

This paper examines a number of recent results in topological graph theory. Invariants such as genus, thickness, skewness, crossing number, and local crossing number are introduced and related to one another. We then deal with topological techniques in the theory of chromatic numbers, and state a very ambitious meta-conjecture which is quite useful in generating true theorems. In closing, we attempt to suggest appropriate directions for further research in topological graph theory, and we give a few results to indicate the richness of the terrain.

AN INTRODUCTION TO TOPOLOGICAL GRAPH THEORY

INTRODUCTION

I shall try in this paper to collect some of the basic concepts and results of topological graph theory. Obviously, this is a somewhat subjective matter, and I make no claims for the completeness of this survey. Rather, I have tried to concentrate on several key results - notably, Euler's Formula and Heawood's Theorem on surfaces other than the plane - which may serve to give a feeling for the subject.

Emphasis is placed on ideas and techniques which are not yet readily accessible in the literature. For example, we omit any mention of the theorems of Kuratowski, Whitney and MacLane which characterize planarity. On the other hand, we outline in some detail the techniques of piecewise - linear embeddings, and discuss the results of Youngs on 2-cell embeddings.

It is assumed that the reader has some knowledge of graph theory (for undefined terms, see Harary's book [33]) and also of elementary pointset topology although this latter stipulation may be removed provided that one is willing to take certain assertions in Section 1 on faith. Our primary goals are to provide information on and to stimulate interest in topological graph theory.

The outline of the paper is as follows. In Section 1, we describe the various orientable and non-orientable surfaces and introduce the idea of embedding a graph in a surface. We discuss proper, or 2-cell, embeddings and the work of Youngs. An elementary but complete proof of Euler's Formula for the plane is given and a necessary condition for proper embeddability is derived. We also show how to realize a graph in 3-space, R^3, and define maximal graphs and triangulations in a surface.

Section 2 contains the definitions of genus, maximum genus, and upper embeddability and gives results on genus and upper embeddability for complete graphs and other important classes. We prove Ringel's result that genus is not increased by contracting a connected subgraph, and we give a detailed treatment of piecewise -

linear embeddings.

In Section 3, a topological proof of the 5 color theorem for planar graphs is generalized to yield some new results. We also show how to obtain several disparate coloring results from a tantalizing-but-false improvement on the Szekeres-Wilf bound on chromatic numbers.

In Section 4, we define the chromatic number $X(S)$ and Heawood number $H(S)$ of a surface S and give a simple proof that $X(S) \leq H(S)$ for $S \neq S_0$, the sphere. As a consequence of our method, we are able to prove part of a result of Dirac that, for certain surfaces S, if G can be embedded in S and $X(G) = X(S)$, then G contains $K_{H(S)}$ as a subgraph.

Section 5 includes a discussion of thickness and results for the standard families and, in addition, it explores the chromatic numbers of graphs with specified thickness. Some variants of thickness, such as book thickness and point-arboricity, are also dealt with.

Section 6 deals with pinched manifolds, or pseudomanifolds, which are obtained from surfaces by introducing certain 0-dimensional singularities. Coloring problems involving thickness, as well as the celebrated empire coloring problem, can be cast in this setting. Further, these results are seen to suggest that the "obstruction" to a geometric coloring problem is always a complete graph.

Finally, in Section 7, we treat crossing numbers and some other topological invariants and obtain some elementary relations among them.

1. EULER'S FORMULA

A closed (open) surface S is a compact (non-compact) topological space in which every point has a neighborhood homeomorphic to the Euclidean plane. Thus, S is just a 2-dimensional closed (open) manifold. We also call a closed surface a surface. If S is a closed surface, then, by the classification of 2-manifolds (see Cairns [15, pp. 15-38]), S consists of a sphere to which handles or crosscaps are attached, where a handle is simply a cylinder minus its ends and a crosscap is a Möbius band which we imagine to be deformed so that its bounding circle is untwisted. The deformation must take place in 4-dimensions since a crosscap cannot be realized in 3-dimensional space without self-intersections. (See Figure 1) To attach a handle to a sphere or any other surface S, remove the interiors of two disjoint closed discs and identify the two bounding circles of a handle with the boundaries of the closed discs. To attach a crosscap, remove the interior of one closed disc and identify the bounding circle of the crosscap with the boundary of the closed circle. (See Figure 2) The result of attaching a handle to the sphere is called the <u>torus</u>; we denote by S_k the result of attaching k handles (k $\geq$ 0) to the sphere. Attaching a crosscap to the sphere yields the <u>projective plane</u> while attaching two crosscaps yields the <u>Klein bottle</u>.

With any closed surface S we may associate an integer e(S) $\leq$ 2, called the <u>Euler characteristic</u> of S, which is given by the formula

$$e(S_k) = 2 - 2k,$$
$$e(S_k') = 2 - 2k,$$
$$e(U_k) = 3 - 2k.$$

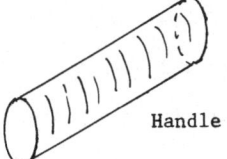

Handle

Cross-cap

<u>Figure 1</u>

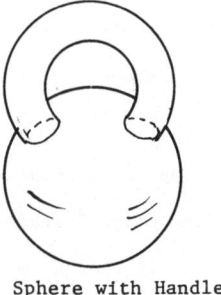

Sphere with Handle
(a torus)

Sphere with Cross-cap
(a projective plane)

<u>Figure 2</u>

where S_k' is obtained from S_{k-1} by attaching two crosscaps and U_k is obtained from S_{k-1} by attaching one crosscap. (So S_1' is the Klein bottle and U_1 is the projective plane.) It can be shown [15] that every surface S can be written uniquely in the form S_k, S_k', or U_k, for some k, and so e(S) is well-defined. One can also use the homology groups of S to define e(S).

Let G be a graph. We do not allow loops or multiple edges, and we require a finite number of vertices. V(G) and E(G) denote, respectively, the sets of vertices and edges. If $|S|$ is the cardinality of a set S, a graph G is said to have order n if $|V(G)| = n$.

Any graph G can be realized as a collection of points and straight-line segments in 3-dimensional euclidean space - that is, we can find a set P of points and L of non-intersecting open straight-line segments such that the endpoints of any element ℓ in L belong to P and there are 1-1 correspondences between V(G) and P and between E(G) and L such that p and q are the endpoints of ℓ if and only if v and w are the vertices joined by e, where p corresponds to v, q to w, and ℓ to e. To obtain the realization, choose a set P of points in 3-space such that P and V(G) are in 1-1 correspondence and such that P is in <u>general position</u>. This means that no 3 points lie on the same line and no 4 points on the same plane. Note that a singleton is in general position and, having a set S with n-1 points in general position, one can choose a new point x in 3-space but not in any line or plane determined by points in S. Thus, $S \cup \{x\}$ is a set of n points in general position. If v and w are joined by an edge e, let e correspond to the open straight-line segment ℓ between the points of P corresponding to v and w. Let L be the set of all such open segments. Two elements of L intersect if and only if their endpoints violate the hypotheses of general position. Similarly, no element of P belongs to an element of L.

This <u>topological realization</u> $|G| = P \cup \bigcup_{\ell \in L} \ell$ of a graph G in 3-space has a well-defined topology as a subspace of R^3. We shall often write G instead of $|G|$. Two graphs G and H are called <u>homeomorphic</u>, denoted $G \approx H$, if they have a common subdivison - i.e., they are isomorphic "up to vertices of degree 2". Clearly, $G \approx H$ if and only if $|G|$ and $|H|$ are homeomorphic as topological spaces. Also,

G is connected if and only if $|G|$ is connected.

Suppose G is a graph and S is a surface. By an <u>embedding</u> f: G → S of G in S we shall mean a topological embedding f (homeomorphism into) of $|G|$ in S. If f: G → S is an embedding, then the image of f, $f(|G|)$, is homeomorphic to $|G|$. Thus, the image of f is just a topological realization of G in S consisting of a set P' of points in S and a set L' of open Jordan arcs (simple curves) in S satisfying the analogous restrictions to those on P and L. We shall often write G for the image of f. When the particular embedding does not matter, we write $G \subset S$ to mean according to context, that G is (or can be) embedded in S. Note that every graph G can certainly be embedded in S_m, where $m = |E(G)|$. G is called <u>planar</u> if $G \subset R^2$. Any embedding of G in R^2 yields an embedding of G in S_0 and vice versa. Thus, G is planar if and only if $G \subset S_0$.

If f: G → S is an embedding, then G (that is, its image under the embedding) is a compact, hence closed, subspace of S. Therefore, S-G is open. Since S is locally connected, the connected components of S-G are open. They are called the <u>regions</u> of the embedding f. Each region is an open surface. If the regions are all <u>open</u> 2-<u>cells</u> (that is, homeomorphic to the euclidean plane), then f is called a <u>proper</u> or 2-<u>cell</u> embedding. We write G < S to mean that G is, or, by the usual abus du language, can be, properly embedded in S. Note that if G is connected, then $G \subset S_0$ if and only if $G < S_0$.

Suppose f: G → S is a proper embedding. How many regions does f have? Euler's famous formula, which we are about to state, shows that this number is finite and that it is completely determined by G and does not depend on f. Call a graph G an (n,m) <u>graph</u> and write G(n,m) if G is connected, has n vertices, and m edges.

<u>Theorem 1.1.</u> If G(n,m) and f: G → S is a proper embedding, then

$$n - m + r = e(S) ,$$

where r is the number of regions.

<u>Proof.</u> We shall give the proof when $S = S_0$, the sphere. A more general proof can be obtained by using the homological characterization of e(S) - (see Spanier [75, p. 173]). We induct on m. If m = 0, G consists of a point so n = 1, m = 0, r = 1

and the theorem holds. Suppose it holds for graphs with m-1 edges. Let $G(m,n)$

be arbitrary $(m \geq 1)$ and $f: G \rightarrow S$ a proper embedding. Call an edge of G a <u>bridge</u>

(or <u>isthmus</u>) if its removal results in a disconnected graph. Clearly, e is a

bridge if and only if e lies on no cycle of G. Hence, if every edge of G is a

bridge, G is a tree. But then $r = 1$ (since G has no cycles) and $m = n - 1$ because

G is a tree. Thus, $n - m + r = n - (n-1) + 1 = 2 = e(S_0)$ so the result holds.

Hence, we may assume that G has an edge e which is not a bridge. Let $G' = G - e$

and let $f': G' \rightarrow S_0$ denote the restriction of f to G'. Then $G'(n,m-1)$ and f' is

a proper embedding with r-1 regions since e must lie on the boundary of 2 regions –

otherwise it would be a bridge. Thus, by the inductive hypotheses, $n - (m-1) +$

$(r-1) = 2$. But $n - m + r = n - (m-1) + (r-1)$.

Suppose $G(n,m)$ and $G < S$. We can use Euler's Formula to derive an ·inequality

relating n, m, and e(S) which generalizes the well-known relation $m \leq 3n - 6$ for

planar graphs.

<u>Corollary 1.2.</u> If $G(n,m)$ and $G < S$, then $m \leq 3(n-e(S))$.

<u>Proof.</u> Let s(R) denote the number of sides (i.e., edges) in the boundary of a

region R with bridges counting twice. Then $2m = \Sigma_R s(R) \geq 3r$ since every region

must have at least 3 sides. Thus, $r \leq \frac{2}{3}m$ so $e(S) = n - m + r \leq n - m + \frac{2}{3}m = n - \frac{m}{3}$.

If $G(n,m)$ and $f: G \rightarrow S$ is an embedding such that every region has 3 sides or,

equivalently, if $m = 3(n-e(S))$, then f (or G) is called a <u>triangulation</u> of S, or

S-<u>triangulation</u>. Note that f must be a proper embedding. Clearly, any S-triangu-

lation G is S-<u>maximal</u> in the sense that if any new edge x is added to G, the aug-

mented graph $G + x$ cannot be properly embedded in S.

The converse of this statement, however, is false for surfaces other than

S_0 or R^2. It is true in the plane – for if G is R^2-maximal (called <u>maximal planar</u>)

and $G < S_0$, then no region with 4 sides or more can have all of its diagonals.

Thus, one of the diagonals could be added within the region without violating

planarity. Since G is maximal planar, this cannot happen so G is a triangulation.

This argument does not work even for the torus since then a k-sided region

$(k \geq 4)$ could have all of its diagonals outside of the region. In fact, the im-

plication is false. For if $K_8 - C_5$ denotes the complete graph K_8 on 8 vertices minus the edges of a 5-cycle C_5, then $K_8 - C_5$ is S_1-maximal, but $K_8 - C_5$ has only

$28 - 5 = 23$ edges, which is strictly less than the $3(8-0) = 24$ edges required by

an S_1-triangulation (see Harary, Kainen, Schwenk, and White [36] and also Duke

and Haggard [24]).

This suggests the following interesting problem:

Question 1.3. By how many edges can an S-maximal graph fail to be a triangulation?

In the sequel, it will be convenient to have an extension of Corollary 1.2.

Certainly, the restriction of connectedness (implicit in the hypothesis G(n,m))

could be removed since we can make G connected by adding extra edges joining its

components. However, the augmented graph might not be properly embedded in S.

Moreover, $G < S$ does not imply that every subgraph of G is properly embedded.

The next result, which resolves these difficulties, is due to Youngs [82].

Basically, it says that if $G \subset S$, then any non-simply-connected region can be

altered by removing handles or crosscaps to produce a simpler surface S' in which

G is properly embedded. The idea goes back at least to König [52, p. 198] and

certainly seems geometrically evident. However, Youngs realized that the proof was

somewhat more delicate than had been previously noted.

Let us say an embedding $G \subset S$ is simplest if, given any other embedding

$G \subset S'$, $e(S) \geq e(S')$. An embedding is maximal if no other embedding has more re-

gions. Youngs proved the following result constructively [82, p. 309].

Theorem 1.4. If $G \subset S$ is an embedding which is not proper, then the surface S can

be modified to produce a surface S* and an embedding $G \subset S*$, where $e(S*) > e(S)$.

Corollary 1.5. If $G \subset S$ is simplest, then $G < S$.

Using this result, Youngs was able to give the following characterization

theorem:

Theorem 1.6. $G \subset S$ is simplest if and only if it is a maximal proper embedding.

Proof. Suppose $G < S$ is simplest and $G < S'$ is a maximal proper embedding. Then

by Euler's Formula

$$n - m + r = e(S), \quad \text{and}$$

$$n - m + r' = e(S'),$$

where r and r' denote, respectively, the number of regions in the embeddings

G < S and G < S'. Since G < S is simplest and G < S' is maximal, we have

$$0 \leq e(S) - e(S') = r - r' \leq 0$$

so e(S) = e(S') and r = r'.

Theorem 1.4 enables us to find a strengthened version of Corollary 1.2.

Theorem 1.7. Let G be any graph with n vertices and m edges and suppose $G \subseteq S$.

Then $m \leq 3(n-e(S))$.

Proof. Obtain a connected graph $\overline{G}$ and embedding $\overline{G} \subseteq S$ from $G \subseteq S$ by adding new

edges joining the components of G. If G is already connected, no new edges are

added. Then $\overline{G}(n,\overline{m})$, where $\overline{m} \geq m$. By Theorem 1.4, we can find a proper embedding

$\overline{G} < S^*$, where $e(S^*) \geq e(S)$. Hence $\overline{m} \leq 3(n-e(S^*))$. But $m \leq \overline{m}$ and $3(n-e(S^*)) \leq$

$3(n-e(S))$ so we are done.

2.   GENUS

We define the genus $\gamma(G)$ of a graph G to be the smallest integer k such that

$G < S_k$. Note that the existence of some integer k such that $G < S_k$ follows from

Corollary 1.5. If G(n,m) and $G < S_k$, then by Corollary 1.2, $m \leq e(n-3(S_k)) =$

$3n - 6 + 6k$ so $k \geq \frac{m-3n+6}{6}$.   Thus

Proposition 2.1. $\gamma(G) \geq \{ \frac{m-3n+6}{6} \}$, where {x} denotes the least integer greater

than or equal to x.

For example, if $G = K_n$, the complete graph on n vertices, then $m = \frac{n(n-1)}{2}$ so

$\frac{m-3n+6}{6} = \frac{(n-3)(n-4)}{12}$.   Hence, $\gamma(K_n) \geq \{ \frac{(n-3)(n-4)}{12} \}$, and a famous result of Ringel

and Youngs [70] asserts that equality does hold. Their proof gives an embedding,

for each n, of K_n in S_k, where $k = \{ \frac{(n-3)(n-4)}{12} \}$. Their elaborate combinatorial

techniques depend on the residue class of n modulo 12.

Since there are trivalent graphs of arbitrarily high genus the inequality of

Proposition 2.1 cannot, in general, be an equality.

The genus of other classes of graphs is known. For example, $\{ \gamma(K_{p,q}) \} =$

$\{ \frac{(p-2)(q-2)}{4} \}$ (Ringel [67])and $\gamma(Q(d)) = 1 + (d-4)2^{d-3}$ (Ringel [68] and Beineke

and Harary [8]), where $K_{p,q}$ denotes the complete bipartite graph and Q(d) denotes

the d-dimensional cube which consists of the points and lines of a d-dimensional hypercube. Since $K_{p,q}$ and $Q(d)$ are bipartite graphs, Proposition 2.1 can be sharpened for them to read $\gamma(G) \geq \{\frac{m-2n+4}{4}\}$. What remains is to show these embeddings are possible.

Another result we should mention here (due to Battle, Harary, Kodama, and Youngs [2]) is that $\gamma(G) = \Sigma_i \gamma(B_i)$, where the B_i are the blocks of G. In particular, then, the genus of a graph is the sum of the genera of its components.

If G is a graph, there is certainly no largest integer k for which $G \subset S_k$ since if $G < S_{\gamma(G)}$, we can add an arbitrary number of handles to any region. On the other hand, there _is_ a largest integer k for which $G < S_k$; call k the maximum genus, $\gamma_M(G)$, of G. For example, if G is a tree or a unicyclic graph (connected with only one cycle), then it is easy to see that $\gamma_M(G) = 0$.

Let $G(n,m)$ be arbitrary and suppose $G < S_k$. By Euler's Formula, $n - m + r = 2 - 2k$ so $k = \frac{m-n+2-r}{2}$. To make k as large as possible, we make r as small as possible. Now k is an integer and $r \geq 1$. Hence, $k \leq [\frac{1}{2}(m-n+1)]$, where [x] is the greatest integer $\leq$ x. Since $\beta(G)$, the cyclomatic number of G, which is just the dimension of the cycle space of G, is given by the formula $\beta(G) = m-n+1$, we obtain a result of Nordhaus, Stewart, and White [61].

Proposition 2.2. $\gamma_M(G) \leq [\frac{\beta(G)}{2}]$

Call G upper embeddable if $\gamma_M(G) = [\frac{\beta(G)}{2}]$. It was shown in [61] that complete graphs are upper embeddable. Ringeisen [63] has shown that complete bipartite and maximal planar graphs are also upper embeddable. This latter result, of course, has the consequence that a planar graph may have arbitrarily large maximum genus. See also Nordhaus [60].

Consider the graph $G = C_r \vee C_s$ obtained from two cycles C_r and C_s by identifying a vertex in C_r with a vertex in C_s. Since $\beta(G) = 2$, $\gamma_M(G) \leq 1$. We claim that $G < S_1$ so that G is upper embeddable. For a torus can be obtained from a square by identifying opposite pairs of sides (see Figure 3). Under the identification, the perimeter of the square becomes two simple closed curves which intersect at a single point x. Hence, we can realize C_r in S_1 using one of these simple closed curves and C_s in S_1 using the other simple closed curve so that the point x corresponds to the vertex which C_r and C_s have in common. The complement of G in

S_1, S_1 - G, consists of a single open 2-cell since S_1 - G can also be obtained by first removing the perimeter of the square and then performing the identifications which do not affect the interior of the square and leave it as an open 2-cell.

This suggests a somewhat more geometric way to interpret Proposition 2.2 - namely, that at least 2 cycles are required to "kill off" each handle of S_k in an embedding of G in S_k in order that the embedding should be proper.

Genus has the very pleasant property that it is not increased by edge contraction. For if G < S_k and e is an edge of G, then we may find a very small closed disc D in S_k intersecting G only in e and its endpoints. (See Figure 4) Now we may shrink D without topologically altering S_k and this has the effect of collapsing e to a point. More generally, Ringel has observed:

<u>Theorem 2.3</u>. If H is a connected subgraph of G, then $\gamma(G/H) \leq \gamma(G)$, where G/H is obtained from G by collapsing H to a point.

This follows from our remark about edge contraction since H can be collapsed stepwise by contracting one edge at a time in a spanning tree.

In general, there is no reason why an embedding of a graph in a manifold should be at all "nice," even locally. For example, some edge could be embedded as a "topologists' sine curve". In particular, we cannot assume that every vertex v of a graph G < S has a small neighborhood in which the edges of G incident with v radiate outward from v like the spokes of a wheel, although we really need some such property to validate our earlier assumption of the existence of a suitably small disc D containing every edge e.

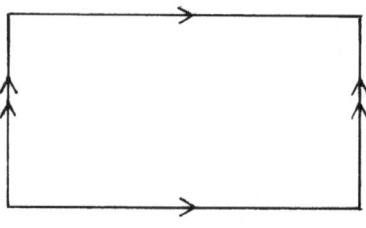

Figure 3

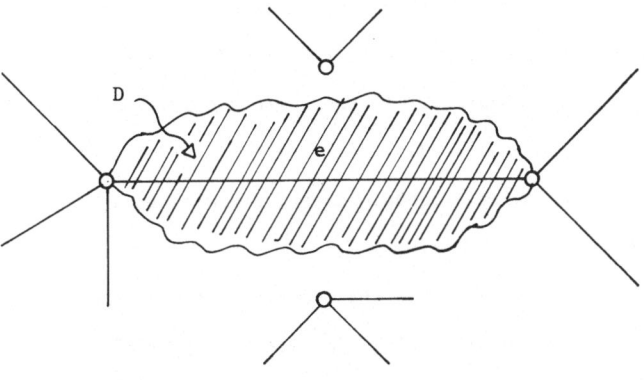

Figure 4

These problems, and others, can be handled by introducing the idea of <u>piece-wise-linear</u> (or PL) embeddings. Let S be an surface. Then S is obtained from the sphere by attaching handles and crosscaps. It is easy to see that, if only handles are attached, they can be deformed so as not to intersect one another and so S can be embedded in R^3. On the other hand, if crosscaps are attached as well, S can be embedded in R^4. Thus, any closed surface can be embedded in a euclidean space of dimension not exceeding 4. We can alter any such embedding (by arbitrarily little) to produce a <u>faceted embedding</u> - that is, we "bend" the surface so that it is divided into "flat" triangles, or <u>facets</u> meeting each other along common edges and at common corners (see Figure 5). Given a faceted embedding of S, we can form a 2-dimensional simplicial complex K (this is just the 2-dimensional analogue of a graph) whose 0-simplices are the corners, 1-simplices the edges, and 2-simplices the facets. Moreover K is linearly embedded in euclidean space and topologically K is just S. We shall denote a faceted embedding by its corresponding complex K. If f: G → S is an embedding and K is any faceted embedding of S, then f can be regarded as an embedding of G in K.

Let K and K' be faceted embeddings of S. We call K' a <u>subdivision</u> of K if, for i = 0,1,2, each i-simplex σ' of K' is contained in some 2-simplex σ of K (see Figure 6). If K is any 2-dimensional simplicial complex, its 0- and 1-simplices determine a graph G(K) called the 1-<u>skeleton</u> of K. Finally, let us choose for each surface S a standard faceted embedding K_S.

Now suppose f: G → K is an embedding, where K is a 2-dimensional complex.
Call f <u>simplicial</u> if f(G) is a subgraph of G(K). If f: G → S is an embedding, we
say that f is <u>piecewise-linear</u> (PL) if there are subdivisions G' of G and K' of K_S
such that the corresponding embedding f': G' → K' is simplicial. For a much more
detailed treatment of simplicial complexes and PL embeddings, see for example
Zeeman [84] or Hudson [41].

The following theorem, which we state but do not prove for graphs in surfaces,
is true in a much more general setting (Spanier [75, p. 128]).

<u>Theorem 2.4</u>. Let f: G → S be an embedding. Then there is an embedding f': G → S
which is PL.

In fact, f' can be chosen arbitrarily close to f, so if f is a proper embedding, f' can be taken to be proper also.

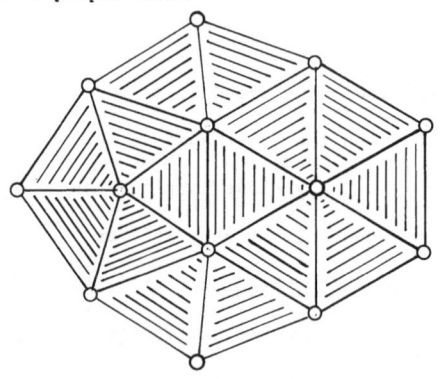

Figure 5. Part of a Faceted Embedding

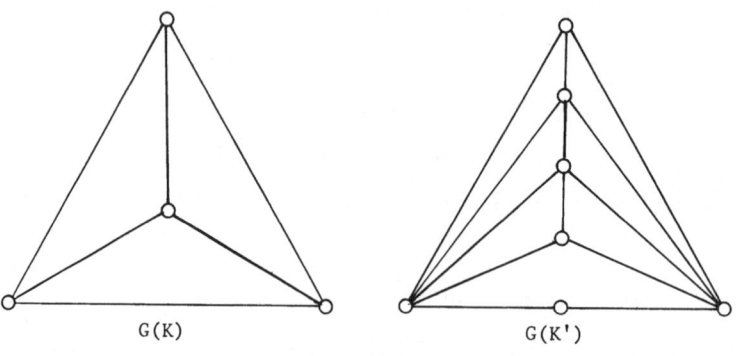

G(K)   G(K')

Figure 6
(The external region should be regarded as a 2-simplex of both K and K'.)

3. CHROMATIC NUMBERS OF GRAPHS

If G is any (n,m) graph, then the average degree d of G is by definition $d = \frac{2m}{n}$. Hence, from Corollary 1.2, if G is planar,

$$d = \frac{2m}{n} \leq \frac{6n-12}{n} = 6 - \frac{12}{n} < 6 .$$

Thus, G must contain a vertex of degree $\leq$ 5.

This helps to suggest the following definition: The Szekeres-Wilf number sw(G) of a graph G is given by

$$sw(G) = \max_{H \subseteq G} \delta(H) ,$$

where H is any subgraph of G and $\delta(H)$ is the minimum degree of H. Of course, it suffices to take the maximum over all induced subgraphs H. Let $\chi(G)$ denote the chromatic number of G (minimum number of colors required to color the vertices of G so that adjacent vertices have different colors). Szekeres and Wilf have proved [76]:

Theorem 3.1. $\chi(G) = 1 + sw(G)$.

The proof is an easy argument by induction on the number of vertices in G. Alternatively, we can derive it as an immediate consequence of the next result. Call G r-critical if $\chi(G) = r$ but $\chi(G-v) = r-1$ for any vertex v ε V(G). Note that if $\chi(G) = r$, then G contains an r-critical subgraph. For more on r-critical graphs, see e.g., Dirac [20] and [21].

Theorem 3.2. If G is r-critical, then $\delta(G) \geq r-1$.

Proof. Suppose G has a vertex v of degree $\leq$ r-2. Color G-v with r-1 colors. Then v is adjacent to at most r-2 different colors so v can be colored with one of the remaining colors. This is an r-1 coloring of G, which is impossible.

Our initial remarks about planar graphs show that if G is planar, sw(G) $\leq$ 5. Hence, by 3.1, we have

Corollary 3.3. If G is planar, $\chi(G) \leq 6$.

This result can be improved upon in several directions. Let us define the skewness $\mu(G)$ of a graph G to be the minimum number of edges whose removal makes G planar. Obviously, removing all the edges would suffice, so skewness is well-defined.

Lemma 3.4. If $\mu(G) \leq 5$, then $sw(G) \leq 5$.

Proof. Let G' be the planar graph obtained from G by removing an appropriate set of $\mu(G)$ edges. Then if G has n vertices and m edges, G' has n vertices and $m - \mu(G)$ edges, so $m - \mu(G) \leq 3n - 6$, or $m \leq 3n - 6 + \mu(G) < 3n$. Hence, if d denotes the average degree, $d = \frac{2m}{n} < 6$. Thus, there must be some vertex of degree ≤ 5. Observing that $\mu(H) \leq \mu(G)$ whenever $H \subseteq G$ completes the proof.

The following result is now immediate.

Theorem 3.5. If $\mu(G) \leq 5$, then $X(G) \leq 6$.

We can strengthen Corollary 3.3 by reducing 6 to 5. This is the famous 5 Color Theorem of Heawood [37]. Heawood originally proved it using a "coloring chain" argument with which Kempe had attempted (unsuccessfully) to prove the 4 Color Conjecture (4CC) - see Saaty [72]. We sketch here a somewhat different proof.

Theorem 3.6. If G is planar, then $X(G) \leq 5$.

Proof. Suppose G is planar and 6-minimal and suppose G is an (n,m) graph. (Thus $X(G) = 6$ but $X(H) < 6$ if $|V(H)| < n$. Minimal implies critical and any critical graph must be connected so there is no loss of generality in assuming G is an (n,m) graph.) By 3.2, $\delta(G) = 5$. Let v be a vertex of degree 5 which is adjacent to $v_1,\ldots,v_5$ and let $e_1,\ldots,e_5$ denote the corresponding edges. If every pair v_i,v_j $(1 \leq i \leq j \leq 5)$ is adjacent, G would contain K_6 as a subgraph. Since K_6 is not planar, some pair, say v_1 and v_2, are not adjacent. Delete e_3, e_4, and e_5 from G and contract e_1 and e_2 to produce a new graph $\overline{G}$. By Theorem 2.3, $\overline{G}$ is planar and hence, since G is minimal, $\overline{G}$ can be 5-colored. This induces a 5-coloring of G-v in which v_1 and v_2 receive the same color which, in turn, induces a 5-coloring of G.

Note that we did not use the non-planarity of K_5, only of K_6, in this proof, which optimists among you may take as propaganda for the 4CC. We also needed $sw(G) \leq 5$ for G planar and that $\overline{G}$, constructed from G by deleting and contracting edges, is planar.

Suppose we replace "G planar" with "$\mu(G) \leq 2$". Then the first two remarks still hold but the third one does not. The problem is that the edges to be deleted from G may have been contracted to form $\overline{G}$ and now constitute a vertex which is not removable. One can nevertheless get around this difficulty by showing that G has at least 8 vertices of degree 5 (see Kainen [42]) and we obtain:

Theorem 3.7. If $\mu(G) \leq 2$, then $\chi(G) \leq 5$.

The theorem implies a result of Hedetneimi that there is no uniquely 5-colorable planar graph. (See Chartrand and Geller [16].) In fact, there is no semi-uniquely 5-colorable planar graph (that is, if G is planar, and v and w ϵ V(G) with v $\neq$ w, then there is a 5-coloring of G assigning different colors to v and w.) See Greenwell [28].

Clearly, sw(G) $\leq$ $\Delta(G)$, the maximum degree of G, so $\chi(G) \leq 1 + \Delta(G)$. The following well-known result due to Brooks [13] improves this inequality by 1 for most graphs.

Theorem 3.8. Let G be connected. Then $\chi(G) \leq \Delta(G)$ unless G is complete or an odd cycle.

It is a pity that an analogous improvement over the Szekeres-Wilf bound, namely $\chi(G) \leq$ sw(G) if G is not complete or an odd cycle, is impossible. For if $G = C_k + K_t$ denotes the join of an odd cycle C_k and K_t, then sw(G) = 2 + t while $\chi(G) = \chi(C_k) + \chi(K_t) = 3 + t$.

The following well-known and elegant theorem is due to Grötzsch [29].

Theorem 3.9. Any planar graph with no 3-cycles can be 3-colored.

We should also include here two results of Ringel [65]. Suppose that a graph G can be drawn in the plane so that no edge crosses more than one other edge. Call such a graph G nearly planar.

Theorem 3.10. If G is nearly planar, then $\chi(G) \leq 7$.

If G is planar with dual graph G*, let H denote the graph whose vertices are those of G and G* and which includes all edges of G and G* together with edges joining each vertex v* in G* to each vertex v̇ in the boundary of v*.

Theorem 3.11.   If G is cubic (regular of degree 3), $\chi(H) \leq 6$.

The results of Kronk [53] and of Kronk and White [54] are similar in flavor to these theorems.

4.   CHROMATIC NUMBERS OF SURFACES

Let S be a surface. We define the <u>chromatic number</u> $\chi(S)$ of S to be the maximum of $\chi(G)$ for $G \subseteq S$. In the same way, we define the <u>Szekeres-Wilf</u> number of S, $sw(S) = \max \{sw(G) | G \subseteq S\}$. Plainly, $\chi(S) \leq 1 + sw(S)$ and so any bound on $sw(S)$ yields a bound on $\chi(S)$.

In order to find a bound on $sw(S)$, we shall find a bound on $d = d(G)$, the average degree of G, for any graph $G \subseteq S$. The following lemma is an easy consequence of Theorem 1.7.

Lemma 4.1.   Suppose $G(n,m)$ and $G \subseteq S$. Then $d \leq 6 - \dfrac{6e(S)}{n}$.

Consider the equation $x^2 - 7x + 6e(S) = 0$. There are two roots $\alpha(S)$ and $\alpha*(S)$ given by

$$\alpha(S) = \frac{7 + \sqrt{49 - 24e(S)}}{2}$$

$$\text{and} \qquad \alpha*(S) = \frac{7 - \sqrt{49 - 24e(S)}}{2} \ .$$

Set $H(S) = [\alpha(S)]$. This is called the <u>Heawood number</u> of the surface S because of the following theorem which is due to Heawood [37].

Theorem 4.2.   Let S be a surface other than the sphere. Then $\chi(S) \leq H(S)$.

If S is the sphere, $H(S) = 4$, and we are dealing with the 4CC.

Proof.   It suffices, by Theorem 3.1, to prove that $sw(G) \leq H(S) - 1$ whenever $G \subseteq S$. When $S = U_1$, the projective plane, then it follows from Lemma 4.1 that $d(G) < G$ whenever $G \subseteq U_1$ and hence $sw(G) \leq 5 = H(U_1) - 1$. For $S \neq U_1$, we are able to prove a slightly stronger result, which finishes the proof of this theorem.

Theorem 4.3.   Let S be a surface other than the sphere or projective plane and let $G \subseteq S$. Then $d(G) \leq \alpha(S) - 1$.

Proof.   If $e(S) = 0$ (i.e., S is the torus of Klein bottle), then $\alpha(S) = 7$ and, by Lemma 4.1, $d(G) \leq 6 = 7 - 1$. Thus, we can assume $e(S) < 0$. Let $G \subseteq S$ be any graph of order n. If $n \geq \alpha(S)$, then

$$d \leq 6 - \frac{6e(S)}{n} \leq 6 - \frac{6e(S)}{\alpha(S)} = \alpha(S) - 1 .$$

On the other hand, if $n < \alpha(S)$, then $d \leq n - 1 < \alpha(S) - 1$.

Remark 4.4. If S is the projective plane, $X(S) \leq H(S) = 6$.

Proof. By Lemma 4.1, if $G \subset S$, $d < 6$ so $sw(G) \leq 5$. Therefore, $sw(S) \leq 5$ and hence $X(S) \leq 6$.

The following result, due to Ringel and Youngs [70] in the orientable case and Ringel [66] in the non-orientable case, shows that the upper bound of Theorem 4.2 is usually exact.

Theorem 4.5. If S is any surface other than the sphere or Klein bottle, then $X(S) = H(S)$. If S is the Klein bottle, $X(S) = 6 = H(S) - 1$.

Proof. The result of Ringel and Youngs, which we mentioned earlier, on the genus of K_n shows that $\gamma(K_{H(S_k)}) = k$ so $X(S_k) \geq H(S_k)$ which establishes equality. Ringel obtained similar results when S is non-orientable and not the Klein bottle. If S is the Klein bottle, $H(S) = 7$ but Franklin [27] noticed that K_7 cannot be embedded in the Klein bottle. This implies that if $G \subset S$ and G is 7-minimal, then there is some vertex v of degree 6 two of whose neighbors are not adjacent. Now we can proceed as in the 5 color theorem to obtain a 6-coloring of G. Hence, $X(S) \leq 6$. Since $K_6 < S$, $X(S) = 6$.

Now let us use Theorem 4.3 to rederive a portion of a result due to Dirac [23]. If $S = S_k$, it is an easy bit of algebra to see that $\alpha(S)$ is integral for $k \equiv 0,3,4$, or 7 (mod 12) and hence, for these residue classes, $\alpha(S) = H(S)$. The theorem is also a consequence of an earlier result of Dirac [22].

Theorem 4.6. Let $k \neq 0$ and suppose $k \equiv 0,3,4$, or 7 (mod 12). If $G \subset S_k$ and $X(G) = H(S_k)$, then G contains K_r as a subgraph, where $r = H(S_k)$.

Proof. Let G' be an r-critical subgraph of G. Then $\delta(G') \geq r-1$ and, by Theorem 4.3, $d(G') \leq \alpha(S_k) - 1 = r-1$. Therefore, G' is regular of degree r-1. By Brooks' Theorem, $G' = K_r$ since $X(G') = r$.

This theorem is almost true for S_0. Dirac [19] has shown that if G is planar and $\chi(G) = 4$, then G has K_4 as a subcontraction. It follows that G contains a subgraph homeomorphic to K_4. However, the wheel graph W_k of Tutte [77] for k odd and greater than 3 has chromatic number 4 but contains no K_4.

Obviously, the preceding result holds for any surface $S \neq U_1$ for which $\alpha(S)$ is integral - that is, whenever $49 - 24e(S)$ is the square of an odd number. For example, when $e(S) = -3$, $\alpha(S) = 9 = H(S)$ and $G \subseteq S$ with $\chi(G) = 9$ must contain K_9 as a subgraph. Thus, Dirac's result holds for many non-orientable surfaces as well.

Question 4.7. Does Dirac's theorem also hold for the projective plane?

5. THICKNESS

If S is a surface and G is a graph, we define the S-thickness of G, $\theta_S(G)$, to be the minimum number t of subgraphs G_i, $1 \leq i \leq t$, of G such that $G = \bigcup G_i$ and $G_i \subseteq S$. Note that one can assume without loss of generality that the G_i are spanning subgraphs of G which are edge disjoint. Hence, if $G(n,m)$, $m \leq \theta_S(G)$ 3 $(n-e(S))$, so we have

Lemma 5.1. If $G(n,m)$, then $\theta_S(G) \geq \frac{m}{3(n-e(S))}$.

Let us write θ_k for θ_S if $S = S_k$, and θ for θ_0. Only θ has been studied extensively. It turns out that, as with genus, the lower bounds on $\theta(G)$ are, in fact, realizable, for most graphs in several important families.

It is an easy consequence of Lemma 5.1 that , for any n, $\theta(K_n) \geq [\frac{n+7}{6}]$. The reverse inequality requires an actual decomposition of K_n into planar subgraphs. For $n \not\equiv 4 \pmod 6$ and $n \neq 9$, this was accomplished by Beineke and Harary [9] and Beineke [4]. When $n \equiv 4 \pmod 6$, the construction was performed by Mayer [57,58] for n = 16, 34,and 40, by Hobbs and Grossman [40] for n = 22, and by Beineke [4] for n = 28. It is known that in the anomalous cases n = 9 and 10, $\theta(K_n)$ is one greater than $[\frac{n+7}{6}]$. See Battle, Harary, and Kodama [1] and Tutte [78]. Thus we have

Theorem 5.2. If $n \neq 9$ or 10, then $\theta(K_n) = [\frac{n+7}{6}]$ for $n \not\equiv 4 \pmod 6$ or $n \equiv 4 \pmod 6$ and $n \leq 45$. Moreover, $\theta(K_9) = 3 = \theta(K_{10})$.

Similar results are known for the thickness of $K_{p,q}$ for suitably restricted p and q (Beineke, Harary and Moon [10] and Beineke [6]).

In particular, if $p = q$, we find

$$\theta(K_{p,p}) = [\tfrac{p+5}{4}]$$

as predicted by Lemma 5.1, again modified for bipartite graphs. If Q_d denotes the d-dimensional cube, Kleinert [50] has shown that

$$\theta(Q_d) = \{\tfrac{d+1}{4}\}.$$

Let us also mention here some recent results on $\theta_S(K_n)$ for $S \neq S_0$. It is ironic, and reminiscent of the situation for chromatic numbers, that Beineke [3], Ringel [69] and Beineke and Harary [7] were able to show:

Theorem 5.3. For all $n \geq 3$, $\theta_1(K_n) = [\tfrac{n+4}{6}]$, $\theta_2(K_n) = [\tfrac{n+3}{6}]$, and $\theta_S(K_n) = [\tfrac{n+5}{6}]$, where S is the projective plane.

Let us digress for a moment to consider a modern-day coloring problem. Suppose that it is desired to simultaneously color a map of the earth and of the moon such that each country on earth is colored the same as that portion of the moon which it has colonized. Countries and lunar colonies are assumed connected; no country has more than one lunar colony. Of course, adjacent countries and adjacent colonies are to be colored differently. What is the smallest number k_0 of colors needed?

Lemma 5.4. Let $X_2 = \sup \{X(G) | \theta(G) \leq 2\}$. Then $k_0 = X_2$.

Proof. Replace both terrestrial and lunar maps by their dual graphs and identify each point on earth with its colony (if any) on the moon. The resulting graph has thickness not exceeding 2.

This suggests the following more general question: Determine $X_r = \sup \{(G) | \theta(G) \leq r\}$. Of course $X_1 = 4$ or 5 depending on the truth or falsity of the 4CC. By Theorem 5.2, $\theta(K_{6r-3}) = r$ for $r \geq 3$ and $\theta(K_8) = 2$. Moreover, for $3 \leq r \leq 7$, $\theta(K_{6r-2}) = r$ and it has been conjectured that this holds for all $r \geq 3$. On the other hand, if $G(n,m)$ and $\theta(G) = r$, then $m \leq 3r(n-2)$ by Lemma 5.1 so $d < 6r$ and $\delta(G) \leq 6r-1$. We obtain, therefore, the following result (see Ringel [66], Wilson [80, p. 84] and Kainen [47]).

Theorem 5.5. (a) $8 \leq X_2 \leq 12$

(b) $6r - 2 \leq X_r \leq 6r$ for $3 \leq r \leq 7$;

(c) $6r - 3 \leq X_r \leq 6r$ for $r > 7$.

Since the lower bound in (a) and (b) is given by the maximum-order complete graph with the appropriate thickness, it seems likely that the lower bound is, in fact, precise. In particular, we submit to the reader the following:

Question 5.6. Does there exist a graph G with thickness 2 and chromatic number greater than 8?

Let us note in passing that the analogous questions for toroidal-thickness θ_1 are completely solved. For if $X_r' = \sup\{X(G) \mid \theta_1(G) \leq r\}$, then by Lemma 5.1, $X_r' \leq 1 + 6r$ and, by Theorem 5.3, $\theta_1(K_{1+6r}) = r$ so we have:

Theorem 5.7. $X_r' = 1 + 6r$.

Moreover, we can easily prove as before that if $\theta_1(G) = r$ and $X(G) = X_r'$, then G contains K_{1+6r} as a subgraph.

There are several interesting variants of thickness we should mention here. Suppose we are interested in covering a graph by forests (acyclic subgraphs); we define the arboricity a(G) of a graph G to be the minimum number of forests whose union is G. The following elegant result is due to Nash-Williams [59].

Theorem 5.8. $a(G) = \max\limits_{2 < p \leq n} \{\frac{m_p}{p-1}\}$, where n is the order of G and m_p denotes the maximum number of edges in any subgraph of G with p vertices.

Since $m_p \leq 3p - 3$ if $\mu(G) \leq 3$, we obtain

Corollary 5.9. If $\mu(G) \leq 3$, then $a(G) \leq 3$.

If instead of covering all the edges, we merely want to cover the vertices but we require that no vertex belong to more than one forest and that the forests be induced (i.e., full) subgraphs of G, we obtain the notion of point-arboricity $\rho(G)$. Specifically then $\rho(G)$ is the minimum number of subsets in a partition of V(G) into disjoint subsets W_i such that the subgraph $G(W_i)$, induced by each W_i, is acyclic. Chartrand and Kronk [17] have shown that every planar graph has point-arboricity ≤ 3. In fact, there is a more general result due to Lick and White [56]. Let $\rho_k(G)$ denote the minimum number of subsets W_i in a partition of V(G) such that for each

W_i, $sw(G(W_i)) \leq k$. Note that $\rho_0(G) = \chi(G)$ and $\rho_1(G) = \rho(G)$. Lick and White proved:

Theorem 5.10. $\rho_k(G) \leq 1 + [\frac{sw(G)}{k+1}]$.

See also Bezhad and Chartrand [12, pp. 58–63].

Since $sw(G) \leq 5$ whenever $\mu(G) \leq 5$, we obtain

Corollary 5.11. $\rho(G) \leq 3$ if $\mu(G) \leq 5$.

As a consequence $\chi(G) \leq 6$ if $\mu(G) \leq 5$ (a fact which we already knew). If $\rho(G)$ were ≤ 2 for all planar G, we could prove the 4CC. Unfortunately, there are planar graphs G with $\rho(G) = 3$.

Let us mention one more form of thickness, the <u>book-thickness</u> of a graph (see Bernhart and Kainen [11]). The book-thickness of G, $bt(G)$, is the minimum number of "pages" (i.e., half-planes) in a "book" (i.e., collection of half-planes all meeting along a common edge or "spine") such that G can be "printed" (drawn) in the book so that all vertices lie on the spine, each edge lies within a single page, and no edges cross. Since the portion of G printed on any particular page is outer-planar, we have a lower bound on $bt(G)$ in terms of $\theta_{op}(G)$, where $\theta_{op}(G)$ denotes the <u>outerplanar thickness</u> of G (minimum number of outerplanar graphs covering G).

Lemma 5.12. $bt(G) \geq \theta_{op}(G)$.

Note that if G is Hamiltonian planar, $\theta_{op}(G) \leq 2$ and, since a forest is outer-planar, $\theta_{op}(G) \leq 3$ for any planar graph G. However, the stronger assertion that $bt(G) \leq 3$ for any planar graph G may well be false. In fact, we conjecture in [11] that there are planar graphs with arbitrarily large book thickness. Note that a graph G has $bt(G) = 1$ if and only if G is outplanar and $bt(G) \leq 2$ if and only if G is a spanning subgraph of a Hamiltonian planar graph. Book thickness has also been investigated by Ollman [62].

For more information on thickness, see the survey articles by Hobbs [39] and Beineke [5].

6.   PINCHED MANIFOLDS

A <u>pinched manifold</u> X is a compact topological space X in which every point x has a neighborhood N_x which is homeomorphic to the union of b_x disjoint copies of

R^2 whose origins have been identified to a single point. The point x is <u>singular</u> if b_x >1, and b_x is called its <u>order</u>; x is <u>normal</u> if b_x = 1. By the compactness of X, there are only a finite number of singular points. Given a surface S and a set $F_1,\ldots,F_r$ of subsets of S with $1 < |F_i| < \infty$, for $1 \leq i \leq r$, identifying together all the points in each F_i one can form a pinched manifold $S(F_1,\ldots,F_r)$ in which the only singular points $x_1,\ldots,x_r$ correspond to $F_1,\ldots,F_r$ and $|F_i|$ is the order of x_i. It is shown in Kainen [43] that every pinched manifold is the underlying topological space of a 2-dimensional simplicial complex which is a pseudo-manifold, and conversely (see Spanier [75, p. 148 and p. 150]).

<u>Lemma 6.1</u>. Every pinched manifold X can be constructed in this way; that is, if X is any pinched manifold, there is a surface S and subsets $F_1,\ldots,F_r$ such that $X = S(F_1,\ldots,F_r)$.

<u>Proof</u>. Each neighborhood N_x contains no singular point other than x. Replacing each neighborhood N_x with the disjoint union of b_x copies of R^2 has the effect of replacing each singular point x with b_x normal points while altering nothing else. Call the resulting surface S(X) and, for each singular point x_i, let $F(x_i)$ be the set of normal points corresponding to x_i. If S = S(X) and $F_i = F(x_i)$, then $X = S(F_1,\ldots,F_r)$.

Let X be a pinched manifold and let G be a graph. If f: G → X is an embedding, we call f <u>regular</u> provided that, for each singular point x ε X, there is a vertex v ε V(G) such that f(v) = x; f is <u>proper</u> if each component of X - f(G) is an open 2-cell. Note that if f is proper, it must certainly be regular.

If X is a pinched manifold, $X = S(F_i,\ldots,F_r)$, we define the <u>Euler character-istic</u> $e(X) = e(S) - \sum_{i=1}^{r} f_i$, where $f_i = |F_i| - 1$. The following analogue to Theorem 1.4 is proved in [43].

<u>Theorem 6.2</u>. Let f: G → X be an embedding which is not proper. Then X can be modified to produce a pinched manifold X* and a proper embedding of G in X*, where $e(X^*) > e(X)$.

Now if f: G → X is a proper embedding, Euler's formula holds; that is, we have,

Theorem 6.3. If $G(n,m)$ and $f: G \to X$ is a proper embedding, then $n - m + r = e(X)$,

where r = number of regions.

Corollary 6.4. If G is any graph of order n and size m edges, and if $f: G \to X$ is any

embedding, then $m \leq 3(n - e(X))$.

Let us introduce some notation. Let X^k denote the pinched manifold

$S_0(\{x_0,\ldots,x_k\})$ and $X_k = S_0(d_1,\ldots,d_k)$, where the d_i are disjoint doubletons

in S_0. For any X, we define $X(X) = \sup \{X(G) | G \subset X\}$. Ringel conjectured in [64]

that $\chi(X_1) = 5$. This was proved by Dewdney [18] who showed that $X(X_k) \leq k + 4$.

The following result which is proved like Theorem 4.2 leads to an improvement of

Dewdney's upper bound.

Theorem 6.5. For any pinched manifold $X \neq S_0$, $X(X) \leq H(X)$, where

$H(X) = [\frac{1}{2}(7 + \sqrt{49 - 24e(X)}]$

Since $e(X_k) = 2 - k$, we obtain (see Kainen [44]):

Corollary 6.6. For $k \geq 1$, $X(X_k) \leq [\frac{7 + \sqrt{1 + 24k}}{2}]$.

This upper bound is strictly smaller than $k + 4$ for $k \geq 6$ and equals $k + 4$

for $k = 3,4,5$. When $k = 1$, we can use the fact that K_6 cannot be embedded in X_1

to prove that $\chi(X_1) \leq 5$ by the same method we used to prove the 5 color theorem.

Since $K_5 \subset X_1$, we have verified Ringel's conjecture. In [47] we improved still

further on Corollary 6.6.

Theorem 6.7. For any k, $X(X_k) \leq 12$.

Proof. Suppose $G \subset X_k$. Then there is a graph $G' \subset S_0$ such that performing the

appropriate "pinches" on S_0 identifies disjoint pairs of vertices of G' to induce

$G \subset X_k$. Since $d(G') < 6$, $d(G) < 12$. Therefore, $X(G) \leq 12$.

We can relate this last theorem to the empire problem of Heawood [38]. Suppose

that in a map some countries have colonies (not adjacent to them) which must be

colored the same as the mother country. If no country has more than one colony,

then taking the dual of the map and identifying the pairs of vertices corresponding

to a country and its colony, we obtain a graph $G \subset X_k$, where k is the number of

countries with colonies. Thus, by Theorem 6.7, 12 colors will always suffice for

this problem. Heawood obtained this result and more generally showed that $6(r + 1)$

colors would do if no country had more than r colonies. Furthermore, he gave an embedding of K_{12} in X_{12}.

Now let us consider the pinched manifolds X^k.

Theorem 6.8. $5 \leq \chi(X^k) \leq 6$ for $k \geq 1$.

Proof. Let $G \subset X^k$. As in the proof of Theorem 6.7, find a graph $G' \subset S_0$ which induces the embedding of G in X^k. If G' is 5-colored, then recolor G' using 6 colors by coloring with a new color all vertices of G' which correspond to the singular point of X^k and not changing any other colors. Thus $\chi(X^k) \leq 6$ and since $K_5 \subset X^k$ for $k \geq 1$, $\chi(X^k) \geq 5$.

Now $K_6 \not\subset X^k$ since $K_5 \not\subset S_0$. This suggests that perhaps $\chi(X^k) = 5$. If the 4CC is true, then by initially 4-coloring G' and then recoloring it with 5 colors in the proof of Theorem 6.8 we could prove $\chi(X^k) = 5$.

The reader may use the notion of pinched manifold to phrase questions of thickness in terms of embedding graphs in pinched manifolds of the form $S(F_1, \ldots, F_r)$, where S is a disjoint union of spheres (this is strictly speaking a generalization of our original notion of pinched manifold since we formerly insisted on surfaces being connected) and the elements of each set F_i belong to separate spheres. See [47] for more on this topic.

Finally, we want to formulate a very attractive conjecture which we have hinted at several times. Let $M(X) = \max \{n | K_n \subset X\}$.

Conjecture 6.9. $\chi(X) = M(X)$ for $X \neq S_0$.

Of course, when $X = S_0$ this is the 4CC. For X an orientable or non-orientable surface, this is true by Theorem 4.5. It also holds for $X = X_1$ and for $X = X_k (k \geq 12)$; it must hold for $X = X^k$ unless the 4CC is false.

We could go still farther. It seems that any "geometric" coloring problem is determined by finding the maximum order of a complete graph satisfying the constraints. For example, Grötzsch's Theorem corresponds to the fact that K_4 is not triangle-free. (Of course, K_3 is also not triangle-free but this is the exception which proves the rule.)

7. CROSSING NUMBERS AND OTHER INVARIANTS

A _drawing_ of a graph G in a surface S is a realization of G in S in which pairs of edges are allowed to cross (at interior points). At most two edges cross at any given point and no edge crosses a vertex. In the same way that a drawing corresponds to a (true) realization, an _immersion_ corresponds to an embedding. Thus, f: G → S is an immersion if and only if it is an embedding except for disjoint pairs of points on edges of G which are identified by f.

Let cr(f) = number of crossings in an immersion f, and let $cr_S(G)$ = min {cr(f)|f: G → S an immersion}. We call $cr_S(G)$ the S-_crossing number_ of G. Set $cr_k = cr_S$ when $S = S_k$ and $cr_0 = cr$. First of all, let us note that $cr_S(G)$ is always finite since cr(G) < ∞. (To prove the latter assertion, it suffices to represent K_n as a polygon with all its diagonals. Hence, $cr(K_n) < ∞$.)

An immersion f: G → S is _minimal_ if cr(f) = $cr_S(G)$; f is _normal_ if, whenever two edges cross, their 4 endpoints are all distinct. Thus, no edge crosses itself and no two edges with a common endpoint cross. The following easy result has been noted by many people.

Lemma 7.1. Any minimal immersion is normal.

Very few results are known for crossing numbers. To try to prove that cr(G)=t, one first gives a construction to show cr(G) ≤ t and then attempts to justify cr(G) ≥ t. The following conjectures have been suggested by appropriate constructions.

Conjecture 7.2. $cr(K_n) = \frac{1}{4} \left\lfloor \frac{n}{2} \right\rfloor \left\lfloor \frac{n-1}{2} \right\rfloor \left\lfloor \frac{n-2}{2} \right\rfloor \left\lfloor \frac{n-3}{2} \right\rfloor$.

This conjecture (and the related construction) is due independently to Guy [30] and Saaty [71]. It is true for n ≤ 10. See also Busacker and Saaty [14].

Conjecture 7.3. $cr(K_{p,q}) = \left\lfloor \frac{p}{2} \right\rfloor \left\lfloor \frac{p-1}{2} \right\rfloor \left\lfloor \frac{q}{2} \right\rfloor \left\lfloor \frac{q-1}{2} \right\rfloor$.

This conjecture was originally a theorem of Zarankiewicz [83] who successfully gave a construction but failed to prove the reverse inequality. Ringel and Kainen independently noticed his error. The conjecture is true for min(p,q) ≤ 5 (Kleitman [51]).

Saaty proved [73] that if $\varsigma r(K_n)$ is a polynomial for n odd and a polynomial for n even (not necessarily the same polynomial), then 7.2 is valid. Kainen [45] showed that asymptotically the Guy-Saaty conjecture is valid provided that Zarankiewicz's Conjecture is valid:

Theorem 7.4. If Conjecture 7.3 holds, then $\lim\limits_{n \to \infty} n^{-4} cr(K_n) = \frac{1}{64}$.

If we restrict ourselves to <u>linear</u> immersions f: G → R^2 (that is, the image of each edge of G is a straight-line in R^2), we obtain the concept of the <u>recti-linear crossing number</u> $\overline{cr}(G)$ of G. Harary and Hill [34] conjectured that $\overline{cr}(K_8) = 19$. This has been verified by Guy [31]. But $cr(K_8) = 18$ so the crossing number and rectilinear crossing number are not the same in general.

Let us pose another problem here for the reader.

Question 7.5. Is $\overline{cr}(K_n) = cr(K_n)$ for $n \equiv 0 \pmod 3$?

It would even be nice to evaluate $\lim\limits_{n \to \infty} \dfrac{cr(K_n)}{\overline{cr}(K_n)}$. Note that the limit is at most 1 (if it exists) since $\overline{cr}(K_n) \geq cr(K_n)$. We should also mention that Zarankiewicz's construction is linear so perhaps $\overline{cr}(K_{p,q}) = cr(K_{p,q})$.

Erdős has pointed out that the problem of determining $\overline{cr}(K_n)$ is precisely the problem of Esther Klein of finding how many convex quadrilaterals are determined by n points in general position in the plane.

If S is a surface and G is a graph, we define the S-<u>skewness</u> $\mu_S(G)$ of G to be the minimum number of edges whose removal permits G to be embedded in S. If G is connected and not a tree, we define $\varepsilon_S(G)$, the <u>Euler deficiency</u> of G on S, to be $\varepsilon_S(G) = m - (\frac{g}{g-2})(n-e(S))$, where n is the order and m the size of G, and g denotes the girth of G (length of shortest cycle).

We obtained the following theorem in [46].

Theorem 7.6. If G is connected and not a tree, then $\varepsilon_S(G) \leq \mu_S(G) \leq cr_S(G)$.

For example, if $S = S_0$ and $G = K_6$, $cr(K_6) \geq \mu(K_6) \geq \varepsilon(K_6) = 15 - 3(6-2) = 3$. But K_6 can be drawn in the plane with 3 crossings by using two concentric triangles. Hence $cr(K_6) = 3 = \mu(K_6)$. However, $\varepsilon(K_7) = 6$ but $cr(K_7) = 9$ so the inequalities are in general not equalities. Moreover, if G is toroidal, $\varepsilon(G) \leq 6$ but Harary, Kainen, and Schwenk [35] showed that there are toroidal graphs with arbitrarily large planar crossing numbers.

We also showed in [46] that the crossing number of a pseudograph Ψ (allowing multiple edges and loops) is related to $cr_S(G)$, where G is the underlying graph of Ψ obtained by deleting all multiple edges and loops.

Specifically,

Theorem 7.7. $cr_S(\Psi) \leq k^2 \, cr_S(G)$, where k is the maximum number of parallel edges joining any pair of vertices. If every pair of vertices is joined by exactly k edges, the inequality becomes equality.

If $G \subset S$, clearly $cr_S(G) = 0$. Thus, $cr_{\gamma(G)}(G) = 0$. In [48] we considered the question of determining $cr_{\gamma(G)-k}(G)$ for G a cube Q_d. This was inspired by the observation that $\varepsilon_{\gamma(Q_d)-k}(Q_d) = 4k$. For k "small" compared to $\gamma(Q_d)$ we could show $cr_{\gamma(Q_d)-k}(Q_d) \leq 8k$. Thus, we have the following theorem which gives information on the crossing number of Q_d in the "stable range". (Recall, from Section 2, that $\gamma(Q_d) = 1 + (d-4)2^{d-3}$.)

Theorem 7.8. For $k \leq 2^{d-4}$, we have $4k \leq cr_{\gamma(Q_d)-k}(Q_d) \leq 8k$.

For S a surface and G a graph, $\lambda_S(G)$, the S-local crossing number of G, is $\min\{\lambda(f) \,|\, f: G \to S$ an immersion$\}$, where $\lambda(f)$ denotes the maximum number of crossings on any edge in $f(G)$. It will be convenient to define the thickness of an immersion $f: G \to S$, $\theta(f)$, to be the minimum number of subgraphs G_i of G such that $f|G_i$ is an embedding and $G = \bigcup G_i$. It is not difficult to show that $\theta_S(G) = \min\{\theta(f) \,|\, f: G \to S$ an immersion$\}$ (see [49], the proof requires use of the homogeneity of a manifold).

Theorem 7.9. $\theta_S(G) \leq 1 + \lambda_S(G)$.

Proof. It suffices to show that, for any immersion $f: G \to S$, $\theta(f) \leq 1 + \lambda(f)$. Define the crossing graph G_f of f as follows: $V(G_f) = E(G)$ and two vertices of G_f are adjacent if and only if the corresponding edges cross. Then $\Delta(G_f) = \lambda(f)$ and $\chi(G_f) = \theta(f)$ so since $\chi(G_f) \leq 1 + \Delta(G_f)$, $\theta(f) \leq 1 + \lambda(f)$.

Guy, Jenkyns, and Schaer [32] have obtained some estimates on the toroidal crossing number of K_n. We should also mention work of Tutte [79] and more recently Levow [55] which gives an algebraic topological formulation of crossing numbers. Levow points out that Tutte's work is an independent rediscovery of the ideas of Shapiro [74] and Wu [81]. See also the survey paper of Guy [31] on crossing numbers and Eggleton's thesis [25].

REFERENCES

1. Battle, J.,Harary, F., and Kodama, Y., Every Planar Graph with Nine Points has a Nonplanar Complement, Bull. Amer. Math. Soc. 68 (1962), 569-571.

2. Battle, J.,Harary, F., Kodama, Y., and Youngs, J. W. T., Additivity of the Genus of a Graph, Bull. Amer. Math. Soc. 68 (1962), 565-568.

3. Beineke, L. W., Minimal Decompositions of Complete Graphs into Subgraphs with Embeddability Properties, Canad. J. Math. 21 (1969), 992-1000.

4. Beineke, L. W., The Decomposition of Complete Graphs into Planar Subgraphs, Graph Theory and Theoretical Physics (F. Harary, Ed.), Academic Press, London, 1967, 139-154.

5. Beineke, L. W., A Survey of Packings and Coverings of Graphs in The Many Facets of Graph Theory (G. Chartrand and S. Kapoor, eds.), Springer, Berlin (1969), 45-53.

6. Beineke, L. W., Complete Bipartite Graphs: Decomposition into Planar Subgraphs, A Seminar in Graph Theory (F. Harary, Ed.), Holt, Rinehart and Winston, New York (1967), 42-53.

7. Beineke, L. W., and Harary, F., Inequalities Involving the Genus of a Graph and its Thickness, Proc. Glasgow Math. Assoc. 7 (1965), 19-21.

8. Beineke, L. W. and Harary,F., The Genus of the n-cube, Canad. J. Math. 17 (1965), 494-496.

9. Beineke L. W. and Harary,F., The Thickness of the Complete Graph, Canad. J. Math. 17 (1965), 850-859.

10. Beineke, L. W., Harary, F. and Moon, J. W., On the Thickness of the Complete Bipartite Graph, Proc. Cambridge Phil. Soc. 60 (1964), 1-5.

11. Bernhart F. and Kainen, P. C., On the Book-thickness of a Graph, to appear.

12. Bezhad, M. and Chartrand, G., Introduction to the Theory of Graphs, Allyn and Bacon, Boston, (1971).

13. Brooks, R. L., On Colouring the Nodes of a Network, Proc. Cambridge Phil. Soc. 37 (1941), 194-197.

14. Busacker, R. G. and Saaty, T. L., Finite Graphs and Networks, McGraw Hill, New York (1965).

15. Cairns, S. S., Introductory Topology, Ronald Press, New York, (1961).

16. Chartrand, G. and Geller, D. P., On uniquely Colorable Planar Graphs, J. Combinatorial Theory 6 (1969), 271-278.

17. Chartrand, G. and Kronk, H. V., The Point-arboricity of Planar Graphs, J. London Math. Soc. 44 (1969), 612-616.

18. Dewdney, A. K., The Chromatic Number of a Class of Pseudo-2 Manifolds, Manuscr. Math, to appear.

19. Dirac, G. A., A Property of 4-chromatic Graphs and Some Remarks on Critical Graphs, J. London Math. Soc. 27 (1952), 85-92.

20. Dirac, G. A., Some Theorems on Abstract Graphs, Proc. London Math. Soc. (Ser. 3) 2 (1952), 69-81.

21. Dirac, G. A., The Structure of k-chromatic Graphs, Fund. Math. 40 (1953), 42-55.

22. Dirac, G. A., A Theorem of R. L. Brooks and A Conjecture of H. Hadwiger, Proc. London Math. Soc. 7 (1957), 161-195.

23. Dirac, G. A., Short Proof of A Map-colour Theorem, Canad. J. Math. 9 (1957), 225-226.

24. Duke, R. A. and Haggard, G., The Genus of Subgraphs of K_8, Israel J. Math. 11 (1972), 452-455.

25. Eggleton, R. B., Crossing Numbers of Graphs, Doctoral Dissertation, University of Calgary (1973).

26. Erdös, P. and Guy, R. K., Crossing Number Problems, Amer. Math. Monthly 80 (1973), 52-58.

27. Franklin, P., A Six Colour Problem, J. Math. Physics 13 (1934), 363-369.

28. Greenwell, D. L., Semi-uniquely Colorable Graphs, Proc. of Second Louisiana Conference on Combinatorics, Graph Theory and Computing Louisiana State University, Baton Rouge (1971), 253-256.

29. Grötzsch, H., Ein Dreifarbensatz fur dreikreisfreie Netze auf der Kugel, Wiss. Z. Martin-Luther Univ. Halle-Wittenberg Math. Naturwiss, Reihe 8 (1958), 109-120.

30. Guy, R. K., A Combinatorial Problem, Nabla (Bull. Malayan Math. Soc.) 7 (1960), 68-72.

31. Guy, R. K., Crossing Numbers of Graphs, Graph Theory and Applications (Y. Alavi, et al., eds.), Springer, Berlin (1972), 111-124.

32. Guy, R. K., Jenkyns, T. A., and Schaer, J., The Toroidal Crossing Number of the Complete Graph, J. Combinatorial Theory 4 (1968), 376-390.

33. Harary, F., Graph Theory, Addison-Wesley, Reading, Mass. (1969).

34. Harary, F., and Hill, A., On the Number of Crossings in a Complete Graph, Proc. Edinburgh Math. Soc. (2) 13 (1962-63), 333-338.

35. Harary, F., Kainen, P. C., and Schwenk, A. J., Toroidal Graphs with Arbitrarily High Crossing Numbers, Nanta Math., to appear.

36. Harary, F., Kainen, P. C., Schwenk, A. J., and White, A. T., A Maximal Toroidal Graph Which is not a Triangulation, Math. Scand., to appear.

37. Heawood, P. J., Map Colour Theorems, Quart. J. Math. 24 (1890), 332-338.

38. Heawood, P. J., On the Four-colour Map Theorem, Quart. J. Math. 29, (1898), 270-285.

39. Hobbs, A. M., A Survey of Thickness, Recent Progress in Combinatorics, (W. T. Tutte, Ed.), Academic Press, New York, (1969), 255-264.

40. Hobbs, A. M. and Grossman, J. W., Thickness and Connectivity in Graphs, J. Res. Nat. Bur. Stand., Section B, to appear.

41. Hudson, J. F. P., Piecewise Linear Topology, Benjamin, New York (1969).

42. Kainen, P. C., A Generalization of the 5 Color Theorem, to appear.

43. Kainen, P. C., Embedding Graphs in Pseudomanifolds, to appear.

44. Kainen, P. C., On the Chromatic Number of Certain 2-complexes in Proc. of the Third Southeastern Conference on Combinatorics, Graph Theory, and Computing, Florida Atlantic University, Boca Raton (1972).

45. Kainen, P. C., On a Problem of P. Erdös, J. Combinatorial Theory 5, (1968), 374-377.

46. Kainen, P. C., A Lower Bound for Crossing Numbers of Graphs with Applications to K_n, $K_{p,q}$ and $Q(d)$, J. Combinatorial Theory 12B (1972).

47. Kainen, P. C., On the Chromatic Number of a Pinched Manifold, to appear.

48. Kainen, P. C., On the Stable Crossing Number of Cubes, Proc. Amer. Math. Soc. 36 (1972), 55-62.

49. Kainen, P. C., Thickness and Coarseness of Graphs, Abh. Math. Sem. Univ. Hamburg. 39 (1973), 88-95.

50. Kleinert, M., Die Dicke des n-dimensionalen Würfel-Graphen, J. Combinatorial Theory 3 (1967), 10-15.

51. Kleitman, D. J., The Crossing Number of $K_{5,n}$, J. Combinatorial Theory 9 (1970), 315-323.

52. König, D., Theorie der endlichen und unendlichen Graphen, Leipzig, 1936, Reprinted by Chelsea, New York, 1950.

53. Kronk, H. V., The Chromatic Number of Triangle-free Graphs, Graph Theory and Applications (Y. Alavi, et al., eds.), Springer, Berlin, (1972), 179-181.

54. Kronk, H. V., and White, A., A 4-color Theorem for Toroidal Graphs, Proc. Amer. Math. Soc., to appear.

55. Levow, R. B., On Tutte's Algebraic Approach to the Theory of Crossing Numbers, in Proc. of the Third Southeastern Conference on Combinatorics, Graph Theory and Computing, Florida Atlantic Univ. Boca Raton. (1972), 315-324.

56. Lick, D. R., and White, A. T., k-Degenerate Graphs, Canad. J. Math. 22 (1970), 1082-1096.

57. Mayer, J., L'epaisseur des graphes complets K_{34} et K_{40}, J. Combinatorial Theory 9 (1970), 162-173.

58. Mayer, J., $\theta(K_{16}) = 3$, J. Combinatorial Theory 13B (1972), 71.

59. Nash-Williams, C. St. J. A., Edge-disjoint Spanning Trees of Finite Graphs, J. London Math. Soc. 36 (1961), 445, 450.

60. Nordhaus, E. A., On the Girth and Genus of a Graph, Graph Theory and Applications (Y. Alavi, et al., eds.), Springer, Berlin, (1972) 207-214.

61. Nordhaus, E. A., Stewart, B. M., and White, A. T., On the Maximum Genus of a Graph, J. Combinatorial Theory, 11 B (1971), 258-267.

62. Ollman, L. T., Book-thickness of Graphs, in Proc. of the Fourth Southeastern Conference on Combinatorics, Graph Theory, and Computing, Florida Atlantic Univ. Boca Raton (1973), to appear.

63. Ringeisen, R. D., Upper and Lower Embeddable Graphs, in Graph Theory and Applications (Y. Alavi, et al., eds.), Springer, Berlin, (1972), 261,268.

64. Ringel, G., Genus of Graphs, in Combinatorial Structures and Their Applications (R. Guy, ed.) Gordon and Breach, New York (1970), 361-366.

65. Ringel, G., Ein Sechsfarbenproblem auf der Kugel, Abh. Math. Sem. Univ. Hamburg 29 (1965), 107-117.

66. Ringel, G., Färbungsprobleme auf Flächen und Graphen, Deutscher Verlag der Wissenschaften, Berlin (1962).

67. Ringel, G., Das Geschlecht des vollständigen paaren Graphen, Abh. Math. Sem. Univ. Hamburg 28 (1965), 139-150.

68. Ringel, G., Über drei kombinatorische Probleme am n-dimensionalen Würfel und Würfelgitter, Abh. Math. Sem. Univ. Hamburg 20 (1955), 10-19.

69. Ringel, G., Die toroidale Dicke des vollständigen Graphen, Math. Z. 87 (1965), 19-26.

70. Ringel, G., and Youngs, J. W. T., Solution of the Heawood Map-coloring Problem, Proc. Nat. Acad. Sci. USA 60 (1968), 438-445.

71. Saaty, T. L., The Minimum Number of Intersections in Complete Graphs, Proc. Nat. Acad. Sci. USA 52 (1964), 688-690.

72. Saaty, T. L., Remarks on the Four Color Problem: the Kempe Catastrophe, Math. Mag. 40 (1967), 31-36.

73. Saaty, T. L., On Polynomials and Crossing Numbers of Complete Graphs, J. Combinatorial Theory 10 (1971), 183-184.

74. Shapiro, A., Obstructions to the Embedding of a Complex in a Euclidean Space: I. The first obstruction, Annals of Math. 66 (1957), 256-269.

75. Spanier, E. H., Algebraic Topology, McGraw Hill, New York (1966).

76. Szekeres, G., and Wilf, H. S., An Inequality, for the Chromatic Number of a Graph, J. Combinatorial Theory 4 (1968), 1-3.

77. Tutte, W. T., A Theory of 3-connected Graphs, Indag. Math. 23 (1961), 441-455.

78. Tutte, W. T., On the Non-biplanar Character of the Complete 9-graph, Canad. Math. Bull. 6 (1963), 319-330.

79. Tutte, W. T., Towards a Theory of Crossing Numbers, J. Combinatorial Theory 8 (1970), 45-53.

80. Wilson, R. J., <u>Introduction to Graph Theory</u>, Academic Press, New York, (1972).

81. Wu, W. T., <u>A Theory of Imbedding, Immersion, and Isotopy of Polytopes in a Euclidean Space</u>, Science Press, Peking (1965).

82. Youngs, J. W. T., Minimum Imbeddings and the Genus of a Graph, <u>J. Math. Mech.</u> 12 (1963), 303-315.

83. Zarankiewicz, K., On a Problem of P. Turán Concerning Graphs, <u>Fund. Math</u> 41 (1954), 137-145.

84. Zeeman, E. C., <u>Seminar on Combinatorial Topology</u>, Institut des Hautes Etudes Scientifiques (1963), mimeographed.

A SURVEY OF FINITE EMBEDDING THEOREMS FOR PARTIAL LATIN SQUARES AND QUASIGROUPS

Charles C. Lindner
Auburn University

ABSTRACT

A latin square is an n × n array such that in each row and column each of
the integers 1,2,3,...,n occurs exactly once. A partial latin square is an n × n
array such that in each row and column each of the integers 1,2,...,n occurs at most
once. An example of a 4 × 4 partial latin square is given below.

An immediate observation shows that the partial latin square P cannot be completed
to a latin square; i.e., the empty cells cannot be filled in with numbers from the
set {1,2,3,4} so that the result is a latin square.
This is because as soon as cell (3,4) is filled in
a contradiction arises. A very obvious question
to ask at this point is whether or not it is
possible to complete P if we are allowed to enlarge
P and introduce additional symbols. That is, does

P =

1	4		
			4
	1	2	
			3

there exist an m × m latin square Q agreeing with P in its upper left hand corner?
(P is said to be embedded in Q). In a by now classic paper, Trevor Evans has
shown that this is always possible. In fact, Evans proved that any n × n partial
latin square could be embedded in some t × t latin square for every t $\geq$ 2n..., the
best possible result of this kind. Evans' paper is now generic. It has become the
starting point for a fascinating collection of problems in the study of latin
squares: the so-called finite embedding problems. It is this collection of problems
that will be surveyed in this set of notes. These notes are reasonably self contained so that certain parts may be a bit pedantic..., hopefully not too much so.
Examples are included at every opportunity to (hopefully) illustrate what is
going on.

A SURVEY OF FINITE EMBEDDING THEOREMS FOR PARTIAL
LATIN SQUARES AND QUASIGROUPS

1. INTRODUCTION

A latin square is an n × n array such that in each row and column each of
the integers 1,2,3,...,n occurs exactly once. We remark that there is nothing
sacred about the integers 1,2,3,...,n and that any set containing n distinct
objects can be used to fill in cells. A quasigroup is an ordered pair (Q,o)
where Q is a set and o is a binary operation Q such that the equations a o x = b
and y o a = b are uniquely solvable for every pair of elements a and b in Q. We
will only be interested in finite latin squares and quasigroups and so we can think
of a quasigroup as a latin square with a headline and a sideline. A partial latin
square is an n × n array such that in each row and column each of the integers
1,2,...,n occurs at most once and a partial quasigroup is a partial latin square
with a headline and a sideline. An example of a 4 × 4 partial latin square is
given below.

$$P = \begin{array}{|c|c|c|c|} \hline 1 & 4 & & \\ \hline & & & 4 \\ \hline & 1 & 2 & \\ \hline & & & 3 \\ \hline \end{array}$$

An immediate observation shows that the partial latin square P cannot be completed
to a latin square; i.e., the empty cells cannot be filled in with numbers from the
set {1,2,3,4} so that the result is a latin square. This is because as soon as
cell (3,4) is filled in a contradiction arises. A very obvious question to ask
at this point is whether or not it is possible to complete P if we are allowed to
enlarge P and introduce additional symbols. That is, does there exist an m × m
latin square Q agreeing with P in its upper left hand corner? (P is said to be
embedded in Q). In his by now classic paper Embedding incomplete latin squares, [4]
Trevor Evans has shown that this is always possible. In fact, Evans proved that

any n × n partial latin square could be embedded in some t × t latin square for
every t ≥ 2n..., the best possible result of this kind. Evans' paper is now
generic. It has become the starting point for a fascinating collection of problems
in the study of latin square: the so-called <u>finite embedding problems</u> for partial
latin squares and quasigroups. It is this collection of problems that will be surveyed in this paper. We will jump back and forth between latin squares and quasigroups using whichever notation and terminology facilitates the topics order
discussion (mostly latin squares). This will cause no difficulty. Examples are
included at every opportunity to (hopefully) illustrate what is going on.

2. THE EVANS CONJECTURE

We begin our study of finite embedding problems in the most logical way: with
latin squares which can be <u>completed</u>; i.e., the empty cells filled in so that the
result is a latin square. As we have already seen this cannot be done in general.
In fact, it is always possible to construct, for n ≥ 2 , an n × n partial latin
square with only n cells occupied which cannot be completed. For example, if n = 4,
the 4 × 4 partial latin square given below cannot be completed.

<u>Example 2.1.</u>

1	2	3	
			4

Quite frequently, of course, a given n × n partial latin square with con-
siderably more than n cells occupied can be completed ..., but in view of the
preceding remarks, the best possible general embedding theorem along these lines
would be that any n × n partial latin squares with at most n - 1 cells occupied
can always be completed. The following conjecture is due to Trevor Evans [4].

The Evans Conjecture. A partial $n \times n$ latin square with at most $n - 1$ cells occupied can always be completed to a latin square.

The following special cases of the Evans Conjecture have been verified. We state each along with an indication of the proof and an example of the construction used in each case.

Theorem 2.2. (C. C. Lindner [11]). Let I be an $n \times n$ partial latin square with $n - 1$ cells occupied. Let r denote the number of rows in which the occupied cells occur and c the number of columns in which the occupied cells occur. If $r \leq [\frac{n}{2}]$ or $c \leq [\frac{n}{2}]$ then I can be completed to a latin square.

Proof. For the time being a (partial) $r \times s$ latin rectangle is meant an $r \times s$ array such that (in some subset of the rs cells of the array) each of the cells is occupied by an integer from the set $1, 2, \ldots, s$ and such that no integer from the set $1, 2, \ldots, s$ occurs in any row or column more than once. In other words, an $r \times s$ (partial) latin rectangle is just the first r rows of some $s \times s$ (partial) latin square. The last statement is justified due to a result of Marshall Hall [8].

We can assume that $r \leq [\frac{n}{2}]$ and so a suitable permutation of the rows of I gives a $r \times n$ partial latin rectangle R. In [11] it is shown that R can be completed to a latin rectangle. This is done by filling in one row at a time of the partial $r \times n$ latin rectangle R as follows. Suppose the first t rows of R have been filled in so that the result is still an $r \times n$ partial latin rectangle. Let $(t+1, i_1), (t+1, i_2), \ldots, (t+1, i_k)$ be the empty cells in row $t + 1$. Denote by S_{i_j}, $j = 1, 2, \ldots, k$, the set of symbols not occuring in row $t + 1$ or column i_j. Then the sets $S_{i_1}, S_{i_2}, \ldots, S_{i_k}$ always have a system of distinct representatives (SDR) [18]. Any SDR will give a completion of row $T + 1$. This technique works only in case $t \leq [\frac{n}{2}] - 1$ so that this technique will not necessarily complete $1 \times m$ partial latin rectangles with $1 > [\frac{m}{2}]$. In [8], Marshall Hall has shown that a suitable number of rows can be added to a latin rectangle so that the result is a latin square. If $\overline{R}$ is the completion of R, then adding $n - r$ rows to $\overline{R}$ gives a latin square which after a suitable permutation of rows gives the desired completion.

Example 2.3.

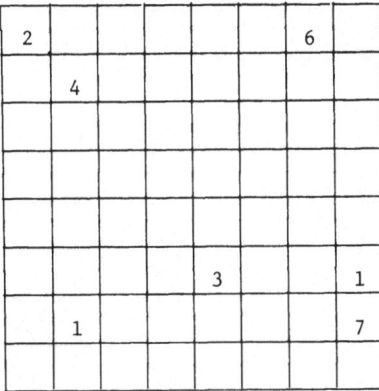

2						6	
	4						
				3			1
	1						7

I

8 × 8 partial latin square with r = 4 and c = 5.

2					6		
	4						
				3			1
	1						7

R

Since r ≤ 4 a suitable permutation of the rows of I transform I into 4 × 8 partial latin rectangle R.

2	5	4	7	8	1	6	3
5	4	3	1	2	6	7	8
7	8	2	6	3	4	5	1
8	1	5	4	6	2	3	7

R̄

Complete R to R̄ by filling in one row at a time via the use of SDR's.

2	5	4	7	8	1	6	3
5	4	3	1	2	6	7	8
7	8	2	6	3	4	5	1
8	1	5	4	6	2	3	7
1	2	7	3	5	8	4	6
3	7	6	2	1	5	8	4
4	6	8	5	7	3	1	2
6	3	1	8	4	7	2	5

R*

Complete R̄ to a 8 × 8 latin square R* by adding 4 rows via the use of SDR's. M. Hall's Theorem [8] guarantees this can be done.

2	5	4	7	8	1	6	3
5	4	3	1	2	6	7	8
3	7	6	2	1	5	8	4
4	6	8	5	7	3	1	2
1	2	7	3	5	8	4	6
7	8	2	6	3	4	5	1
8	1	5	4	6	2	3	7
6	3	1	8	4	7	2	5

L

Permute the rows of R*
to obtain a latin square
L agreeing with I. This
gives the desired com-
pletion.

We remark that in the above example a great deal of technique is necessary to go from R to $\bar{R}$ and from $\bar{R}$ to R*. We will not go into this here. We are only interested in the fact that the procedures described can in fact be carried out.

Theorem 2.4. (J. Marica and J. Schönheim [15]). An $n \times n$ partial latin square with $n-1$ cells occupied can be completed to a latin square if the occupied cells are in different rows and different columns.

Proof. The proof is based on the following result due to Marshall Hall [9] that if A is an abelian group of order n and $b_1,\ldots,b_n$ are any n elements of A such that $b_1+b_2+\ldots+b_n = 0$, then there is a permutation

$$\begin{pmatrix} a_1 & a_2 & \cdots & a_n \\ c_1 & c_2 & \cdots & c_n \end{pmatrix}$$

of the elements of A so that $a_i + c_i = b_i$, $i = 1,2,\ldots,n$. Since the occupied cells are in different rows and columns they can all be considered to be on the main diagonal. Let $b_1,b_2,\ldots,b_{n-1}$ be the (not necessarily distinct) elements in the occupied cells. If we identify the distinct b_i's with distinct elements of an abelian group of order n and set $b_n = - (b_1 + b_2 + \ldots + b_{n-1})$, then $b_1 + b_2 + \ldots + b_n = 0$. Hence there is a permutation of A so that $a_i + c_i = b_i$, $i = 1, 2, \ldots, n$. Now defining a latin square L by filling in cell (i,j) with $a_i + c_j$ gives the required completion.

Example 2.5.

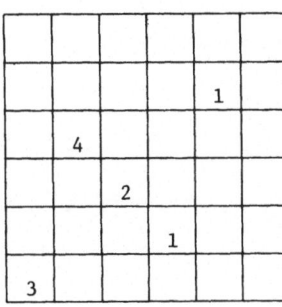

6 × 6 partial latin square with no two occupied cells in the same row or column (based on 0,1,2,3,4,5)

I

Permute occupied cells onto the main diagonal in any order.

$\overline{I}$

Take A to be the additive group of integers modulo 6.

$$b_1 = 4$$
$$b_2 = 2$$
$$b_3 = 1$$
$$b_4 = 1$$
$$b_5 = 3$$

$$b_1+b_2+b_3+b_4+b_5 = 5 \ (\text{mod } 6) \ .$$

Choose $b_6 = 1$ so that now $b_1+b_2+b_3+b_4+b_5+b_6 = 0$. M. Hall's Theorem [9] guarantees the existence of a permutation on A

$$\begin{pmatrix} a_1 & a_2 & a_3 & a_4 & a_5 & a_6 \\ c_1 & c_2 & c_3 & c_4 & c_5 & c_6 \end{pmatrix}$$

such that

$$a_1 + c_1 = b_1 = 4$$
$$a_2 + c_2 = b_2 = 2$$
$$a_3 + c_3 = b_3 = 1$$
$$a_4 + c_4 = b_4 = 1$$
$$a_5 + c_5 = b_5 = 3$$
$$a_6 + c_6 = b_6 = 1$$

One such permutation is given by

$$\begin{pmatrix} 0 & 1 & 2 & 4 & 3 & 5 \\ 4 & 1 & 5 & 3 & 0 & 2 \end{pmatrix} \ .$$

4	1	5	3	0	2
5	2	0	4	1	3
0	3	1	5	2	4
2	5	3	1	4	0
1	4	2	0	3	5
3	0	4	2	5	1

I*

Cell (i,j) is occupied by $a_i + c_j$ (mod 6) where

$$a_1 = 0 \qquad\qquad c_1 = 4$$
$$a_2 = 1 \qquad\qquad c_2 = 1$$
$$a_3 = 2 \qquad\qquad c_3 = 5$$
$$\text{and}$$
$$a_4 = 4 \qquad\qquad c_4 = 3$$
$$a_5 = 3 \qquad\qquad c_5 = 0$$
$$a_6 = 5 \qquad\qquad c_6 = 2$$

5	3	0	2	4	1
2	0	3	5	1	4
0	4	1	3	5	2
1	5	2	4	0	3
4	2	5	1	3	0
3	1	4	0	2	5

L

Permute the rows and columns of I*
to obtain a latin square L agreeing
with I. This gives the desired
completion.

As in the previous example, technique is involved which has been omitted, in particular, the construction of the permutation used to construct I* .

We close this section with a very clever observation due to Allan Cruse which
can be used to extend these results. First we will need some preliminary remarks
(which will also be useful to later sections).

Let L be a (partial) latin square and $\mathcal{L}$ the set of triples associated with
L : i.e., $(x,y,z) \in \mathcal{L}$ if and only if cell (x,y) of L is occupied by z. $\mathcal{L}'$ is
called a <u>conjugate</u> of $\mathcal{L}$ if for some $\alpha \in S_3$, $(x_1, x_2, x_3) \in \mathcal{L}'$ if and only if
$(x_{1\alpha}, x_{2\alpha}, x_{3\alpha}) \in \mathcal{L}$ [19]. Two latin squares are said to be conjugate if their
associated triples have this property. The proof of the following theorem is tri-
vial.

Theorem 2.6. Let L be a partial latin square. If L can be completed then so can any conjugate of L.

Corollary 2.7. Let I be a partial latin square of order n with at most n - 1 cells occupied. Then I can be completed provided that at most $[\frac{n}{2}]$ symbols are used or n - 1 distinct symbols are used and the occupied cells are in different rows or different columns.

This last result is obtained by noting that in the first case I is the conjugate of a partial latin square having occupied cells in at most $[\frac{n}{2}]$ rows or columns whereas in the second case I is the conjugate of a partial latin square in which no two of the occupied cells are in the same row or column.

Example 2.8. Let P be the 8 × 8 partial latin square given below.

P =

	1						
7		2					
		6					
				1			
6						7	

Now the number of occupied rows is 5 and the number of occupied columns is 6,both > 4 and so Theorem 2.2 is not applicable. Since the occupied cells are not pair-wise in different rows and columns Theorem 2.4 cannot be used for a completion. However since only the 4 symbols 1, 2, 6 and 7 are used, Corollary 2.7 can be used for a completion as follows. The set of triples $\mathcal{P}$ associated with P is $\mathcal{P} = \{(1,2,1),$ $(7,6,1)$, $(2,4,2)$, $(5,3,6)$, $(8,1,6)$, $(2,1,7)$, $(8,7,7)\}$. Applying the permutation (1 2 3) to $\mathcal{P}$ gives the conjugate collection $\mathcal{P}' = \{(1,1,2)$, $(1,7,6)$, $(2,2,4)$, $(6,5,3)$, $(6,8,1)$, $(7,2,1)$, $(7,8,7)\}$. The partial latin square associated with $\mathcal{P}'$ is just the partial latin square I given in Example 2.3. Let L be the completion of I given in Example 2.3. Applying the permutation (1 3 2), the inverse of

(1 2 3), to the collection of triples associated with L gives a conjugate collection

of triples whose associated latin square L* is a completion of P.

5	1	3	4	2	8	6	7
7	5	8	2	1	4	3	6
8	6	2	1	7	3	5	4
2	3	5	7	4	6	1	8
3	2	6	8	5	7	4	1
1	7	4	6	3	2	8	5
4	8	7	5	6	1	2	3
6	4	1	3	5	8	7	2

Completion of P
obtained by applying
the permutation (1 2 3)
to the latin square L
given in Example 2.3.

L*

At present the Evans Conjecture remains OPEN.

3. THE EVANS EMBEDDING THEOREM

As we have already seen in sections 1 and 2, it is not always possible to fill

in the cells of a partial latin square so that the result is a latin square. Two

problems arise quite naturally with respect to partial latin squares which cannot

be completed. The first rather obvious problem is whether a partial latin square

which cannot be completed can be finitely embedded at all; i.e., embedded in a

finite latin square. If the answer to this problem is yes, the second problem is

to determine (if possible) the smallest possible general embedding theorem. That

is to determine for each positive integer n the smallest positive integer N such

that any n × n partial latin square can be embedded in some t × t latin square

where t ≤ N. Both of these problems have been successfully handled by Trevor Evans

[4]. We will need a few preliminary notions before stating and proving Evans'

Theorem.

We begin by giving a more general defition of a latin rectangle. Recall that

in section 2 we defined an r × s latin rectangle to be (courtesy of M. Hall [8])

the first r rows of some s × s latin square. We will now agree that an r × s latin rectangle based on $1,2,\ldots,n$ is an $r \times s$ array such that each cell is occupied by one of the integers $1,2,\ldots,n$ and such that each of these integers occurs in each row and column at most once. If we omit the expression "based on $1,2,\ldots,n$" it will be understood that the $r \times s$ latin rectangle is based on $1,2,\ldots,s$ which is the definition of latin rectangle given in section 2.

Example 3.1.

$$R = \begin{array}{|c|c|c|c|c|}\hline 1 & 4 & 7 & 8 & 3 \\ \hline 2 & 9 & 1 & 3 & 4 \\ \hline 5 & 2 & 6 & 1 & 2 \\ \hline \end{array}$$

R is a 3×5 latin rectangle based on $1,2,3,4,5,6,7,8,9$. It is also a 3×5 latin rectangle based on any set of positive integers containing $1,2,3,4,5,6,7,8,9$.

An $r \times s$ latin rectangle R is said to be embedded in the $n \times n$ latin square L provided that the $r \times s$ upper left hand corner of L is R.

The following theorem is due to H. J. Ryser and gives necessary and sufficient conditions for an $r \times s$ latin rectangle to be embedded in an $n \times n$ latin square.

Theorem 3.2. (H. J. Ryser [17]). Let R be an $r \times s$ latin rectangle based on $1,2,\ldots,n$. Let $N(i)$ denote the number of times that the integer i occurs in R. Then R can be embedded in an $n \times n$ latin square if and only if $N(i) \geq r + s - n$ for every $i = 1,2,\ldots,n$.

Example 3.3.

(1)

$$R = \begin{array}{|c|c|c|}\hline 2 & 4 & 1 \\ \hline 3 & 1 & 2 \\ \hline \end{array}$$ 2×3 latin rectangle based on $1,2,3,4$.

R can be embedded in a 4×4 latin square since $2 + 3 - 4 = 1$ and $N(1) = 2 \geq 1$, $N(2) = 2 \geq 1$, $N(3) = 1 \geq 1$, and $N(4) = 1 \geq 1$.

(2)

$$T = \begin{array}{|c|c|c|}\hline 1 & 2 & 3 \\ \hline 3 & 1 & 2 \\ \hline 2 & 4 & 1 \\ \hline \end{array}$$ 3×3 latin rectangle based on $1,2,3,4,5$.

T cannot be embedded in a 5 × 5 latin square since 3 + 3 - 5 = 1 and N(5) = 0 < 1.

(3)

Q =

2	1	4	6
3	5	1	7
6	4	9	3

3 × 4 latin rectangle based on 1,2,3,4,5,6,7,8,9,10 .

Q can be embedded in a 10 × 10 latin square since 3 + 4 - 10 = -3 and N(i) $\geq$ 0 $\geq$ -3 for every i = 1,2,3,4,5,6,7,8,9,10.

We are now in a position to state and prove the Evans Embedding Theorem. (This theorem was also obtained independently by S. K. Stein).

Theorem 3.4. (T. Evans [4]). A partial n × n latin square can be embedded in a t × t latin square for every t $\geq$ 2n.

Proof. Let P be an n × n partial latin square based on 1,2,...,n and t a positive integer $\geq$ 2n. Write t = k + n and let K be a k × k latin square based on n + 1, n + 2, ..., n + k = t. Fill in any unoccupied cells of P with the entry in the corresponding cell in the n × n upper left hand corner of K. This turns the partial latin square P into an n × n latin rectangle $\overline{P}$ based on 1,2,...,n + k = t. By Ryser's Theorem $\overline{P}$ can be embedded in a t × t latin square T if and only if N(i) $\geq$ n + n - t for all i = 1,2,...,t. Since t $\geq$ 2n, n + n - t $\leq$ 0 and the condition N(i) $\geq$ n + n - t is trivially satisfied completing the proof.

The following example illustrates the construction used in this proof.

Example 3.5.

1	2	3	
	3	2	4
	4	1	
			1

4 × 4 partial latin square based on 1,2,3,4 .

P

5	6	7	8	9
6	7	8	9	5
7	8	9	5	6
8	9	5	6	7
9	5	6	7	8

K

5 × 5 latin square
based on 5,6,7,8,9 .

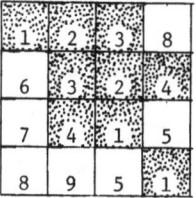

1	2	3	8
6	3	2	4
7	4	1	5
8	9	5	1

$\overline{P}$

4 × 4 latin rectangle
based on 1,2,3,4,5,6,7,
8,9 obtained from P by
filling in the empty cells
with the entries in the
corresponding cells in K .

1	2	3	8	5	6	9	7	4
6	3	2	4	1	7	5	8	9
7	4	1	5	6	9	2	3	8
8	9	5	1	2	4	3	6	7
3	5	6	7	4	8	1	9	2
4	8	7	9	3	2	6	5	1
2	6	8	3	9	1	7	4	5
5	7	9	2	8	3	4	1	6
9	1	4	6	7	5	8	2	3

T

9 × 9 latin square
with P embedded in
the upper left hand
corner.

This example illustrates the embedding of a 4 × 4 partial latin square in a 9 × 9
latin square. Going from the 4 × 4 latin rectangle $\overline{P}$ to the latin square T takes
lots of paper and <u>lots of technique</u> which we will not go into here. What is important is that Ryser's Theorem guarantees that it can be done.

We close this section by remarking that Evans' Theorem is the <u>best possible</u>
embedding theorem of its kind for partial latin squares because a latin square is
also a partial latin square and (just like a group) cannot be embedded in a latin

square less than twice its size. The reader with a vivid imagination can even construct for every n ≥ 4 a "honest-to-gosh" partial latin square which cannot be embedded in a latin square less than twice its size. Those without vivid imaginations see [4]. Of course in many cases a _particular_ n × n partial latin square can be embedded in a latin square less than twice its size. However in view of the preceding remarks, an n × n partial latin square cannot in general be embedded in a latin square smaller than 2n × 2n.

4. STEINER TRIPLE SYSTEMS AND LATIN SQUARES

There are many properties that a latin square can have in addition to being a latin square. Such properties as being idempotent, commutative, or idempotent and commutative (for example). We can ask the same questions for these latin squares as we did for latin squares in general. For example, can a _finite_ partial commutative latin square be embedded in a _finite_ commutative latin square, and if so, what is the best possible general embedding theorem with respect to the size of the containing square. In order to look at these more exotic embedding problems (as they were originally solved) we must first introduce some results on (partial) Steiner triple systems. In particular a very important theorem due to Christine Treash [20] on the finite embeddability of finite partial Steiner triple systems. This may look at first as though it is off of the subject but as we shall see in the sections which follow, it is exactly what is necessary.

A _Steiner triple system_ is a pair (S,t) where S is a finite set and t is a collection of three element subsets of S such that every pair of elements of S occur in exactly one of the triples of t. It is well known that the spectrum for Steiner triple systems is the set of all n ≡ 1 or 3 (mod 6) [16]. There is a very useful connection (to put it mildly) between Steiner triple systems and latin squares due to R. H. Bruck [1]. The connection is as follows. Let (Q,q) be a Steiner triple system and define a latin square based on Q as follows.

(1) cell (x,x) is occupied by x, and

(2) cell (x,y), x ≠ y, is occupied by z if and only if {x,y,z} ε q.

The resulting latin square called a <u>Steiner latin square</u> (when considered as a
quasigroup) satisfies the following identities.

$$\text{Steiner identities}\begin{cases}x^2 = x \\ x(xy) = y \\ (yx)x = y\end{cases}\begin{matrix}\text{idempotent identity} \\ \\ \text{symmetric identities}\end{matrix}$$

In terms of latin squares the identity x^2 = x says that cell (x,x) is occupied by
x, whereas x(xy) = y says that whenever cell (x,y) is occupied by z then cell
(x,z) is occupied by y (analogously for (yx)x = y). It is important to note that
the symmetric identities imply the <u>commutative</u> law and so a Steiner latin square
is commutative. Conversely, any latin square based on the set Q satisfying the
Steiner identities gives rise to a Steiner triple system (Q,q) by placing the
3-element subset {x,y,z} in q if and only if cell (x,y) is occupied by z. A Steiner
triple system (Q,q) and Steiner latin square based on Q are said to be associated
provided that {x,y,z} ε q if and only if cell (x,y) is occupied by z.

Example 4.1.

S = {1,2,3,4,5,6,7}

t = {{1,2,4}, {1,3,7}, {1,5,6},

    {2,3,5}, {2,6,7}, {3,4,6},

    {4,5,7}}

(S,t) is a <u>Steiner triple system</u>

1	4	7	2	6	5	3
4	2	5	1	3	7	6
7	5	3	6	2	4	1
2	1	6	4	7	3	5
6	3	2	7	5	1	4
5	7	4	3	1	6	2
3	6	1	5	4	2	7

Steiner latin square
associated with (S,t)

A <u>partial Steiner triple system</u> is a pair (Q,q) where Q is a finite set and q is a
collection of three element subsets of Q such that every pair of elements of Q
occur in at most one of the triples of q. Whereas the spectrum for Steiner triple
systems is the set of all positive integers n ≡ 1 or 3 (mod 6) there is no cardinality restriction on Q when (Q,q) is partial. A partial Steiner triple system is
equivalent to a partial Steiner latin square; i.e., a partial latin square such
that

(1)   cell (x,x) is occupied by x, all x, and

(2)   if cell (x,y) is occupied by z then cell (y,z) is

occupied by x, cell (z,x) is occupied by y, cell

(y,x) is occupied by z, cell (x,z) is occupied by

y, and finally cell (z,y) is occupied by x.

Example 4.2.

   $Q = \{1,2,3,4,5\}$

   $q = \{\{1,2,4\}, \{2,3,5\}\}$

   (Q,q) is a __partial__ Steiner triple system

1	4		2	
4	2	5	1	3
	5	3		2
2	1		4	
	3	2		5

__partial__ Steiner latin square
associated with (Q,q)

The partial Steiner triple system (P,p) is said to be __embedded__ in the Steiner triple system (V,v) if (not too surprisingly) $P \subseteq V$ and $p \subseteq v$. The partial Steiner triple system (Q,q) of Example 4.2 is embedded in the Steiner triple system (S,t) of Example 4.1. The clever observer will note that the partial Steiner latin square associated with (Q,q) is also embedded in the Steiner latin square associated with (S,t). The question arises immediately as to whether __every__ partial Steiner triple system (partial Steiner latin square) can be finitely embedded in a Steiner triple system (Steiner latin square). The question was answered in the affirmative by Christine Treash [20].

Theorem 4.2. (C. Treash [20]). A finite partial Steiner triple system can be finitely embedded.

   Treash's construction is __difficult__ and __large__. Since the proof is so very tedious we will not even indicate a proof here. The technique used in the proof is not necessary for what follows. A simple proof of Treash's Theorem would be very desirable ... also a much smaller embedding. Treash's results guarantee that a partial Steiner triple system (Q,q) of order $|Q| = n$ can be embedded in a Steiner triple system (S,t) of order $|S| < 2^{2n}$. The best possible result would <u>probably</u> be <u>close to</u> 2k + 1 where k is the smallest positive integer greater than n which is ≡ 1 or 3 (mod 6). Both of these problems are far from a solution.

The use of Treash's Theorem in order to obtaining embedding theorems for certain kinds of latin squares gives very large containing squares. As we shall subsequently see, different techniques can be used to improve the size of the containing square. But now we turn our attention to the use of Treash's Theorem in order to solve an interesting (at least to the author) collection of finite embedding problems.

5. EMBEDDING IDEMPOTENT LATIN SQUARES

By a partial <u>idempotent</u> latin square is meant a partial square with the additional requirement that cell (x,x) is <u>occupied</u> by x.

<u>Example 5.1.</u>

1	4			
4	2	1		
		3	2	
5			4	
				5

<u>partial 5 x 5 idempotent latin square</u>

The problem as to whether every partial <u>idempotent</u> latin square can be finitely embedded in an <u>idempotent</u> latin square is due to S. K. Stein. The solution to this problem was obtained by the author in [12] via the use of Treash's Theorem. Subsequently, A. J. W. Hilton [10] obtained a much smaller embedding using completely elementary methods. Hilton's method will be dealt with in a later section. We concern ourselves now with the original embedding solution (because the technique is quite useful in looking at similar problems).

Let P be a n × n partial idempotent latin square based on the set N = {1,2,3, ...,n}. If the only occupied cells are on the main diagonal then P can be completed since it is well known that there is an idempotent latin square of every size. We consider the case where there are t occupied cells of P off the main diagonal, say (a_1,b_1), (a_2,b_2), ..., (a_t,b_t) where cell (a_i,b_i) is occupied by c_i. In [12] it is shown that it is possible to construct a finite partial Steiner triple system (S,T) with the following properties:

    (1) $N \subseteq S$,

    (2)   T contains four triples of the form $\{a_i,b_i,x\}$, $\{a_i,c_i,y\}$,

           $\{b_i,c_i,z\}$, and $\{x,y,z\}$, where none of x, y, or z belong to N,

    (3)   if $x \in S \setminus N$, T contains exactly one triple of the form

           $\{a,b,x\}$, where a and b belong to N.

Now embed (S,T) in a finite Steiner triple system $(\overline{S},\overline{T})$ and let I be the Steiner latin square associated with $(\overline{S},\overline{T})$. We now transform I into a latin square containing P in its upper left hand corner. This is accomplished as follows. For each $i = 1,2,...,t$, let $T(a_i,b_i) = \{\{a_i,b_i,x\}, \{a_i,c_i,y\}, \{b_i,c_i,z\}, \{x,y,z\}\}$ and $R(a_i,b_i) = \{(a_i,b_i), (a_i,y), (z,b_i), (z,y)\}$. Now in I the cells (a_i,b_i) and (z,y) are occupied by x and the cells (a_i,y) and (z,b_i) are occupied by c_i. If the x's and the c_i's are interchanged in these four cells the result is a latin square with cell (a_i,b_i) occupied by c_i. In [12] it is shown that, for $i \neq j$, $R(a_i,b_i) \cap R(a_j,b_j) = \emptyset$ and so the above procedure can be performed on each of the <u>rectangles</u> $R(a_i,b_i)$ simultaneously. This, of course transforms I into a latin square Q with P embedded in the upper left hand corner. Q is clearly idempotent since none of the cells in any of the rectangles $R(a_i,b_i)$ belongs to the main diagonal. This gives the following theorem.

<u>Theorem 5.2</u>. (C. C. Lindner [12]). A finite partial idempotent latin square can be embedded in a finite idempotent latin square.

    Because Theorem 5.2 is based on Treash's Theorem the containing square can possibly be almost as large as 2^{2n}. A different construction due to A. J. W. Hilton (which we will look at in a later section) gives a containing square no larger than 4n [10]. This is still not the best possible result. The best possible general result would be that an n x n partial idempotent latin square can be embedded in an idempotent latin square no larger than 2n + 1. This problem is still open. As the following example illustrates we can frequently find containing squares smaller than those guaranteed by Theorem 5.2 or even Hilton's result.

Example 5.3.

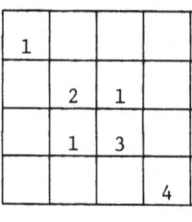

Partial 4 × 4 idempotent latin square. Note that P cannot be completed.

P

Let (S,T) be the partial Steiner triple system given by

$S = \{1,2,3,4,5,6,7\}$

$T = \{\{2,3,5\}, \{2,1,6\}, \{3,1,7\}, \{5,6,7\}\}$.

It is easily seen that (S,T) has properties (1), (2), and (3) stated in the proof

of Theorem 5.2. Let $(\overline{S},\overline{T})$ be the Steiner triple system given by

$\overline{S} = \{1,2,3,4,5,6,7\}$

$\overline{T} = \{\{2,3,5\}, \{2,1,6\}, \{3,1,7\}, \{5,6,7\}, \{1,4,5\}, \{2,4,7\}, \{3,4,6\}\}$.

Then (S,T), we have

$T(2,3) = \{\{2,3,5\}, \{2,1,6\}, \{3,1,7\}, \{5,6,7\}\}$

$R(2,3) = \{(2,3), (2,6), (7,3), (7,6)\}$

$T(3,2) = \{\{3,2,5\}, \{3,1,7\}, \{2,1,6\}, \{5,7,6\}\}$

$R(3,2) = \{(3,2), (3,7), (6,2), (6,7)\}$

1	6	7	5	4	2	3
6	2	5	7	3	1	4
7	5	3	6	2	4	1
5	7	6	4	1	3	2
4	3	2	1	5	7	6
2	1	4	3	7	6	5
3	4	1	2	6	5	7

Steiner latin square associated with $(\overline{S},\overline{T})$ with the rectangles R(2,3) and R(3,2) marked.

I

1	6	7	5	4	2	3
6	2	1	7	3	5	4
7	1	3	6	2	4	5
5	7	6	4	1	3	2
4	3	2	1	5	7	6
2	5	4	3	7	6	1
3	4	5	2	6	1	7

Latin square Q with P embedded in the upper left hand corner obtained from I by interchanging the entries in the rectangles R(2,3) and R(3,2).

Q

Remarks. The partial latin square P given in the above example is commutative. There are two reasons for this. The first is that this gives the simplest non-trivial example of a partial idempotent latin square which cannot be completed (so that the example does not become overly tedious) and secondly it will serve as an example for results obtained in the next section (on embedding partial commutative latin squares). The technique used in Theorem 5.2 to embed partial idempotent latin squares has nothing to do with the partial latin square being commutative. Finally, the partial Steiner triple system (S,T) has the same size as $(\overline{S}, \overline{T})$; i.e. $|S| = |\overline{S}|$. Quite obviously, Treash's Theorem was not used to obtain the embedding. The construction used in Theorem 5.2 uses only the fact that the partial Steiner triple system (S,T) can be finitely embedded. The means by which the embedding is obtained is incidental. Treash's Theorem guarantees a finite embedding can be obtained.

6. EMBEDDING COMMUTATIVE LATIN SQUARES

Shortly after the problem of embedding partial idempotent latin squares was solved the following two additional questions were raised by S. K. Stein.

   (1) Can a finite partial idempotent commutative latin square be embedded in a finite idempotent commutative latin square?, and

   (2) Can a finite partial commutative latin square be embedded in a finite commutative latin square?

Stein conjectured when he raised these two questions that the answer to (1) was probably very easy while the answer to (2) was probably less obvious. He was right on both counts.

Theorem 6.1. (C. C. Lindner [14]). A finite partial idempotent commutative latin square can be embedded in a finite idempotent commutative latin square.

Proof. The same construction used in the proof of Theorem 5.2 gives the desired completion. This is really not too surprising since a bit of reflection results in seeing that $T(x,y) = T(y,x)$ and therefore for every pair of occupied cells (x,y) and (y,x) in the partial square P that the rectangles $R(x,y)$ and $R(y,x)$ have the property that $(a,b) \in R(x,y)$ if and only if $(b,a) \in R(y,x)$. Since Steiner latin squares are commutative interchanging symbols in these rectangles does not affect commutativity.

We remark that although the cells in $R(x,y)$ and $R(y,x)$ are symmetrically placed about the main diagonal when the partial latin square P is commutative this is obviously not the case if P is not commutative. As promised, Example 5.3 serves as an example of the technique used to embedd partial idempotent commutative latin squares.

Having dispensed with the easy part of this section we proceed with the less obvious part.

Theorem 6.2. (C. C. Lindner [12]). A finite partial commutative latin square can be embedded in a finite commutative latin square.

Proof. The proof involves drawing the right kind of pictures. Let P be an $n \times n$ partial commutative latin square based on $1,2,\ldots,n$. Let I be the $3n \times 3n$ partial idempotent latin square based on $1,2,\ldots,3n$ given below.

The blocks (1,2), (1,3), (2,1), and (3,1) are empty and only the main diagonal of blocks (1,1) and (2,2) are occupied. In block (3,3) is P with the main diagonal of P removed. The blocks (2,3) and (3,2) are empty except for the projection of the occupied cells in P on the main diagonal (if any) onto their main diagonals. Now I is a finite partial idempotent commutative latin square and so by Theorem 6.1 can be embedded in a finite idempotent commutative latin square K based on $1,2,\ldots,k$.

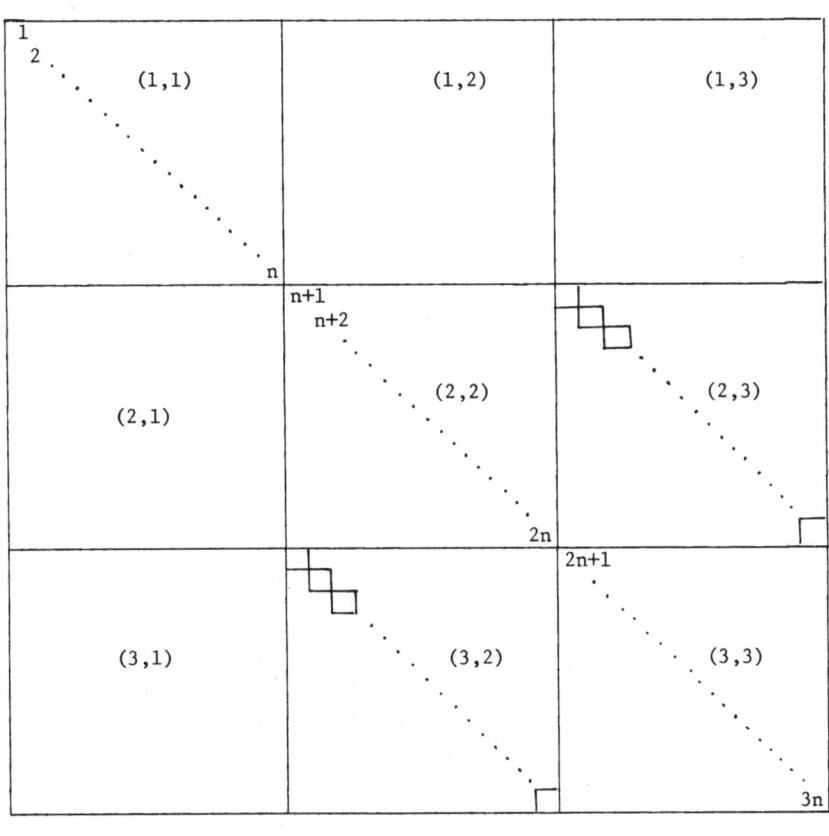

I

Now define a new latin square K(e) based on a new symbol e along with 1,2,...,k from K by "stripping" the main diagonal of K as shown in the accompanying diagram.

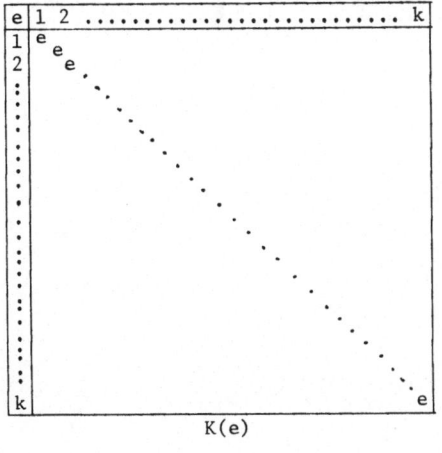

K(e)

The big block is just K with the entries on the main diagonal replaced by e (and the main diagonal projected onto the 1 × k and k × 1 blocks along the headline and sideline).

If cell (i,i) in P was occupied with c_i, then in $K(e)$ cells $(2n + i + 1, n + i + 1)$

and $(n + i + 1, 2n + i + 1)$ are occupied by c_i. Let $R(i,i) = \{(2n + i + 1, n+i+1),$

$(n + i + 1, 2n + i + 1), (2n + i + 1, 2n + i + 1), (n + i + 1, n + i + 1)\}$. Clearly,

if $i \neq j$ and cells (i,i) and (j,j) are occupied in P then $R(i,i) \cap R(j,j) = \emptyset$.

Hence interchanging the e's and the c_i's in each of these squares gives a copy of

P somewhere in $K(e)$. A suitable permutation of rows and columns now places P in

the upper left hand corner of a commutative latin square.

Example 6.3.

3	2	
2	1	
		3

P

Partial 3 × 3 commutative latin square. Note that P is not idempotent and cannot be completed.

1								
	2							
		3						
			4			3		
				5			1	
					6			3
			3			7	2	
				1		2	8	
					3			9

I

Idempotent commutative partial latin square with P less its main diagonal in the lower right hand corner and the main diagonal of P projected onto the diagonals of the big blocks (2,3) and (3,2).

1	9	2	5	8	7	4	3	6
9	2	1	7	3	8	5	6	4
2	1	3	8	7	9	6	4	5
5	7	8	4	6	2	3	9	1
8	3	7	6	5	4	9	1	2
7	8	9	2	4	6	1	5	3
4	5	6	3	9	1	7	2	8
3	6	4	9	1	5	2	8	7
6	4	5	1	2	3	8	7	9

K

I is embedded in K.
In fact I is actually
completed to K.

e	1	2	3	4	5	6	7	8	9
1	e	9	2	5	8	7	4	3	6
2	9	e	1	7	3	8	5	6	4
3	2	1	e	8	7	9	6	4	5
4	5	7	8	e	6	2	3	9	1
5	8	3	7	6	e	4	9	1	2
6	7	8	9	2	4	e	1	5	3
7	4	5	6	3	9	1	e	2	8
8	3	6	4	9	1	5	2	e	7
9	6	4	5	1	2	3	8	7	e

K(e)

K(e) is obtained from
K by projecting the
main diagonal onto
the headline and
sideline of K and
filling in the empty
cells with e. Note
that

$R(1,1) = \{(5,5),(5,8),(8,5),(8,8)\}$

$R(2,2) = \{(6,6),(6,9),(9,6),(9,9)\}$

$R(3,3) = \{(7,7),(7,10),(10,7),(10,10)\}$.

e	1	2	3	4	5	6	7	8	9
1	e	9	2	5	8	7	4	3	6
2	9	e	1	7	3	8	5	6	4
3	2	1	e	8	7	9	6	4	5
4	5	7	8	3	6	2	e	9	1
5	8	3	7	6	1	4	9	e	2
6	7	8	9	2	4	3	1	5	e
7	4	5	6	e	9	1	3	2	8
8	3	6	4	9	e	5	2	1	7
9	6	4	5	1	2	e	8	7	3

$\overline{K(e)}$

$\overline{K(e)}$ obtained from $K(e)$ by interchanging the entries in the squares $R(1,1)$, $R(2,2)$, and $R(3,3)$. P is now in the lower right hand corner of $\overline{K(e)}$.

3	2	8	7	4	5	6	e	9	1
2	1	7	8	3	6	4	9	e	5
8	7	3	9	6	4	5	1	2	e
7	8	9	e	1	2	3	4	5	6
4	3	6	1	e	9	2	5	8	7
5	6	4	2	9	e	1	7	3	8
6	4	5	3	2	1	e	8	7	9
e	9	1	4	5	7	8	3	6	2
9	e	2	5	8	3	7	6	1	4
1	5	e	6	7	8	9	2	4	3

C

A suitable permutation of rows and columns of $\overline{K(e)}$ embeds P in the upper left hand corner of C.

Some remarks are in order. In the proof of Theorem 6.1 (as in the proof of Theorem 5.2) Treash's Theorem guarantees a finite embedding. Any means of obtaining the embedding is legal. It is often possible to obtain (as has been previously mentioned) much smaller embeddings for a given partial Steiner triple system than the results guaranteed by Treash's Theorem. In fact, for every fairly small partial Steiner triple system of order n that the author has looked at, it has always been possible to embed this system by trial and error (without too much difficulty) in a Steiner triple system of order less than or equal to 2k + 1 where k is the smallest positive integer greater than or equal to n which is $\equiv$ 1 or 3 (mod 6). This is very tantilizing. However (as has also been previously noted) there is as yet no better general embedding theorem for partial Steiner triple systems than Treash's Theorem. In the proof of Theorem 6.2 any means of embedding the partial idempotent commutative latin square I in a finite idempotent commutative latin square K will do. As in Example 6.3 this can frequently be done by trial and error (lots of error) so that the resulting square is much smaller than the size guaranteed by Theorem 6.1. As a final comment (this has also been mentioned several times) we remark that regardless of how one goes about any of the embedding theorems looked at so far, a terrific amount of technique is involved most of which has been omitted.

7. CONJUGATE QUASIGROUPS AND IDENTITIES

We now dramatically switch from latin squares to quasigroup notation and terminology. We restate briefly in terms of quasigroups the results of sections 4, 5, and 6.

<u>Section 4</u>. <u>A Steiner quasigroup</u> is a quasigroup satisfying the identities

$$
\text{Steiner identities} \quad
\begin{cases}
x^2 = x & \text{idempotent identity} \\
\left.\begin{array}{l} x(xy) = y \\ (yx)x = y \end{array}\right\} & \text{symmetric identities}
\end{cases}
$$

A partial Steiner quasigroup is a partial idempotent quasigroup (Q,o), q o q = q
for all q ε Q, such that whenever p ≠ q ε Q and p o q is defined then so are q o p,
p o (p o q), (q o p) o p, q o (p o q), and (q o p) o q and furthermore p o q = q o p,
p o (p o q) = (q o p) o p = q, and q o (p o q) = (q o p) o q = p. In terms of
quasigroups Treash's Theorem says that a finite partial Steiner quasigroup can be
embedded in a finite Steiner quasigroup.

Section 5. A finite partial idempotent quasigroup can be embedded in a finite
idempotent quasigroup.

Section 6. A finite partial commutative (idempotent commutative) quasigroup can
be embedded in a finite commutative (idempotent commutative) quasigroup.

We now reintroduce the use of the conjugates of a quasigroup in a slightly
more algebraic setting. Let (Q, $\otimes$) be any (partial) quasigroup. Define six
(partial) binary operations $\otimes(1,2,3)$, $\otimes(1,3,2)$, $\otimes(2,1,3)$, $\otimes(2,3,1)$,
$\otimes(3,1,2)$, and $\otimes(3,2,1)$ on Q as follows: a $\otimes$ b = c if and only if

$$a \otimes (1,2,3)b = c$$
$$a \otimes (1,3,2)c = b$$
$$b \otimes (2,1,3)a = c$$
$$b \otimes (2,3,1)c = a$$
$$c \otimes (3,1,2)a = b$$
$$c \otimes (3,2,1)b = a$$

The six (not necessarily distinct) (partial) quasigroups (Q, $\otimes(i,j,k)$) are called
the conjugates of (Q,$\otimes$) [19]. Two quasigroup identities $w_1(x,y) = v_1(x,y)$ and
$w_2(x,y) = v_2(x,y)$ are said to be conjugate provided that if (Q,$\otimes$) is a quasigroup satisfying one of them, then at least one conjugate of (Q,$\otimes$) satisfies
the other.

Theorem 7.1. (S. K. Stein [19]). The quasigroup identities xy = yx, x(xy) = y,
and (yx)x = y are conjugates.

By a partial x(xy) = y ((yx)x = y) quasigroup is meant a partial quasigroup
(Q,o) such that whenever p o q is defined then so is p o (p o q) ((p o q) o q) and
furthermore p o (p o q) = q ((p o q) o q = p).

Corollary 7.2. Let $(Q,\otimes)$ be a partial $x(xy) = y$ $((yx)x = y$ quasigroup. Then at least one conjugate of $(Q,\otimes)$ is a partial commutative quasigroup.

Proof. Let $(Q,\otimes)$ be a partial $x(xy) = y$ quasigroup. Claim: $(Q,\otimes(3,2,1)$ is a partial commutative quasigroup. So let $x,y \in Q$ and suppose that $x \otimes(3,2,1)y = z$. Then $z \otimes(1,2,3)y = x$. Since $\otimes(1,2,3) = \otimes$ and Q satisfies $x(xy) = y$, $z \otimes y = x$ implies $z \otimes x = y$; i.e., $z \otimes(1,2,3)x = y$. Hence $y \otimes(3,2,1)x = z$ and so $(Q,\otimes(3,2,1))$ is a partial commutative latin square as claimed.

Corollary 7.3. Let $(Q,\otimes)$ be a finite partial $x(xy) = y$ $((yx)x = y)$ quasigroup. Then $(Q,\otimes)$ can be embedded in a finite $x(xy) = y$ $((yx)x = y)$ quasigroup.

Proof. Let $(Q,\otimes)$ be a finite partial $x(xy) = y$ quasigroup. Then $(Q,\otimes(3,2,1))$ is a finite partial commutative quasigroup. Embed $(Q,\otimes(3,2,1))$ in the finite commutative quasigroup $(V,\odot)$. Then $(V,\odot(3,2,1))$ is a $x(xy) = y$ quasigroup and $(Q,\otimes)$ is embedded in $(V,\odot(3,2,1))$.

Example 7.4.

$\otimes$	1	2	3
1		2	
2	2	1	
3	1		3

$(Q,\otimes)$

Partial $x(xy) = y$ quasigroup. Note that not only can $(Q,\otimes)$ not be completed to a $x(xy) = y$ quasigroup but that $(Q,\otimes)$ cannot be completed to any quasigroup.

$\otimes(3,2,1)$	1	2	3
1	3	2	
2	2	1	
3			3

$(Q,\otimes(3,2,1))$

Partial commutative quasigroup obtained from $(Q,\otimes)$ by conjugation.

The partial commutative quasigroup $(Q,\otimes(3,2,1))$ is the partial commutative quasigroup in Example 6.3 (after adding a headline and a sideline). Let $(V,\odot)$ be quasigroup in Example 6.3 in which $(Q,\otimes(3,2,1))$ is embedded. Then $(V,\odot(3,2,1))$ is a partial $x(xy) = y$ quasigroup and $(Q,\otimes)$ is embedded in $(V,\odot(3,2,1))$.

⊙ (3,2,1)	1	2	3	4	5	6	7	8	9	e
1	e	2	8	5	4	7	6	3	3	1
2	2	1	9	6	7	4	5	e	3	8
3	1	5	3	7	2	9	4	8	6	e
4	5	7	6	8	1	3	2	4	e	9
5	6	e	7	9	8	1	3	5	4	2
6	7	6	5	e	3	2	1	9	8	4
7	4	3	2	1	e	8	9	6	7	5
8	3	4	1	2	9	e	8	7	5	6
9	9	8	4	3	6	5	e	2	1	7
e	8	9	e	4	5	6	7	1	2	3

$(V, \odot(3,2,1))$ is a $x(xy) = y$ quasigroup obtained from $(V, \odot)$ by conjugation and $(Q, \otimes)$ is embedded in $(V, \odot(3,2,1))$.

$$(V, \odot(3,2,1))$$

Let I_1, I_2, I_3, I_4, I_5, I_6, I_7, I_8 denote the following collection of quasigroup identities.

$$I_1 = \{x^2 = x\}$$
$$I_2 = \{xy = yx\}$$
$$I_3 = \{x(xy) = y\}$$
$$I_4 = \{(yx)x = y\}$$
$$I_5 = \{x^2 = x, \ xy = yx\}$$
$$I_6 = \{x^2 = x, \ x(xy) = y\}$$
$$I_7 = \{x^2 = x, \ (yx)x = y\}$$
$$I_8 = \{x^2 = x, \ x(xy), \ (yx)x = y\} \ .$$

In view of the preceding results along with the observation that the idempotent identity $x^2 = x$ is invariant under conjugation we have the following theorem.

<u>Theorem 7.5</u>. A finite partial I_k quasigroup can be embedded in a finite I_k quasigroup for every $k \ \varepsilon \ \{1,2,3,4,5,6,7,8\}$.

Remarks. As we shall see in the next two sections there are some simpler methods
for the finite embeddings and some methods (not necessarily simpler) that give
smaller embeddings than those obtained in Theorems 5.2, 6.1, and 6.2. Therefore
some justification for their extensive use so far is in order.

(1) The constructions based on Treash's Theorem (that a finite
     partial Steiner triple system can be embedded in a finite
     Steiner triple system) are the original constructions used
     to solve the embedding problems mentioned so far. They are
     of interest in themselves. They give a unified approach to
     all of the problems considered in this set of notes.

(2) The problem of whether or not finite partial totally symmetric
     quasigroups (quasigroups satisfying the two identities x(xy) = y
     and (yx)x = y) can be finitely embedded in a totally symmetric
     quasigroup is still unsolved. Possibly the constructions used
     so far can be used to settle this problem. There seems (at
     least to the author) to be no way for the methods of Hilton
     and Cruse to gain a solution.

(3) Although not mentioned in this set of notes, the construction
     given in Theorem 5.2 to embed partial idempotent quasigroups
     can be used to show that finite partial quasigroups satisfying
     the identities x^2 = x and x(yx) = y can be finitely embedded
     in a quasigroup satisfying these identities [13]. This cannot
     be obtained via the methods of either Hilton or Cruse.

(4) These constructions may prove useful in dealing with some of
     the problems listed at the end of this set of notes ... they
     certainly cannot hurt.

8. HILTON'S METHOD

Shortly after the problem of finitely embedding partial idempotent latin squares was solved, A. J. W. Hilton [10] gave a different solution. Hilton's construction is simpler and gives a smaller containing square. Hilton's techniques can also be used to embed partial commutative latin squares which also gives a better bound than the original solution.

Theorem 8.1.   (A. J. W. Hilton [10]). A partial $n \times n$ idempotent latin square can be embedded in an idempotent latin square no larger than $4n \times 4n$.

Proof. Let P be a partial $n \times n$ idempotent latin square based on $1,2,...,n$. By Evans' Theorem P can be embedded in a $2n \times 2n$ latin square A based on $1,2,...,2n$.

A =

A is based on $1,2,...,$, $2n$ and the main diagonal of the block P is $1,2,...,n$ in that order.

Now let B, C, and D be the $2n \times 2n$ latin squares as defined below.

B =

B is based on $1,2,...,2n$ and the main diagonal of the block B_1 is $n+1,n+2,...,2n$ in that order.

C is based on $2n+1$, $2n+2$, ..., $4n$ and the main diagonal of the block C_4 is $2n+1, 2n+2, ..., 2n$ in that order.

D is based on $2n+1$, $2n+2$, ..., $4n$ and the main diagonal of the block D_4 is $3n+1, 3n+2, ..., 4n$ in that order.

Let I be the $4n \times 4n$ latin square based on $1,2,...,4n$ defined below. Then I is idempotent and contains P in the upper left hand corner.

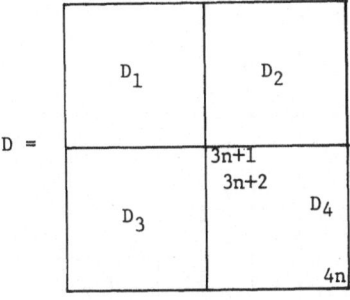

P is embedded in the idempotent latin square I.

Example 8.2.

$$P = \begin{array}{|c|c|c|c|}
\hline
1 & 4 & 2 & \\
\hline
 & 2 & & 3 \\
\hline
 & & 3 & \\
\hline
 & 1 & & 4 \\
\hline
\end{array}$$

Partial 4 × 4 idempotent latin square. Note that P <u>cannot</u> be completed.

A =

1	4	2	8	5	6	7	3
6	2	8	3	1	4	5	7
7	8	3	6	4	1	2	5
8	1	6	4	7	5	3	2
5	6	7	1	2	3	4	8
2	3	5	7	6	8	1	4
3	7	4	5	8	2	6	1
4	5	1	2	3	7	8	6

P is embedded in the upper left hand corner of A. T. Evans' Theorem guarantees this can be done.

B =

5	8	6	7	1	2	3	4
7	6	8	5	2	3	4	1
8	5	7	6	3	4	1	2
6	7	5	8	4	1	2	3
1	2	3	4	5	6	7	8
2	3	4	1	6	7	8	5
3	4	1	2	7	8	5	6
4	1	2	3	8	5	6	7

$$C = \begin{array}{|cccc|cccc|}
\hline
9 & 10 & 11 & 12 & 13 & 14 & 15 & 16 \\
10 & 11 & 12 & 9 & 14 & 15 & 16 & 13 \\
11 & 12 & 9 & 10 & 15 & 16 & 13 & 14 \\
12 & 9 & 10 & 11 & 16 & 13 & 14 & 15 \\
\hline
13 & 14 & 15 & 16 & 9 & 12 & 10 & 11 \\
14 & 15 & 16 & 13 & 11 & 10 & 12 & 9 \\
15 & 16 & 13 & 14 & 12 & 9 & 11 & 10 \\
16 & 13 & 14 & 15 & 10 & 11 & 9 & 12 \\
\hline
\end{array}
\qquad
D = \begin{array}{|cccc|cccc|}
\hline
13 & 14 & 15 & 16 & 9 & 10 & 11 & 12 \\
14 & 15 & 16 & 13 & 10 & 11 & 12 & 9 \\
15 & 16 & 13 & 14 & 11 & 12 & 9 & 10 \\
16 & 13 & 14 & 15 & 12 & 9 & 10 & 11 \\
\hline
9 & 10 & 11 & 12 & 13 & 16 & 14 & 15 \\
10 & 11 & 12 & 9 & 15 & 14 & 16 & 13 \\
11 & 12 & 9 & 10 & 16 & 13 & 15 & 14 \\
12 & 9 & 10 & 11 & 14 & 15 & 13 & 16 \\
\hline
\end{array}$$

1	4	2	8	13	14	15	16	5	6	7	3	9	10	11	12
6	2	8	3	14	15	16	13	1	4	5	7	10	11	12	9
7	8	3	6	15	16	13	14	4	1	2	5	11	12	9	10
8	1	6	4	16	13	14	15	7	5	3	2	12	9	10	11
9	10	11	12	5	8	6	7	13	14	15	16	1	2	3	4
10	11	12	9	7	6	8	5	14	15	16	13	2	3	4	1
11	12	9	10	8	5	7	6	15	16	13	14	3	4	1	2
12	9	10	11	6	7	5	8	16	13	14	15	4	1	2	3
13	14	15	16	1	2	3	4	9	12	10	11	5	6	7	8
14	15	16	13	2	3	4	1	11	10	12	9	6	7	8	5
15	16	13	14	3	4	1	2	12	9	11	10	7	8	5	6
16	13	14	15	4	1	2	3	10	11	9	12	8	5	6	7
5	6	7	1	9	10	11	12	2	3	4	8	13	16	14	15
2	3	5	7	10	11	12	9	6	8	1	4	15	14	16	13
3	7	4	5	11	12	9	10	8	2	6	1	16	13	15	14
4	5	1	2	12	9	10	11	3	7	8	6	14	15	13	16

P is embedded in the upper left hand corner of I.

I

As always, a lot of technique is involved in the above example. Hilton's result is the best general result to date on embedding partial idempotent latin squares. It is still not the best possible result. The best possible general embedding theorem would be that a partial idempotent latin square of size n can always be in an idempotent latin square of size no larger than 2n + 1.

We close this section by using Hilton's method to embed partial commutative latin squares.

Theorem 8.3. A partial n × n commutative latin square can be embedded in a commutative latin square no larger than 4n × 4n.

Proof. Let P be a partial n × n commutative latin square based on 1,2,...,n. Let T be a n × n commutative latin square based on n+1,n+2,...,2n. Let T(P) be the n × n commutative latin rectangle based on 1,2,...,2n obtained from P by filling in the unoccupied cells with the entries in the corresponding cells in T. Then T(P) is an n × n commutative latin rectangle based on 1,2,...,2n. By Ryser's Theorem T(P) can be embedded in a 2n × 2n latin square based on 1,2,...,2n. Let A be such an embedding and partition A as shown in the accompanying diagram.

$$
A = \begin{array}{|c|c|}
\hline
T(P) & A_1 \\
\hline
A_2 & A_3 \\
\hline
\end{array}
$$

Now let M be a n × n commutative latin square based on 2n+1,2n+2,...,3n and K a n × n commutative latin square based on 3n+1,3n+2,...,4n. Let N be the 4n × 4n commutative latin square based on 1,2,...,4n defined below, where A_1^T, A_2^T, and A_3^T denote the transposes of A_1, A_2, and A_3 respectively.

$$N = $$

M	K	T(P)	A_1
K	M	A_2	A_3
T(P)	A_2^T	M	K
A_1^T	A_3^T	K	M

Now transform N into the latin square N(P) given below.

$$N(P) = $$

T(P)	K	M	A_1
K	M	A_2	A_3
M	A_2^T	T(P)	K
A_1^T	A_3^T	K	M

N(P) is a 4n × 4n commutative latin square which contains P in the upper left hand corner.

Remarks. Since the identities xy = yx, x(xy) = y, and (yx)x = y are conjugate, Theorem 8.3 also shows that a partial x(xy) = y ((yx)x = y) quasigroup of order n can be embedded in a x(xy) = y ((yx)x = y) quasigroup of order 4n. This is still not the best possible result along these lines. The best possible result would be 2n and this has recently been obtained by Allan Cruse [2]. Section 9 will be devoted to Cruse's method of embedding partial commutative and idempotent commutative latin squares.

9. CRUSE'S METHOD

In this final section we give Allan Cruse's constructions for finitely embedding partial commutative and partial idempotent commutative latin squares [2]. Cruse's constructions yield the best possible general embeddings for these types

of latin squares. However, Cruse's constructions are technically difficult (much
more so than Theorem 8.3 and possibly as difficult as Treash's Theorem). Cruse's
embedding theorems follow almost immediately from the following analogue of Ryser's
Theorem for commutative latin rectangles. We will not indicate a proof here.

Theorem 9.1. (A. Cruse [2]). Let R be $r \times r$ commutative latin rectangle based on
$1,2,\ldots,n$. Let $N(i)$ denote the number of times that the integer i occurs in R.
Then R can be embedded in an $n \times n$ commutative latin square if and only if
$N(i) \geq 2r - n$ for every $i = 1,2,\ldots,n$ and at least r of the $N(i)$ have the same
parity as n.

Example 9.2.

$$R = \begin{array}{|c|c|c|} \hline 1 & 2 & 4 \\ \hline 2 & 5 & 1 \\ \hline 4 & 1 & 5 \\ \hline \end{array}$$

3×3 commutative latin rectangle
based on $1,2,3,4,5,6,7$.

In order for R to be embeddable in a 7×7 commutative latin square (1) $N(i) \geq$
$2 \cdot 3 - 7 = -1$ for each i, and (2) at least 3 of the $N(i)$ must be odd (have the
same parity as 7). Condition (1) is trivially satisfied but condition (2) is not
since only $N(1) = 3$ is odd. Hence R cannot be embedded in a 7×7 commutative
latin square. If we change the entry in cell (2,2) of R to 6 then the resulting
3×3 commutative latin rectangle R can be embedded in a 7×7 commutative latin
square since condition (1) is still trivially satisfied, but now condition (2) is
satisfied since $N(1) = 3$, $N(5) = 1$, and $N(6) = 1$ are odd.

We remark that Cruse's construction involves adding an "edge" at a time to
the commutative latin rectangle (via the use of SDR's) until it is extended to
the desired commutative latin square. As always, a lot of technique is involved.

Theorem 9.3. (A. Cruse [2]). A partial $n \times n$ commutative latin square can be
embedded in a $t \times t$ commutative latin square for every even $t \geq 2n$.

Proof. The proof is obtained by slightly modifying Evans' proof of the embeddability of partial latin squares which are not necessarily commutative. Let P be
an $n \times n$ partial commutative latin square based on $1,2,\ldots,n$ and let $t \geq 2n$ be an

even positive integer. Write t = k + n and let K be a k x k commutative latin

square based on n+1,n+2,...,n+k = t. Using K to fill in the unoccupied cells in

P turns P into an n x n commutative latin rectangle $\overline{P}$ based on 1,2,...,t. Condition

(1) satisfied since 2n − t ≤ 0. To see that condition (2) is satisfied note that

t is even and that at least n of 1,2,...,t do not occur on the main diagonal of P

and so N(i) is even for these numbers. Hence $\overline{P}$ can be embedded in a t × t com-

mutative latin square by Theorem 9.1.

Corollary 9.4. (A. Cruse [2]). A partial n × n idempotent commutative latin

square can be embedded in an idempotent commutative latin square of size 2n + 1

(or any larger odd number).

Proof. The proof is completely analogous to the proof of Theorem 9.3 except that

now K is taken to be an idempotent commutative latin square of size either t − n

or t − n − 1 (depending on which of these numbers is odd).

We close this section with an example of the construction used in Theorem 9.3.

Example 9.5.

P =

1			
	4	3	
	3	2	4
		4	1

4 x. 4 partial commutative latin square which cannot be completed to a commutative latin square or any latin square for that matter.

K =

5	6	7	8
6	7	8	5
7	8	5	6
8	5	6	7

4 × 4 commutative latin square based on 5,6,7,8.

$\overline{P}$ =

1	6	7	8
6	4	3	5
7	3	2	4
8	5	4	1

4 × 4 commutative latin rectangle based on 1,2,3,4,5,6,7,8 obtained from P by filling in the empty cells with the corresponding entries in K.

$\overline{P}$ satisfies the conditions of Theorem 9.1 in order to be embedded in an 8 × 8 commutative latin square. As mentioned after the statement of Theorem 9.1 this is done by adding an edge at a time via the use of SDR's.

1	6	7	8	4
6	4	3	5	1
7	3	2	4	6
8	5	4	1	2
4	1	6	2	3

M_1

1	6	7	8	4	2
6	4	3	5	1	7
7	3	2	4	6	5
8	5	4	1	2	6
4	1	6	2	3	8
2	7	5	6	8	3

M_2

1	6	7	8	4	2	3
6	4	3	5	1	7	8
7	3	2	4	6	5	1
8	5	4	1	2	6	7
4	1	6	2	3	8	5
2	7	5	6	8	3	4
3	8	1	7	5	4	2

M_3

1	6	7	8	4	2	3	5
6	4	3	5	1	7	8	2
7	3	2	4	6	5	1	8
8	5	4	1	2	6	7	3
4	1	6	2	3	8	5	7
2	7	5	6	8	3	4	1
3	8	1	7	5	4	2	6
5	2	8	3	7	1	6	4

M_4

The last latin square on the right M_4 is a commutative latin square with P embedded in the upper left hand corner.

Great care must be taken in adding an edge at a time in expanding $\overline{P}$ to M_4. The "right" SDR must be chosen in adding each edge or it may become necessary to start over. (The use of SDR's in proving Theorem 2.2 involves no such restriction; i.e., as soon as a row is completed it is never necessary to undo anything). Eventually, Theorem 9.1 guarantees that after a finite length of time and a lot of effort we can add the necessary edges.

10. CONCLUDING REMARKS

In this set of notes I have tried to give a fairly comprehensive survey of finite embedding theorems for a certain collection of partial latin squares. Quite obviously there are embedding theorems not mentioned. For example, partial idempotent $x(yx) = y$ quasigroups can be finitely embedded [13]. However a look at the problems in the next section along with the references should give a fairly accurate picture of the state of finite embedding theorems for partial latin squares and quasigroups. Next, I should mention the connection between the finite embedding theorems obtained in this paper and universal algebra. This is due to Trevor Evans. If I_k is any one of the eight collections of quasigroup identities defined in section 7 then the fact that any partial I_k quasigroup can be embedded in a finite I_k quasigroup can be used to show (via [3], [5], [6], and [7]) that the finitely presented quasigroups in the quasigroup varieties I_k are residually finite, Hopfian, and have a solvable word problem. The reader is referred to [12] and [14] for the appropriate details. Finally, I have purposely omitted all but the simplest proofs in order not to become bogged down in details and therefore lose sight of the main objectives of this set of notes (whatever they are !!). This may give the impression that much of what has been discussed is really not very difficult. An even cursory look at the papers on which these notes are based will quickly dispel any such ideas.

We close out these notes with some problems (next section).

11. PROBLEMS

(1) The Evans' Conjecture remains open [4].

(2) Give a simple proof of Treash's Theorem (that a finite partial Steiner triple system can be embedded in a finite Steiner triple system [20]).

(3)   Treash's Theorem guarantees that a partial Steiner
       triple system of order n can be embedded in a Steiner
       triple system no larger than 2^{2n} [20]. Find a smaller
       general embedding theorem. The best possible result
       is probably 2k + 1 where k is the smallest positive
       integer greater than or equal to n which is $\equiv$ 1 or 3
       (mod 6).

(4)   The best general embedding theorem (to date) for partial
       idempotent latin squares due to A. J. W. Hilton [10]
       guarantees that a partial idempotent latin square of size
       n can be embedded in an idempotent latin square of size
       4n.   The best possible result would be 2n + 1. Can this
       be obtained?

(5)   A quasigroup satisfying the identities x(xy) = y and
       (yx)x = y is called totally symmetric. If the idempotent
       identity x^2 = x is added we have the Steiner identities.
       Although Treash's Theorem guarantees that a finite partial
       Steiner quasigroup can be embedded in a finite Steiner
       quasigroup, things are different if the idempotent identity
       is dropped. In fact the problem as to whether a finite
       partial totally symmetric quasigroup can be embedded in a
       finite totally symmetric quasigroup is open [14]. This
       problem seems closely related to the other embedding theorems
       surveyed in this set of notes. However all attempts by the
       author to gain a solution using the techniques used in this
       paper have failed.

(6)   Allan Cruse [2] has shown that a partial n × n commutative
      (idempotent commutative) latin square can be embedded in a
      2n × 2n commutative $(2n + 1) \times (2n + 1)$ idempotent commutative)
      latin square. These are the best possible results. However
      both are based on Cruse's analogue of Ryser's Theorem [17].
      A simple construction would be desirable.

(7)   In [13] it is shown that a finite partial idempotent $x(yx) = y$
      quasigroup can be finitely embedded. The identities $x^2 = x$
      and $x(yx) = y$ are consequences of the Steiner identities; i.e.,
      every quasigroup which satisfies the Steiner identities also
      satisfies the identities $x^2 = x$, $x(yx) = y$. Let I be a collec-
      tion of identities (including $x^2 = x$) which are consequences
      of the Steiner identities. Under what conditions can a finite
      partial I quasigroup be embedded in a finite I quasigroup?
      This is probably very hard ... and is a good place to stop.

Postscript.   Since the writing of this paper problem (5) has been solved by
R. Chaffer, M. Egger, R. St. Andre, and D. Smith. In this same paper they also
show that partial $x(yx) = y$ quasigroups (not necessarily idempotent) can be finitely
embedded. The techniques used in both proofs are based on the construction given
in Theorem 5.2 in order to embed partial idempotent latin squares.

REFERENCES

1. Bruck, R. H., What is a Loop? *Studies in Modern Algebra* (A. A. Albert, ed.), Prentice Hall, Englewood-Cliffs, N. J., (1963).

2. Cruse, Allen, On Embedding Incomplete Commutative Latin Squares, *J. Combinatorial Theory*, (to appear).

3. Evans, Trevor, Embeddability and the Word Problem, *J. London Math. Soc.*, 28 (1953), 76-80.

4. Evans, Trevor, Embedding Incomplete Latin Squares, *Amer. Math. Monthly*, 67 (1960), 958-961.

5. Evans, Trevor, Finitely Presented Loops, Lattices, etc. are Hopfian, *J. London Math. Soc.*, 44 (1969), 551-552.

6. Evans, Trevor, Some Connections Between Residual Finiteness, Finite Embeddability and the Word Problems, *J. London Math. Soc.* (2), 1 (1969), 399-403.

7. Evans, Trevor, The Word Problem for Abstract Algebras, *J. London Math. Soc.*, 26 (1971), 64-71.

8. Hall, Marshall, An Existence Theorem for Latin Squares, *Bull. Amer. Math. Soc.*, (1945), 387-388.

9. Hall, Marshall, A Combinational Problem on Abelian Groups, *Proc. Amer. Math. Soc.*, 3 (1952), 584-587.

10. Hilton, A. J. W., Embedding an Incomplete Diagonal Latin Square in a Complete Diagonal Latin Square, *J. Combinatorial Theory*, 15A (1973), 121-128.

11. Lindner, C. C., On Completing Latin Rectangles, *Canad. Math. Bull.*, 13 (1970), 65-68.

12. Lindner, C. C., Embedding Partial Idempotent Latin Squares, *J. Combinatorial Theory*, 10A (1971), 240-245.

13. Lindner, C. C., Finite Partial Cyclic Triple Systems Can be Finitely Embedded, *Algebra Universalis* 1 (1971), 93-96.

14. Lindner, C. C., Finite Embedding Theorems for Partial Latin Squares, Quasigroups, and Loops, *J. Combinatorial Theory*, 13A (1972), 339-345.

15. Marica, J. and Schönheim, Incomplete Diagonals of Latin Squares, *Canad. Math. Bull.*, 12 (1969), 235.

16. Moore, E. H., Concerning Triple Systems, *Math. Ann.*, 43 (1893), 271-285.

17. Ryser, H. J., A Combinatorial Theorem with an Application to Latin Rectangles, *Proc. Amer. Math. Soc.*, 2 (1951), 550-552.

18. Ryser, H. J., Combinational Mathematics, The Carus Mathematical Monographs, No. XIV, Math. Assoc. Amer., (1963).

19. Stein, S. K., On the Foundations of Quasigroups, *Trans. Amer. Math. Soc.*, 85 (1957), 228-256.

20. Treash, C., The Completion of Finite Incomplete Steiner Triple Systems with Application to Loop Theory, *J. Combinatorial Theory*, 10A (1971), 259-265.

COMPUTING THE CHARACTERISTIC POLYNOMIAL OF A GRAPH

Allen J. Schwenk*
University of Michigan

ABSTRACT

How can one actually compute the eigenvalues of a graph? In principal, there
are three methods. Namely, (1) we can search for p orthogonal eigenvectors,
(2) we can determine the first p moments by counting closed walks and then find
the spectrum from the moments, or (3) we can use certain subgraphs to determine the
coefficients of the characteristic polynomial and then find its roots.

In practice, however, all of these approaches may prove to be too tedious. If
the spectrum is a natural concept, then we should expect to find simple relations
between the spectra of related graphs. This is indeed the case. In Sections 2 and
3, we present four theorems which are structural results linking the spectrum of a
graph to the spectra of certain subgraphs. Then in Section 4, we find the spectra
of graphs formed by certain binary operations. In the final section, we apply these
results to obtain the spectra of several known families of graphs: complete graphs,
complete bigraphs, cubes, cycles, wheels, paths, ladders, and möbius ladders.

*Research supported in part by grant 73-2502 from the Air Force Office of Scienti-
fic Research. This paper was part of the author's doctoral dissertation "The Spectrum of a Graph," The University of Michigan, 1973.

COMPUTING THE CHARACTERISTIC POLYNOMIAL OF A GRAPH

1. INTRODUCTION

In general, we adhere to the terminology in Harary's book [6] with the following additions. When we refer to the <u>characteristic polynomial</u> of graph G, written $\phi(G;x) = \sum_{n=0}^{p} a_n x^{p-n}$, we really mean the characteristic polynomial of the adjacency matrix of G. For brevity, write $\phi(G;x) = \phi(G) = \phi(x)$ provided the omitted variable is obvious from the context. The roots of $\phi(G)$ are the <u>eigenvalues</u> of G, denoted λ_1, λ_2, ..., λ_p. These p roots (which may include multiple roots) comprise the <u>spectrum</u> of G. Two graphs are <u>cospectral</u> if they have the same spectrum. We let $V(G)$ denote the points of G. In particular, if H is a subgraph of G, then $G - V(H)$ is the graph remaining when the points of H are removed from G.

Many investigators have interpreted some or all of the coefficients of $\phi(G;x)$ in terms of graphical structure. Collatz and Sinogowitz [1] noted that $a_1 = 0$, that $-a_2$ is the number of lines of G, and $-a_3/2$ is the number of triangles in G. Harary [5] examined the determinant of the adjacency matrix, or in other words, a_p. His result is easily generalized to interpret all of the coefficients. This has been done (Theorem 1) independently by Sachs [9] and Spialter [10].

Let $c(G)$ and $k(G)$ be the numbers of cycles and components in G. In a <u>mutation graph</u>, each component is either a cycle or K_2; these graphs were introduced in Harary [5]. Each permutation with no fixed points can be viewed as a one-to-one mapping taking each point of the underlying mutation graph to one of its neighbors. Note that a given mutation graph G corresponds to $2^{c(G)}$ permutations, since each cycle may be mapped in either of two directions. Finally, let G_n be the set of those n-point subgraphs of G which are mutations.

<u>Theorem 1</u>. For $n \geq 1$, $a_n = \sum_{M \varepsilon G_n} (-1)^{k(M)} (2)^{c(M)}$.

A detailed proof of this theorem is presented in the works of Sachs [9] and Spialter [10].

2. THE SPECTRUM OF G MINUS A POINT OR A LINE

Our first two results display respectively the relations between the characteristic polynomial of graph G and the polynomials of G minus one point and G minus one line.

It is both convenient and consistent to define the characteristic polynomial of the empty graph to be $\phi(K_o) = 1$.

__Theorem 2.__ Let v be a point of graph G and let $\mathcal{C}(v)$ be the collection of cycles containing v. Then $\phi(G)$ satisfies

(1) $\quad \phi(G) = x\phi(G-v) - \sum\limits_{u \text{ adj } v} \phi(G-v-u) - 2 \sum\limits_{Z \epsilon \mathcal{C}(v)} \phi(G-V(Z)).$

__Proof.__ Recall that $\phi(G) = \Sigma a_i x^{p-i}$, and Theorem 1 expresses a_i in terms of the i-point mutation subgraphs. We now present a one-to-one correspondence between those mutations contributing to a_i on the left and those contributing to one of the terms on the right. Namely, letting M be an i-point mutation of G, we have three possibilities:

(A) If $v \notin M$, let M' be the same mutation, only now viewed as a subgraph of $G - v$.

(B) If $v \epsilon K_2 \subset M$, let M' be $M - V(K_2)$ viewed as a subgraph of $G - V(K_2)$.

(C) If $v \epsilon C_n \subset M$, let M' be $M - V(C_n)$ viewed as a subgraph of $G - V(C_n)$.

It is easy to see that this does indeed establish a one-to-one correspondence. Moreover, if M contributes an amount m toward the coefficient of x^{p-1} on the left, we observe that on the right M' also contributes m in each case as we now demonstrate:

(A) Since M' = M, we see that M' contributes m to the coefficient of x^{p-1-i} in $\phi(G-v)$, and thus supplies m toward the coefficient of x^{p-1} in $x\phi(G-v)$.

(B) In this case M' = M - V(K_2), so M' contributes

(2) $\quad (-1)^{k(M')} (2)^{c(M')} = -(-1)^{k(M)} (2)^{c(M)} = -m$ to $x^{p-2-(i-2)} = x^{p-i}$ in

$\phi(G-v-u)$. Hence, it supplies $+mx^{p-i}$ to $-\phi(G-v-u)$.

(C)   Finally, here we have $M' = M - V(C_n)$, so M' contributes

(3) $(-1)^{k(M')} (2)^{c(M')} = -\frac{1}{2} (-1)^{k(M)} (2)^{c(M)} = -m/2$ to $x^{p-n-(i-n)} = x^{p-i}$

in $\phi(G-V(C_n))$. Thus, M' contributes $+mx^{p-i}$ in $-2\phi(G-V(C_n))$.

Thus, the contribution of each mutation M to the left side is matched by a corresponding contribution on the right side by the mutation M'. This completes the proof of the theorem.

Let us consider a special case of this theorem when v is an endpoint and hence lies on no cycles. Then equation (1) becomes Theorem 2 of Harary, King, Mowshowitz, and Read [7].

Corollary 2a. If v is an endpoint of graph G and u is the point adjacent to v, then

(4) $\phi(G) = x\phi(G-v) - \phi(G-v-u).$

Another simplification of (1) occurs when v is a cutpoint. The coalescence of two rooted graphs G,r and H,s, denoted $G \cdot H$, is the graph formed by identifying the two roots. Thus, the new root t becomes a cutpoint joining G to H.

Corollary 2b. If G and H are two rooted graphs with roots r and s, then the characteristic polynomial of the coalescence $G \cdot H$ is

(5) $\phi(G \cdot H) = \phi(G)\phi(H-s) + \phi(G-r)\phi(H) - x\phi(G-r)\phi(H-s).$

Proof. We apply Theorem 2 to $G \cdot H$ with the coalesced point t as the point v to get

(6) $\phi(G \cdot H) = x\phi(G \cdot H-t) - \sum_{u \text{ adj } t} \phi(G \cdot H-t-u) - 2 \sum_{z \varepsilon \mathscr{C}(t)} \phi(G \cdot H-V(Z)).$

Next we observe that removing t from $G \cdot H$ yields the same graph as the union of $G - r$ and $H - s$. But it is trivial that the characteristic polynomial of the union is just the product of the component polynomials. Thus, we may replace $x\phi(G \cdot H-t)$ by $x\phi(G-r)\phi(H-s)$ in (6). Since t is a cutpoint in $G \cdot H$, each cycle containing t is either a cycle of G containing r or a cycle of H containing s. Similarly, each point adjacent to t is either in G or in H. Thus, we may split up the sums in (6) according as u (or Z) is in G or H, that is,

(7) $\phi(G \cdot H) = x\phi(G-r)\phi(H-s) - \underset{u \text{ adj } r}{\Sigma} \phi(G \cdot H-t-u) - \underset{u \text{ adj } s}{\Sigma} \phi(G \cdot H-t-u)$

$-2 \underset{Z \varepsilon \mathcal{C}(r)}{\Sigma} \phi(G \cdot H-V(Z)) - 2 \underset{Z \varepsilon \mathcal{C}(s)}{\Sigma} \phi(G \cdot H-V(Z)).$

But now each of the graphs obtained by removing some points from G · H can be re-written as a union of disjoint graphs so that here too the characteristic polynomial can be replaced by the product of the component polynomials. Thus we now have

(8) $\phi(G \cdot H) = x\phi(G-r)\phi(H-s) - \underset{u \text{ adj } r}{\Sigma} \phi(G-r-u)\phi(H-s)$

$- \underset{u \text{ adj } s}{\Sigma} \phi(G-r)\phi(H-s-u) - 2 \underset{Z \varepsilon \mathcal{C}(r)}{\Sigma} \phi(G-V(Z))\phi(H-s)$

$-2 \underset{Z \varepsilon \mathcal{C}(s)}{\Sigma} \phi(G-r)\phi(H-V(Z))$

Next, we group the first and third sums together and factor out $\phi(H-s)$, and semi-larly, we factor $\phi(G-r)$ out of the second and fourth sums to obtain

(9) $\phi(G \cdot H) = x\phi(G-r)\phi(H-s) - \phi(H-s)[\underset{u \text{ adj } r}{\Sigma} \phi(G-r-u) + 2 \underset{Z \varepsilon \mathcal{C}(r)}{\Sigma} \phi(G-V(Z))]$

$-\phi(G-r)[\underset{u \text{ adj } s}{\Sigma} \phi(H-s-u) + 2 \underset{Z \varepsilon \mathcal{C}(s)}{\Sigma} \phi(H-V(Z))].$

But now the quantities in brackets can be replaced by applying Theorem 2 just to G and just to H respectively, yielding

(10) $\phi(G \cdot H) = x\phi(G-r)\phi(H-s) - \phi(H-s)[x\phi(G-r)-\phi(G)] - \phi(G-r)[x\phi(H-s)-\phi(H)],$

which simplifies to equation (5).

As an example, we apply Theorem 2 to the cycle C_n to get

(11) $\phi(C_n) = x\phi(P_{n-1}) - 2\phi(P_{n-2}) - 2 \cdot$

Thus, the characteristic polynomial of the cycle can be found from the polynomials of paths, or conversely, if the polynomials of the cycles are known, we may use (11) to find the polynomials of the paths recursively. On the other hand, applying Corollary 2a to the path P_n yields

(12) $\phi(P_n) = x\phi(P_{n-1}) - \phi(P_{n-2})$,

which can be used to find the characteristic polynomial of the path directly.

The path P_{2n-1} can be thought of as the coalescence of two copies of P_n rooted at an endpoint, so from Corollary 2b we have

(13) $\phi(P_{2n-1}) = 2\phi(P_{n-1}) \phi(P_{n-1}) - x\phi^2(P_{n-1})$,

so that

$$(14) \qquad \phi(P_{2n-1}) = \phi(P_{n-1})(2\phi(P_n) - x\phi(P_{n-1})).$$

Thus, we conclude that every eigenvalue of P_{n-1} is also an eigenvalue of P_{2n-1}. This will be evident when we present the spectra of paths later in this paper. In fact, we observe that Corollary 2b implies that whenever $\phi(G-r) = \phi(H-s)$, then $\phi(G-r)$ divides $\phi(G \cdot H)$.

Since the dependence of $\phi(G \cdot H)$ upon G involves only $\phi(G)$ and $\phi(G-r)$, we are led to define the following concept. Two rooted graphs G_1, r_1 and G_2, r_2 are called <u>cospectrally rooted</u> if, not only are they cospectral as graphs, but also $\phi(G_1-r_1) = \phi(G_2-r_2)$. We now have the following useful result.

<u>Corollary 2c.</u>   If G_1 and G_2 are cospectrally rooted and H is any rooted graph, then $\phi(G_1 \cdot H) = \phi(G_2 \cdot H)$.

In general, a cospectral pair of connected graphs cannot be used to build a cospectral pair of larger connected graphs, but this corollary asserts that a cospectrally rooted pair <u>can</u> be used to build larger cospectral pairs. Moreover, if we form the successive coalescence of n copies of G_1 or G_2, the spectrum of the graph formed is independent of the number of copies of G_1. Thus, this construction can be used to provide n+1 cospectral connected graphs, because the polynomial with n graphs G_i in the coalescence, $\phi(G_1 \cdot \ldots \cdot G_1 \cdot G_2 \cdot \ldots \cdot G_2)$, is independent of the number of occurrences of G_1 in it. Figure 1 displays the two smallest pairs of cospectrally rooted graphs and trees. Using these in Corollary 2c, we can construct n+1 co-spectral connected graphs on 6n + 1 points, and n+1 cospectral trees on 8n + 1 points.

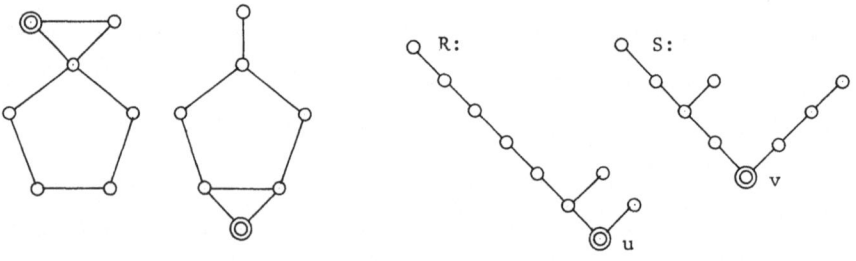

cospectrally rooted graphs   cospectrally rooted trees

<u>Figure 1.   The Two Smallest Pairs of Cospectrally Rooted Graphs and Trees</u>

The next theorem is a similar result relating $\phi(G)$ to the characteristic polynomial of G minus one line. The method of proof is identical to that of Theorem 2 and is omitted.

<u>Theorem 3.</u> Let uv be a line of G and let $\mathcal{C}$(uv) be the set of all cycles containing this line. The characteristic polynomial of G satisfies

(15) $\phi(G) = \phi(G-uv) - \phi(G-u-v) - 2 \sum_{Z\varepsilon\,\mathcal{C}(uv)} \phi(G-V(Z))$.

The next result follows directly from Theorem 3.

<u>Corollary 2.2a</u>   Let G, r and H, s be two disjoint rooted graphs. The characteristic polynomial of the graph formed by joining the roots with a new line rs is

(16) $\phi(G \cup H \cup rs) = \phi(G)\phi(H) - \phi(G-r)\phi(H-s)$.

Notice that if G and H are cospectrally rooted, then equation (16) becomes

(17) $\phi(G \cup H \cup rs) = \phi^2(G) - \phi^2(G-r)$.

In particular, this equation is satisfied when G and H are two isomorphic rooted graphs. For example, the path P_{2n} can be viewed as the result of joining two copies of P_n by a new line joining corresponding endpoints. Thus, by (17) we have

(18) $\phi(P_{2n}) = \phi^2(P_n) - \phi^2(P_{n-1})$.

Theorem 3 is closely related to Theorem 2 as can be seen by applying Theorem 3 to the cycle to get

(19) $\phi(C_n) = \phi(P_n) - \phi(P_{n-2}) - 2$.

This formula can also be obtained by use of equations (11) and (12) which resulted from two applications of Theorem 2.

3.   SOME SPECIAL SPECTRA

The automorphism group of G induces a partition of its points into orbits. If u and v are in the same orbit, say θ_1, then certainly for any orbit θ_j, point u must have as many neighbors in θ_j as v does. Thus, denoting the set of points adjacent to v as N(v), we have

(20) $|N(u) \cup \theta_j| = |N(v) \cup \theta_j|$.

A partition $V_1 \cup V_2 \cup ... \cup V_m$ of V(G) is <u>equitable</u> if for each i and for all u,v ε V_i, $|N(u) \cap V_j| = |N(v) \cap V_j|$ for all j. The following theorem is implicit within the proof of a result of Mowshowitz [8].

<u>Theorem 4.</u> If $V_1, V_2, \ldots V_m$ is an equitable partition of graph G with $t_{ij} = |N(v) \cap V_j|$ for $v \in V_i$, and T is the matrix (t_{ij}), then $\phi(T)$ divides $\phi(G)$.

<u>Sketch of Proof</u>. For each eigenvector $\alpha = (a_1, \ldots, a_m)$ of T, we find a corresponding eigenvector β of G by setting the v_i coordinate of β equal to a_j where $v_i \in V_j$. Vector β is easily seen to be an eigenvector of the adjacency matrix A of G. Moreover, any linear dependence among the vectors so formed must provide an analogous dependence among the eigenvectors of T, and so we conclude that the vectors so found are linearly independent. Thus, $\phi(T)$ divides $\phi(G)$.

As mentioned above, most applications of this theorem use the automorphism group to partition the points into the orbits, but in fact other partitions are possible and useful, as in the proof of the next corollary. For example, the partition of V(G) into singletons is always equitable; however, in this case the matrix T is trivially seen to be A itself, and the conclusion of the theorem becomes a tautology. At the other extreme, if G is regular, we may take an equitable partition with one orbit. Then T is the 1 x 1 matrix (r), so $\phi(T) = x - r$, and the theorem asserts that $x - r$ divides $\phi(G)$, as is well known.

We now wish to apply this result to find a common formula for the characteristic polynomials of the coalescence graphs $G_1 \cdot \ldots \cdot G_1 \cdot G_2 \cdot \ldots \cdot G_2$ mentioned after Corollary 2c. Given the rooted graph G, r, let $G * n$ be the graph formed by adding n-1 new points each adjacent to the same points as r.

<u>Corollary 4a</u>. If $G_1, G_2, \ldots, G_n$ are n cospectrally rooted graphs, then

(21) $x^{n-1} \phi(G_1 \cdot G_2 \cdot \ldots \cdot G_n) = \phi^{n-1}(G_1 - r_1) \phi(G_1 * n)$

<u>Proof</u>. We have already seen that in (21), the graph $G_1 \cdot G_2 \cdot \ldots \cdot G_n$ may be replaced by $G_1 \cdot G_1 \cdot \ldots \cdot G_1$, which we now denote $n * G_1$. (This notation is intentionally similar to $G_1 * n$ because these two graphs play rather symmetric roles in the discussion that follows. It is hoped that no confusion will result.) If G_1 has p points, we define a partition of $n * G_1$ with p parts by letting orbit V_1 be the coalesced root, and letting each orbit $V_2, \ldots, V_p$ be the n copies of point v_i for each point other than the root in G_1. This is an equitable partition, so we proceed to examine the structure of the matrix T of Theorem 4. Since each copy of G_1 in $n * G_1$ has only one point from each orbit, we see that the entries of T are

given by

(22)
$$t_{1j} = \begin{cases} n & \text{if } r_1 \text{ adj } v_j \text{ in } G_1 \\ 0 & \text{otherwise} \end{cases}$$

$$t_{j1} = \begin{cases} 1 & \text{if } r_1 \text{ adj } v_j \text{ in } G_1 \\ 0 & \text{otherwise} \end{cases}$$

and, for $i, j > 1$,

$$t_{ij} = \begin{cases} 1 & \text{if } v_i \text{ adj } v_j \text{ in } G_1 \\ 0 & \text{otherwise .} \end{cases}$$

Now T provides p eigenvectors α_i for G, each with the property that all points in the same orbit have equal coordinates in α_i. In addition, each eigenvector β of $G_1 - r_1$ provides $n-1$ eigenvectors γ_i of $n * G_1$ by setting γ_i equal to β on the first copy of G_1, and equal to $-\beta$ on the $(i+1)$'st copy of G_1, and $\gamma_i = 0$ on all other copies. These vectors γ_i are orthogonal to the eigenvectors obtained from T because for each orbit V_j, the coordinates of vector γ_i are given by $+\beta$ on one point, $-\beta$ on another, and zero on the remaining $n-2$ points. Thus, this procedure provides $(n-1)(p-1)$ eigenvectors of $n * G_1$, which, together with the p eigenvectors of T account for all $np - n + 1$ eigenvectors of $n * G_1$. Thus,

(23)
$$\phi(n * G_1) = \phi^{n-1}(G_1 - r_1)\phi(T).$$

Now what about $\phi(G_1 * n)$, the graph formed by replacing the root by n identical points? We define an equitable partition having these n points in V_1, and the remaining orbits are singletons. Letting S be the corresponding matrix for this partition, we find that S is precisely T', the transpose of T. (This is the relation between $G_1 * n$ and $n * G_1$. Namely, since $S = T'$, we have $\phi(S) = \phi(T)$ and so $G_1 * n$ and $n * G_1$ have p common eigenvalues, the eigenvalues of T.) Thus, we know p eigenvalues of $G_1 * n$, so we proceed to find the remaining $n-1$ eigenvalues. These are all 0, as can be seen by setting the coordinate of one point in V_1 to be $+1$, of one of the $n-1$ remaining points to be -1, and setting all other coordinates to be 0. This procedure constructs the $n-1$ linearly independent eigenvectors with eigenvalue 0. Consequently,

(24)
$$\phi(G_1 * n) = x^{n-1}\phi(T),$$

so we may substitute for $\phi(T)$ in (23) to get

(25) $x^{n-1} \phi(n * G_1) = \phi^{n-1}(G_1 - r_1) \phi(G_1 * n)$.

Replacing $n * G_1$, by $G_1 \cdot \ldots \cdot G_1 \cdot G_2 \cdot \ldots \cdot G_2$, we obtain equation (21), completing the proof of the corollary.

Consider the tree $Y_n = P_{n-1} * 2$ where the path was planted, as shown in Figure 2. Then Corollary 4a can be used to find the spectrum of Y_n from the spectra of paths because

(26) $x \phi(P_{n-1} \cdot P_{n-1}) = \phi(P_{n-2}) \phi(Y_n)$,

which implies

(27) $\phi(Y_n) = x \phi(P_{2n-3}) / \phi(P_{n-2})$.

Looking ahead to (71) to find the spectra of paths and substituting into (27), we find that

(28) $\phi(Y_n) = x \prod_{j=1}^{n-1} (x - 2 \cos \pi(2j-1)/(2n-2))$.

P_5:

Y_6:

Figure 2.   A Rooted Path and the Corresponding Graph $Y_n = P_{n-1} * 2$

The first four structural theorems are sufficiently general to be applied to any graph, however, the results may not be simple enough to be efficient. The next three theorems only apply to graphs with very specific structure, but when they do apply, they provide exceedingly useful results. The first of these results uses the unique homogeneous polynomial of degree p in two variables $\psi(G;x,y)$ such that $\psi(G;x,1) = \phi(G;x)$. Of course each term of this new polynomial is obtained from the term $a_n x^{p-n}$ in $\phi(G;x)$ by replacing it by $a_n x^{p-n} y^n$.

Theorem 5.   Let G_o be a given p-point graph and let H,r be a rooted n-point graph. Let G be the graph formed by coalescing one copy of H at each point of G_o.   Then

(29) $$\phi(G) = \psi(G_0; \phi(H), \phi(H-r)) .$$

<u>Proof.</u> The left side of (29) can be written as $\Sigma a_m x^{pn-m}$ since G has pn points.

Now by Theorem 1, we have

(30) $$a_m = \sum_{M \epsilon G_m} (-1)^{k(M)} (2)^{c(M)} .$$

For each m-point mutation M contributing to a_m, we shall find a corresponding contribution to the coefficient of x^{pn-m} on the right side of (29). To do this, we first write $\phi(G_0) = \Sigma b_i x^{p-i}$, so that

(31) $$\psi(G_0; \phi(H), \phi(H-r)) = \sum_{j=0}^{p} b_j \phi^{p-j}(H) \phi^j (H-r) .$$

Now each component of mutation M of G either lies in one of the p copies of H or else it lies in G_0. Thus, we may partition M in a natural way into p+1 mutations $M = M_0 \cup M_1 \cup \ldots \cup M_p$ where M_0 is a mutation of G_0, and each remaining M_i is a mutation of the i'th copy of H. We let k_i and c_i be the numbers of components and of cycles in M_i. Then clearly $k(M) = \Sigma k_i$ and $c(M) = \Sigma c_i$. Now let $v_1, \ldots, v_p$ be the points of G_0. If $v_i \epsilon M_0$, then certainly $v_i \notin M_i$, and so, whenever $v_i \epsilon M_0$, we choose to view M_i not as a mutation of H, but rather as a mutation of H-r. Now M_0 contributes $(-1)^{k_0}(2)^{c_0}$ to $b_{|M_0|}$, and if $v_i \epsilon M_0$, then M_i contributes $(-1)^{k_i}(2)^{c_i} x^{n-1-|M_i|}$ to the corresponding factor of $\phi(H-r)$, and finally, if $v_i \notin M_0$, then M_i contributes $(-1)^{k_i}(2)^{c_i} x^{n-|M_i|}$ to the appropriate factor of $\phi(H)$. Thus, the contribution of the M_i's to the right side is the product of these individual terms which is

(32) $$(-1)^{\Sigma k_i}(2)^{\Sigma c_i} x^{pn-\Sigma|M_i|} = (-1)^{k(M)} (2)^{c(M)} x^{pn-m}.$$

But this is identical to the contribution of M to the left side as seen from equation (30). Moreover, M and the partition $M_0 \cup M_1 \cup \ldots \cup M_p$ clearly form a one-to-one correspondence between mutations of G and ordered (p+1)-tuples of mutations of G_0 and of p copies of H. On invoking this equality of contributions for every mutation, (29) is verified.

Notice that if we apply this theorem with $H = K_1$, then $H-r = K_0$ and $G = G_0$ and (29) asserts

(33) $$\phi(G_0) = \psi(G_0; \phi(K_1), \phi(K_0)).$$

But $\phi(K_1) = x$, and we have previously defined $\phi(K_0) = 1$ just before Theorem 2. Thus

$$(34) \qquad\qquad \phi(G_0) = \psi(G_0;x,1),$$

which agrees as it must with the definition of ψ, and so we have further evidence that our choice for $\phi(K_0)$ was correct.

Since the graph G is formed by p coalescences, we may replace each copy of H in light of Corollary 2c by a cospectrally rooted graph H' and still have the same spectrum. Thus we have proved the next statement.

<u>Corollary 5a.</u> Let G_0 be a given p-point graph and let H_i, r_i be a collection of p mutually cospectrally rooted graphs. Let G be the graph formed by coalescing H_i at v_i for each point v_i of G_0. Then

$$(35) \qquad \phi(G) = \psi(G_0;\phi(H_1),\phi(H_1-r_1)).$$

This corollary is very useful for constructing large collections of cospectral graphs. For example, if H_1,r_1 and H_2,r_2 are two cospectrally rooted graphs, then all of the 2^p graphs formed by coalescing either H_1 or H_2 at each point of G_0 are cospectral. Of course, some of these may be isomorphic, but we can prevent that from happening by selecting G_0 to be an identity graph. It is easy to see that if two of the graphs so formed are isomorphic, then this isomorphism restricted to G_0 is a nontrivial automorphism, contrary to the stipulation that G_0 is an identity graph. Thus, using the cospectrally rooted pairs of Figure 1, we can construct 2^n cospectral connected graphs on 7n points for each $n \geq 6$. Similarly, we construct 2^n cospectral trees on 9n points for each $n \geq 7$ since there exists an identity tree for each $n \geq 7$.

4. SPECTRA OF GRAPHS FORMED BY BINARY OPERATIONS

Several well known binary operations on graphs provide valuable information for analyzing graphical structure. (See Harary [6, pp. 21-23]). We wish to express the characteristic polynomial of the graph formed by a binary operation in terms of the polynomials of the two constituent graphs. In some cases, we only achieve this goal when some regularity assumptions are made about these two graphs.

Let G and H be graphs with point sets $V(G) = \{u_1,u_2,\ldots\}$ and $V(H) = \{v_1,v_2\ldots\}$ and line sets $X(G)$ and $X(H)$. The <u>union</u> of G and H is defined as usual by

(36) $V(G \cup H) = V(G) \cup V(H)$

           $X(G \cup H) = X(G) \cup X(H).$

Assuming $V(G) \cap V(H) = \phi$, the adjacency matrix of $G \cup H$ has two diagonal blocks while the off diagonal blocks are zero. Thus, we see that

(37) $\phi(G \cup H) = \phi(G)\phi(H)$ if $V(G) \cap V(H) = \phi.$

The union is seldom used to describe the structure when G and H have points in common, so equation (37) is adequate.

In four common binary operations having $V(G) \times V(H)$ as the point set, adjacency is defined by

Operation	Symbol	$(u_1,v_1) \text{adj} (u_2,v_2)$ if:
Cartesian Product	$G \times H$	$u_1 = u_2$ and v_1 adj v_2
		or $v_1 = v_2$ and u_1 adj u_2
Conjunction	$G \wedge H$	u_1 adj u_2 and v_1 adj v_2
Strong product	$G * H$	$d(u_1,u_2) \le 1$ and
		$d(v_1,v_2) \le 1$
Composition	$G[H]$	u_1 adj u_2 or
		$u_1 = u_2$ and v_1 adj v_2

The symbol for the strong product was chosen because

(38) $G * H = (G \times H) \cup (G \wedge H).$

Note that composition is the only "nonabelian" operation here, that is, in general

(39) $G[H] \neq H[G].$

The next theorem extends a result of Cvetković [2].

Theorem 6.   Let $\phi(G) = \Pi(x-\lambda_i)$ and $\phi(H) = \Pi(x-\mu_j)$. The characteristic polynomials of the cartesian product, conjunction, and strong product are

(40) $\phi(G \times H) = \Pi\Pi(x - \lambda_i - \mu_j)$

(41) $\phi(G \wedge H) = \Pi\Pi(x - \lambda_i \mu_j)$

(42) $\phi(G * H) = \Pi\Pi(x - \lambda_i\mu_j - \lambda_i - \mu_j).$

Proof.   The method of proof is identical for all three equations, and so we shall just present the proof of (42). Let $p = |V(G)|$ and $s = |V(H)|$, and let $\alpha = (a_1, a_2, \ldots, a_p)$ and $\beta = (b_1, b_2, \ldots, b_s)$ be eigenvectors of G and H with eigenvalues λ and μ.

Let $\alpha \times \beta$ be the vector of length ps whose (u_i, v_j) coordinate is $a_i b_j$. We show that $\alpha \times \beta$ is an eigenvector of $G * H$ with eigenvalue $\lambda\mu + \lambda + \mu$. Let us find the (u_1, v_1) coordinate of the action of $\alpha \times \beta$ on A, the adjacency matrix of $G * H$. This coordinate has the value

(43)
$$S_{11} = \Sigma a_i b_j$$

where the sum is over all (i, j) such that $(u_i, v_j) \text{adj} (u_1, v_1)$. But, in light of the definition of adjacency in the strong product, S_{11} can be decomposed into three sums:

(44)
$$S_{11} = \sum_{\substack{u_i \text{ adj } u_1 \\ v_j \text{ adj } v_1}} a_i b_j + \sum_{\substack{u_i \text{ adj } u_1 \\ v_j = v_1}} a_i b_1 + \sum_{\substack{v_j \text{ adj } v_1 \\ u_i = u_1}} a_1 b_j.$$

Factoring each sum yields

(45)
$$S_{11} = (\sum_{u_i \text{ adj } u_1} a_i)(\sum_{v_j \text{ adj } v_1} b_j) + b_1(\sum_{u_i \text{ adj } u_1} a_i) + a_1(\sum_{v_j \text{ adj } v_1} b_j).$$

Now since α is an eigenvector of G, $\Sigma a_i = \lambda a_1$, and similarly $\Sigma b_j = \mu b_1$, so we see that

(46)
$$S_{11} = \lambda a_1 \mu b_1 + b_1 \lambda a_1 + a_1 \mu b_1 = (\lambda\mu + \lambda + \mu) a_1 b_1.$$

Applying this argument, given here for (u_1, v_1), on an arbitrary coordinate (u_i, v_j) we find that $S_{ij} = (\lambda\mu + \lambda + \mu) a_i b_j$. Thus $\alpha \times \beta$ is an eigenvector of $G*H$ with eigenvalue $\lambda\mu + \lambda + \mu$. Since this argument applies to any pair of eigenvectors α_i and β_j, we have found ps eigenvectors of $G * H$. If they are linearly independent, we are through; but, in fact, if the α_i's have been chosen to be orthogonal, and similarly for the β_j's, then these constructed eigenvectors are also orthogonal which, of course, implies linear independence. To see this let $\alpha_i \times \beta_j$ and $\alpha_k \times \beta_\ell$ be two eigenvectors of $G * H$. Let $\alpha_i(m)$ denote the m'th coordinate of α_i. Then

(47)
$$(\alpha_i \times \beta_j) \cdot (\alpha_k \times \beta_\ell) = \sum_m \sum_n \alpha_i(m) \beta_j(n) \alpha_k(m) \beta_\ell(n)$$

$$= (\sum_m \alpha_i(m) \alpha_k(m))(\sum_n \beta_j(n) \beta_\ell(n))$$

$$= (\alpha_i \cdot \alpha_k)(\beta_j \cdot \beta_\ell),$$

the last two factors being scalar products. Now either $i \neq k$ so that $\alpha_i \cdot \alpha_k = 0$, or $j \neq \ell$ and $\beta_j \cdot \beta_\ell = 0$. In either case, we have shown that $\alpha_i \times \beta_j$ and $\alpha_k \times \beta_\ell$ are orthogonal, completing the proof of (42).

This theorem provides another useful tool for constructing cospectral graphs, for if G_1, G_2 and H_1, H_2 are two cospectral pairs, then certainly $G_1 \times H_1$ and $G_2 \times H_2$ are also cospectral. Similarly their conjunctions and strong products are cospectral. For example, to prove that there exist cospectral n-connected graphs, let G_1 and G_2 be any connected cospectral pair. Then $G_1 \times K_n$ and $G_2 \times K_n$ are cospectral and n-connected.

The _join_ of two graphs G + H is the graph formed by joining every point of G to every point of H. Or, using complementation, we may write, following Zykov [11],

$$(48) \qquad\qquad G + H = \overline{\overline{G} \cup \overline{H}}.$$

The join and the composition may be viewed as special cases of a more general operation which we call "generalized composition". If G is labeled and has p points, then the graph $G[H_1, H_2, \ldots, H_p]$ is formed by taking the disjoint graphs $H_1, H_2, \ldots, H_p$ and then joining every point of H_i to every point of H_j whenever u_i adj u_j in G. Thus the ordinary composition is just

$$(49) \qquad\qquad G[H] = G[H, H, \ldots, H],$$

and the join is given by

$$(50) \qquad\qquad G + H = K_2[G, H].$$

In general, it does not appear likely that the characteristic polynomial of the generalized composition can always be expressed in terms of the polynomials of $G, H_1, H_2, \ldots, H_p$, but if all the H_i's are regular we have the following result.

__Theorem 7.__ If $H_1, H_2, \ldots, H_p$ are all regular, then $V(H_1) \cup V(H_2) \cup \ldots \cup V(H_p)$ is an equitable partition of $G[H_1, H_2, \ldots, H_p]$. Let T denote the matrix associated with this partition. Then the characteristic polynomial of the generalized composition is

$$(51) \qquad \phi(G[H_1, H_2, \ldots, H_p]) = \phi(T) \, \Pi \phi(H_i)/(x - r_i) \ .$$

__Proof.__ The partition chosen is clearly equitable, for $v \in V(H_i)$ is adjacent to either all or none of the points in $V(H_j)$ according as u_i is or is not adjacent to u_j in G. By Theorem 4, $\phi(T)$ divides $\phi(G[H_1, H_2, \ldots, H_p])$ and each eigenvector is constant on each set $V(H_i)$.

The largest eigenvector of each H_i is $(1, 1, \ldots, 1)$. Each of the remaining $p_i - 1$ eigenvectors of H_i is orthogonal to this vector, so its coordinates sum to zero.

Consequently, it is easily seen to yield an eigenvector of $G[H_1,H_2,\ldots,H_p]$ if we set all the other coordinates to be zero. This provides

$$\sum_{i=1}^{p} (p_i-1) = (\sum_{i=1}^{p} p_i) - p$$

eigenvectors which, together with the p eigenvectors of T, will form a complete set of eigenvectors unless they are linearly dependent. But, since each pair is orthogonal, they are in fact linearly independent. We divide $\phi(H_i)$ by $(x - r_i)$ to remove the factor corresponding to eigenvector $(1,1,\ldots,1)$. Multiplying by $\phi(T)$ completes the proof.

For convenience, we specialize this result to the ordinary composition and to the join.

<u>Corollary 7a.</u> If H is regular of degree r with s points and G has p points, then

$$(52) \qquad \phi(G[H]) = s^p \; \phi(G;\frac{x-r}{s} \,) \; [\phi(H)/(x - r)]^p.$$

<u>Proof.</u> According to Theorem 7, we have an equitable partition of the ps points of G[H] into p parts, each of size s. Consequently, the matrix T has r's on the diagonal, and entries s exactly where A(G) has 1's, that is,

$$(53) \qquad\qquad T = sA(G) + rI.$$

But then,

$$(54) \qquad\qquad \phi(T) = \det((x-r)I - sA(G))$$
$$= s^p \det((\frac{x-r}{s})I - A(G))$$
$$= s^p \; \phi(G;\frac{x-r}{s}).$$

Thus, Theorem 7 implies formula (52).

<u>Corollary 7b.</u> If H_1 and H_2 are regular graphs of degrees r_1 and r_2 and with p_1 and p_2 points, then the characteristic polynomial of their join is

$$(55) \qquad \phi(H_1+H_2)=(x^2-(r_1+r_2)x+r_1r_2-p_1p_2) \frac{\phi(H_1)\phi(H_2)}{(x-r_1)(x-r_2)} \;.$$

<u>Proof.</u> According to equation (50), we need to evaluate $\phi(K_2[H_1,H_2])$. Applying Theorem 7, we obtain the matrix

$$(56) \qquad\qquad T = \begin{pmatrix} r_1 & p_2 \\ p_1 & r_2 \end{pmatrix}$$

whose characteristic polynomial is

$$(57) \qquad\qquad \phi(T) = (x^2 - (r_1+r_2)x + r_1r_2 - p_1p_2),$$

which, together with equation (51) implies (55).

5. THE SPECTRA OF EIGHT WELL-KNOWN FAMILIES OF GRAPHS

Let us now apply the structural results of this paper to obtain the spectra of various common families of graphs. Most of these spectra have been presented before (see Collatz and Sinogowitz [1], Cvetković [3], Doob [4], and Harary, King, Mowshowitz, and Read [7], but we derive them here to demonstrate the usefulness of these structural theorems. The spectra of wheels, ladders, and mobius ladders are believed to be new. The characteristic polynomial of the complete graph is

$$(58) \qquad \phi(K_p) = (x - p + 1)(x + 1)^{p-1}.$$

To demonstrate this, we first note that $\phi(K_1) = x$ satisfies (58), and we proceed by induction. Assume (58) has been verified for all integers less than p. Let us write write $K_p = K_n + K_{p-n}$ for n in the range $1 \le n \le p - 1$. We now apply Corollary 7b with $r_1 = n-1$, $p_1 = n$, $r_2 = p-n-1$, $p_2 = p-n$ to obtain

$$(59) \qquad \phi(K_n + K_{p-n}) = (x^2 - (p-2)x + 1 - p)(x+1)^{n-1}(x+1)^{p-n-1}$$
$$= (x-p+1)(x+1)(x+1)^{p-2},$$

and so we have proved (58) by induction. Similarly, since $\phi(\overline{K}_p) = x^p$ and $K_{m,n} = \overline{K}_m + \overline{K}_n$, we may apply Corollary 7b with $r_1 = r_2 = 0$ and $p_1 = m$, $p_2 = n$ to obtain

$$(60) \qquad \phi(K_{m,n}) = (x^2 - mn)x^{m+n-2}.$$

The cubes constitute the next family of interest. They can be defined inductively by $Q_1 = K_2$ and $Q_n = Q_{n-1} \times_n K_2$. We claim that

$$(61) \qquad \phi(Q_n) = \prod_{i=0}^{n} (x - n + 2i)^{\binom{n}{i}}.$$

When n = 1, equation (61) reduces to $\phi(K_2) = (x-1)(x+1)$ which agrees with (58). Assuming this formula has been demonstrated for n-1 and applying equation (40) of Theorem 6, we see that

$$(62) \qquad \phi(Q_{n-1} \times K_2) = \prod_{i=0}^{n-1} (x-n+1+2i+1)^{\binom{n-1}{i}}(x-n+1+2i-1)^{\binom{n-1}{i}}$$
$$= \prod_{i=0}^{n-1} (x-n+2(i+1))^{\binom{n-1}{i}}(x-n+2i)^{\binom{n-1}{i}}$$
$$= \prod_{i=0}^{n} (x-n+2i)^{\binom{n-1}{i}+\binom{n-1}{i-1}},$$

which reduces to equation (61).

We shall now find the spectrum of the cycle $C_n = v_1 v_2 \cdots v_n v_1$ by an entirely different approach, namely, we shall specify n linearly independent eigenvectors. Let $\varepsilon = \varepsilon(j)$ be 0 if $j < n/2$ and 1 if $j \geq n/2$. Set the v_i coordinate of eigenvector α_j denoted by $\alpha_j(v_i)$ to be $\cos 2\pi j(i - \varepsilon(j))/n$. (This is the method of pulling the solution out of the hat and then verifying it.) Using a trigonometric identity we see that the v_i coordinate of the adjacency matrix $A(C_n)$ acting on α_j is

$$(63) \qquad (A\alpha_j)(v_i) = \cos 2\pi j(i+1-\varepsilon)/n + \cos 2\pi j(i-1-\varepsilon)/n$$

$$= 2 \cos 2\pi J/n \cos 2\pi j(i-\varepsilon)/n$$

$$= 2 \cos 2\pi j/n \; \alpha_j(v_i).$$

Since this holds for each coordinate v_i, we have shown that α_j is indeed an eigenvector whose eigenvalue is $2 \cos 2\pi j/n$. The only possible linear dependence can occur among vectors with the same eigenvalue. But the only time $2 \cos 2\pi j/n = 2 \cos 2\pi k/n$ for $1 \leq j < k \leq n$ is when $k = n - j$. Thus, the only possible linear dependence is of the form $a\alpha_j + b\alpha_{n-j} = 0$. Without loss of generality, we assume $a = 1$, so that

$$(64) \qquad \alpha_j = -b\alpha_{n-j},$$

and

$$(65) \qquad \alpha_j \cdot \alpha_j = b^2 \alpha_{n-j} \cdot \alpha_{n-j}.$$

Now $j < k = n - j$ implies that $\varepsilon(j) = 0$ and $\varepsilon(n-j) = 1$. Consequently, the v_i coordinate of α_{n-j} is

$$(66) \qquad \alpha_{n-j}(v_i) = \cos 2\pi(n-j)(i-1)/n$$

$$= \cos 2\pi j(i-1)/n$$

$$= \alpha_j(v_{i-1}).$$

That is, α_{n-j} is just α_j rotated $2\pi/n$ radians. Therefore $\alpha_j \cdot \alpha_j = \alpha_{n-j} \cdot \alpha_{n-j}$, so (65) implies that $b = \pm 1$. We now see that the v_n coordinate of equation (64) becomes

$$(67) \qquad \cos 2\pi j = \pm \cos 2\pi(n-j)(n-1)/n$$

$$1 = \pm \cos 2\pi j/n.$$

But this is impossible for $1 \leq j < n/2$, so we conclude there is no linear dependence among the n eigenvectors constructed. Thus,

$$(68) \qquad \phi(C_n) = \prod_{j=1}^{n} (x - 2 \cos 2\pi j/n).$$

Since the <u>wheel</u> is defined as $W_n = K_1 + C_{n-1}$, we apply Corollary 7b to conclude that

(69)
$$\phi(W_n) = (x^2 - 2x - n + 1) \prod_{j=1}^{n-2} (x - 2 \cos 2\pi j/(n-1))$$

$$= (x - 1 - \sqrt{n})(x - 1 + \sqrt{n}) \prod_{j=1}^{n-2} (x - 2 \cos 2\pi j/(n-1)) .$$

Now consider the path $P_n = v_1 v_2 \cdots v_n$. For each eigenvector α of P_n we construct a corresponding eigenvector β of $C_{2n+2} = u_1 u_2 \cdots u_{2n+2} u_1$ by setting

(70)
$$\beta(u_i) = \begin{cases} \alpha(v_i) & \text{if } i \leq n \\ 0 & \text{if } i = n+1 \text{ or } 2n+2 \\ -\alpha(v_{i-n-1}) & \text{if } n+2 \leq i \leq 2n+1. \end{cases}$$

Thus $\phi(P_n)$ divides $\phi(C_{2n+2})$. But P_n has diameter n-1, and so, as observed in Doob [4], its n eigenvalues must be distinct. Thus, the spectrum of P_n consists of n of the n+2 <u>distinct</u> eigenvalues of C_{2n+2}. Now C_{2n+2} has two simple eigenvalues (namely ± 2) and n paired eigenvalues (2 cos $2\pi j/(2n+2)$ for $1 \leq j \leq n$). But +2 and -2 cannot possibly be eigenvalues of the path because either of these values leads to a contradiction in the last coordinate when we try to form an eigenvector. Consequently,

(71)
$$\phi(P_n) = \prod_{j=1}^{n} (x - 2 \cos \pi j/(n+1)).$$

The <u>ladder</u> is defined as $P_n \times K_2$. Using (71) and equation (40) of Theorem 6 we obtain as an immediate corollary:

(72) $\phi(P_n \times K_2) = \displaystyle\prod_{j=1}^{n} (x - 1 - 2 \cos 2\pi j/(n+1))(x + 1 - 2 \cos 2\pi j/(n+1)).$

We conclude this paper by finding the spectra of <u>möbius ladders</u>, defined as follows: M_{2n} is obtained from $C_{2n} = v_1 v_2 \cdots v_{2n} v_1$ by adding the n additional lines $v_i v_{i+n}$, and M_{2n+1} is C_{2n+1} with the 2n+1 additional lines $v_i v_{i+n}$. We define eigenvectors α_j just as we did for the cycles C_{2n} and C_{2n+1}, only now we find that the eigenvector α_j of möbius ladder M_{2n} has the eigenvalue

(73) $\lambda = 2 \cos \pi j/n + \cos \pi j$

$$= (-1)^j + 2 \cos \pi j/n,$$

and so

$$(74) \qquad \phi(M_{2n}) = \prod_{j=1}^{2n} (x - (-1)^j - 2 \cos \pi j/n) \ .$$

Similarly, the eigenvector α_j of the möbius ladder M_{2n+1} has the eigenvalue

$$(75) \qquad \lambda = 2 \cos 2\pi j/(2n+1) + 2 \cos 2\pi jn/(2n+1)$$

$$= 4 \cos \pi j(n+1)/(2n+1) \cos \pi j(n-1)/(2n+1) \ .$$

Thus, its spectrum is

$$(76) \qquad \phi(M_{2n+1}) = \prod_{j=1}^{2n+1} (x - 4 \cos \frac{\pi j(n+1)}{(2n+1)} \cos \frac{\pi j(n-1)}{(2n+1)}) \ .$$

Acknowledgement. The author is grateful to F. Harary for his helpful comments.

REFERENCES

1. Collatz, L., and Sinogowitz, U., "Spectra of finite graphs", Abh. Math. Sem. Univ. Hamburg, 21 (1957) 64-77.

2. Cvetković, D. M., "Spectrum of the graph of n-tuples", Publ. Elektrotehn. Fak. Univ. Beograd, Ser. Mat. Fiz., Nos. 274-301 (1969) 91-95.

3. Cvetković, D. M., "Graphs and their spectra", Publ. Elektrotehn. Fak. Univ. Beograd, Ser. Mat. Fiz., Nos. 354-356 (1971) 1-50.

4. Doob, M., "Graphs with a small number of distinct eigenvalues", Annals of the New York Academy of Sciences, 175 (1970) 104-110.

5. Harary, F., "The determinant of the adjacency matrix of a graph", SIAM Rev., 4 (1962) 202-210.

6. Harary, F., Graph Theory. Addison-Wesley, Reading, 1969.

7. Harary, F., King, C., Mowshowitz, A., and Read, R.C., "Cospectral graphs and digraphs", Bull. London Math. Soc. 3 (1971) 321-328.

8. Mowshowitz, A., "The adjacency matrix and the group of a graph", New Directions in the Theory of Graphs (F. Harary, ed.). Academic Press, New York, 1973.

9. Sachs, H., "Beziehungen zwischen den in einem Graphen enthaltenen Kreisen und seinem charakteristischen Polynom.", Publ. Math. Debrecen 11 (1964) 119-134.

10. Spialter, L., "The atom connectivity matrix and its characteristic polynomial", J. Chem. Documentation 4 (1964) 261-274.

11. Zykov, A. A., "On some properties of linear complexes, (Russian). Mat. Sbornik 24 (1949) 163-188. Amer. Math. Soc. Translation N. 79, 1952.

FINDING AN INDEPENDENT SET IN A PLANAR GRAPH

Michael O. Albertson
Smith College

ABSTRACT

A subset H of the vertices of a graph is independent if no two vertices
in H are adjacent. The Erdös-Vizing Problem suggests that a planar graph has
an independent set of vertices that contains at least 1/4 of the vertices of
the graph. The purpose of this paper is to give an algorithm that produces
an independent set in a planar graph that contains more than 2/9 of the vertices
of the graph.

FINDING AN INDEPENDENT SET IN A PLANAR GRAPH

A subset H of the vertices of a graph G is independent if no two vertices in H are adjacent. The Four Color Conjecture implies the Erdös-Vizing Problem, namely that it is possible to find an independent set of the vertices of a planar graph that contains at least 1/4 of the vertices of the graph. The purpose of this paper is to give an explicit algorithm for finding an independent set of vertices in a planar graph which contains more than 2/9 of the vertices of the graph. We do not contend that the algorithm is a good one, merely that it works.

We assume G is a triangulation. If not, introducing edges to make it so cannot make the task easier. The strategy of the algorithm will be to reduce the question of finding an independent set in G to that of finding an independent set in G', a graph on fewer vertices. Suppose G has N vertices and G' has N-q vertices.

Suppose H' is independent in G' with

$$|H'| > \frac{2}{9}(N-q) .$$

If there exists H independent in G with

$$|H| = |H'| + p \quad \text{and} \quad p \geq \frac{2}{9} q ,$$

then H will be a required independent set, namely

$$|H| > \frac{2}{9} N .$$

If we reduce the number of vertices in G enough, there will be an obvious four-coloring of G. Taking the vertices that have the most frequently used color enables one to build an independent set of the appropriate size. The algorithm consists of a sequence of reductions. Every time a new triangulation appears, begin with Step 1. If Step 1 is impossible, go on to Step 2, etc.

<u>Step 1</u>. Look for a vertex of degree 3. Suppose v_1 is a vertex of degree 3 with neighbors v_2, v_3, and v_4. Let $G' = G - v_1 - v_2 - v_3 - v_4$. If H' is independent in G'

then $H' + v_1$ will be independent in G.

Step 2. Look for a vertex of degree 4. Suppose v_1 is a vertex of degree 4 with neighbors v_2, ..., v_5. Since G is planar, some pair of v_2, ..., v_5 (say v_2 and v_4) are not adjacent. Let G' be $G - v_1 - v_3 - v_4 - v_5$ with v_2 adjacent to all vertices in G that are in G' and are adjacent to v_3, v_4, or v_5. Suppose H' is independent in G'. If v_2 is not in H' then $H' + v_1$ is independent in G. If v_2 is in H', then $H' + v_4$ is independent in G.

Step 3. Look for a 3-cycle that is not a face boundary. Suppose v_1, v_2, v_3 is a 3-cycle with N_1 interior vertices and N_2 exterior vertices. Assume $N_1 \equiv j \pmod 9$ and $N_2 \equiv k \pmod 9$. We consider the ordered pairs (j,k) as separate cases. Luckily only three techniques suffice to reduce every case. Assume $j \geq k$.

Case a. The ordered pair (j, k) is not $(3,3)$, $(4,3)$, $(4,4)$, $(7,4)$, $(8,3)$, $(8,4)$, $(8,7)$, or $(8,8)$. Let G_1 be the induced subgraph on the vertices interior to v_1, v_2, v_3 and G_2 be the induced subgraph on the vertices exterior to v_1, v_2, v_3. For $i = 1, 2$ we can find H_i independent in G_1 with $|H_i| > \frac{2}{9} N_i$. Thus $H = H_1 + H_2$ will be independent in G.

Subcase 1. $0 \leq k \leq j \leq 4$.

Since $|H_1|$ is an integer greater than $2N_1/9$, we have

$$|H_1| \geq \frac{2}{9} N_1 + \frac{9-2j}{9} .$$

Similarly,

$$|H_2| \geq \frac{2}{9} N_2 + \frac{9-2k}{9} .$$

Therefore

$$|H| \geq \frac{2}{9} (N_1 + N_2 + 3) + \frac{12-2(j+k)}{9} .$$

If $j+k \leq 5$ we have $|H| > \frac{2}{9} N$.

Subcase 2. $0 \leq k \leq 4$, $5 \leq j \leq 8$.

As above we have,

$$|H_1| \geq \frac{2}{9} N_1 + \frac{18-2j}{9} \quad .$$

Similarly,

$$|H_2| \geq \frac{2}{9} N_2 + \frac{9-2k}{9} \quad .$$

Therefore

$$|H| \geq \frac{2}{9} (N_1 + N_2 + 3) + \frac{21-2(j+k)}{9} \quad .$$

If $j+k \leq 10$ we have $|H| > \frac{2}{9} N$.

Subcase 3. $5 \leq k \leq j \leq 8$.

As above we have,

$$|H_1| \geq \frac{2}{9} N_1 + \frac{18-2j}{9} \quad .$$

Similarly,

$$|H_2| \geq \frac{2}{9} N_2 + \frac{18-2k}{9} \quad .$$

Therefore

$$|H| \geq \frac{2}{9} (N_1 + N_2 + 3) + \frac{30-2(j+k)}{9} \quad .$$

If $j+k \leq 14$ we have $|H| > \frac{2}{9} N$.

Case b. The ordered pair (j,k) is one of $(4,3)$, $(4,4)$, $(7,4)$, $(8,3)$, $(8,4)$, $(8,7)$, or $(8,8)$. Let G_3 be G_1 together with a vertex u adjacent to all vertices in G_3 that are adjacent to v_1 , v_2 , v_3 in G. Let G_4 be G_2 together with v_1 , v_2 , v_3 and all induced edges. For $i = 3,4$, we can find H_i independent in G_i with $|H_i| > \frac{2}{9} |G_i|$. If u is in H_3 then $H = H_3 + H_4 - u$ is independent in G . Now let u be not in H_3 . At most one of v_1 , v_2 , v_3 is in H_4 . If v_i is in H_4 then $H = H_3 + H_4 - v_i$ is independent in G . In either case $|H| = |H_3| + H_4 - v_i$ is independent in G . In either case $|H| = |H_3| + |H_4| - 1$.

Subcase 1. $j = 8$.

We know

$$|H_3| \geq \frac{2}{9} (N_1 + 1) + 9/9 \quad .$$

Similarly

$$|H_4| \geq \frac{2}{9} (N_2 + 3) \quad .$$

Therefore

$$|H| \geq \frac{2}{9} (N_1 + N_2 + 3) + \frac{2}{9} + 9/9 - 1 > \frac{2}{9} N \quad .$$

Subcase 2. $j = 4$.

As above

$$|H_3| \geq \frac{2}{9} (N_1 + 1) + 8/9$$

and

$$|H_4| \geq \frac{2}{9} (N_2 + 3) \quad .$$

Thus

$$|H| \geq \frac{2}{9} (N_1 + N_2 + 3) + \frac{2}{9} + 8/9 - 1 > \frac{2}{9} N \quad .$$

The same construction with the roles of j and k reversed takes care of the pair $(7, 4)$.

Case c. Suppose $N_1 \equiv N_2 \equiv 3 \pmod 9$. For $1 \leq p < q \leq 3$ let G_{pq} be G_4 in the previous case with vertices v_p and v_q identified. For $1 \leq p < q \leq 3$ let E_{pq} be G_1 together with v_1, v_2, v_3 and their induced edges with vertices v_p and v_q identified. Denote this new vertex by v_{pq}. For $1 \leq p < q \leq 3$ we can find F_{pq} independent in E_{pq} with

$$|F_{pq}| > \frac{2}{9} |E_{pq}|$$

which implies

$$|F_{pq}| \geq \frac{2}{9} (N_2 + 2) + 8/9 \quad .$$

Similarly we can find H_{pq} independent in G_{pq} with

$$|H_{pq}| > \frac{2}{9} |G_{pq}|$$

which implies

$$|H_{pq}| \geq \frac{2}{9} (N_1 + 2) + 8/9 \quad .$$

From F_{pq} and H_{pq} we will find H indpendent in G with

$$|H| = |H_{pq}| + |F_{pq}| - 1 \geq \frac{2}{9}(N_1 + N_2 + 3) + \frac{18}{9} - 1 > 2/9\ N$$

Consider G_{12} and E_{12}. If v_{12} or v_3 is not in H_{12} suppose v_{12} is in F_{12}. $H = H_{12} + F_{12} - v_{12}$ will be independent in G of the appropriate size. Thus we may assume one of v_{12} or v_3 appears in H_{12} and F_{12}. If the same vertex appears in both, then $H = H_{12} + F_{12}$ will be independent where v_{12} goes to either v_1 or v_2. Thus we may assume v_{12} is in E_{12} and v_3 is in H_{12}. Now consider G_{13}. If v_{13} or v_2 is not in H_{13} then we are done as above. If v_{13} is in H_{13}, let $H = (F_{12} - v_{12}) + (H_{13} - v_{13}) + v_1$. If v_2 is in H_{13}, let $H = (F_{12} - v_{12}) + (H_{13} - v_2) + v_2$.

$\underline{\text{Step 4}}$. Look for a pair of triangles v_1, v_2, v_3 and v_2, v_3, v_4 with deg $v_1 = 5$ and deg $v_4 \leq 6$. (Recall that every vertex has degree ≥ 5.) The configuration is shown in Figure 1. There cannot exist edges $v_5 v_7$ or $v_8 v_{10}$ or $v_8 v_{11}$. since these would form 3-cycles. It also must be the case that v_5 is distinct from v_8, v_9, v_{10}, and v_{11} since if $v_5 = v_j$ for some j, $9 \leq j \leq 11$, then v_5, v_2, v_4 is a 3-cycle, which is not a face boundary. Similarly v_7 must be distinct from v_8, ..., v_{11}. Also v_8 and v_{11} (or v_{10} if deg $v_4 = 5$) are distinct from v_5, v_6, v_7. Construct G' from G by deleting v_1, v_4, v_5, ..., v_{11} and connecting v_2 to all neighbors of v_5, v_6, v_7, and connecting v_3 to all neighbors of v_8, ..., v_{11}. We can find H' independent in G' with $|H'| > \frac{2}{9}(N-9)$. If neither v_2 nor v_3 is in H' then $H' + v_1 + v_4$ is independent in G. If v_2 is in H' then $H = H' - v_2 + v_4 + v_5 + v_7$ is independent in G. If v_3 is in H' then $H = H' - v_3 + v_1 + v_8 + v_{11}$ (or v_{10} if deg $v_4 = 5$). In each of these cases $|H| > 2/9\ N$.

We have given an algorithm to construct an independent set of the vertices of a triangulation. It is not obvious why every triangulation must satisfy the condition of one of the steps. It is shown in [1] that the conditions of steps 1, 2 or 4 are required to be in a triangulation, i.e., every triangulation contains a vertex of degree less than five or the configuration in Figure 1. This is analogous to a result of Kotzig on Eulerian polyhedra [4].

It seems possible that an algorithm exists to arrive at an independent set H with $|H|/N > 2/9$. One suspects for $|H|/N$ close to 1/4 the task would be difficult, since almost no progress has been made on the Erdös-Vizing problem.

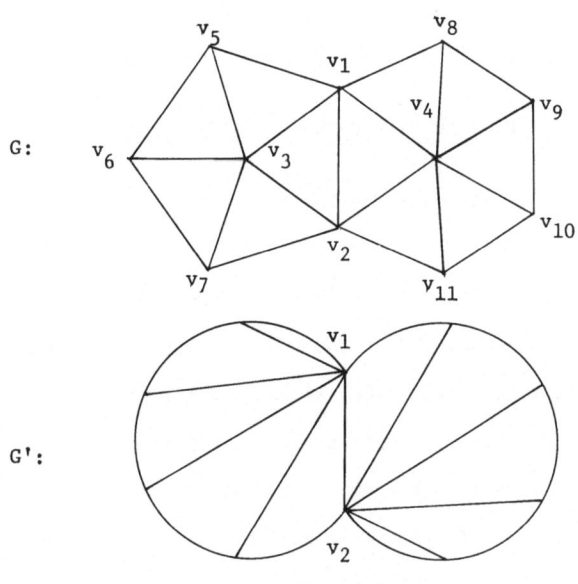

Figure 1

Note: I am indebted to the referee for bringing Kotzig's work to my attention.

REFERENCES

1. Albertson, M. O., "A Lower Bound for the Independence Number of a Planar Graph", J. Combinatorial Theory, submitted.

2. Berge, C., Graphes et Hypergraphes, Dunod, Paris, 1970.

3. Haken, W., "An Existence Theorem for Planar Maps", J. Combinatorial Theory 14B, 1973.

4. Kotzig, A., "Contribution to the Theory of Eulerian Polyhedra", Mat. Fyz. Casopis 5 (1955) 101-113.

A CLASS OF STARTER INDUCED 1-FACTORIZATIONS

B. A. Anderson
Arizona State University

ABSTRACT

A technique of Mullin and Nemeth for constructing strong starters in certain Abelian groups of prime power order is considered. These strong starters induce 1-factorizations of complete graphs. It is easy to show that these 1-factorizations possess enough symmetry to insure that if $\{F_i, F_j\}$ and $\{F_k, F_m\}$ are pairs of distinct 1-factors from such a 1-factorization, then the cycle structures of $F_i \cup F_j$ and $F_k \cup F_m$ are identical. The method is applied to construct what is apparently the first example of a 1-factorization of the complete graph on 28 points, K_{28}, such that the union of every two distinct 1-factors is a Hamiltonian circuit. This problem, generalized to K_{2n}, has arisen in several contexts and the above result strengthens the conjecture of various authors that such 1-factorizations exist on all K_{2n}. The first unsettled case is now K_{36}.

A CLASS OF STARTER INDUCED 1-FACTORIZATIONS

1. INTRODUCTION

Suppose K_{2n} denotes the complete graph on 2n points and let F be a 1-factorization of it. Call F <u>perfect</u> iff whenever F_i and F_j are distinct 1-factors of F, $F_i \cup F_j$ is a Hamiltonian circuit. For background on the currently available methods of constructing perfect 1-factorizations, see [1], [4], [6], [7] and especially [3].

The purpose of this note is to outline a process for constructing 1-factorizations that might be refined to yield a construction of perfect 1-factorizations on an infinite class of complete graphs, many of which are, at present, not known to possess perfect 1-factorizations. The method has so far been applied to find new solutions for several already settled cases and the first solution on K_{28}.

It will become apparent that many interesting questions concerning the construction to be presented are not considered. This article contains the material presented at the conference and should be considered as only a preliminary report on the method. It is planned to discuss the matter more fully at a later time.

Several concepts and results will be introduced to clarify the terminology used in the theorems.

<u>Definition 1.1.</u> Suppose t is a positive integer, r = 2t + 1 and G is an Abelian group of order r written additively. Then S is a <u>starter</u> in G iff

$$S = \{\{x_1, y_1\}, \ldots, \{x_t, y_t\}\}$$

such that every non-zero element of G occurs as

      (i)   an element of some pair of S

and   (ii) a difference of some pair of S.

If, in addition

      (iii) $1 \leq i \leq t$ implies $x_i + y_i \neq 0$,

      (iv) $1 \leq i, j \leq t, i \neq j$ implies $x_i + y_i \neq x_j + y_j$,

S is said to be a <u>strong starter</u>.

Consult [5], [9] for information on strong starters. Starters induce 1-factorizations in a very nice way. Suppose G is an Abelian group of order r = 2t + 1 and S is a starter in G. Label the points of K_{r+1} by ∞ and the elements of G. If g ∈ G, define ∞ + g = g + ∞ = ∞ . Let S* = S $\cup$ {∞,0}. Clearly S* is a 1-factor of K_{r+1} . The 1-factorization F induced by S is then the family of translates of S* by elements of G. That is, if g ∈ G

$$S* + g = \{∞,g\} \cup \{\{x_i + g, y_i + g\}: 1 \leq i \leq t\}.$$

Hence

$$F = \{S* + g: g \in G\}.$$

It is easy to show [9, p. 46] that F is a 1-factorization of K_{r+1} . One can use the notation (G,S) → F.

The following result is the basis of the construction.

Theorem 1.2. (Mullin-Nemeth [8]) Suppose p is a prime and m is a positive integer such that 3 < $p^m \equiv$ 3(mod 4). Thus p^m = 2t + 1, t odd. Let x be primitive in GF[p^m] and define

$$S(x) = \{\{x^0,x^1\},\{x^2,x^3\},...,\{x^{2t-2}, x^{2t-1}\}\} .$$

Then S(x) is a strong starter for the additive group of GF[p^m] . We will use the notation F(x) to denote the 1-factorization induced by S(x).

Now, let S_{2n} be the symmetric group on the vertices of K_{2n} . Clearly if π ∈ S_{2n} , then π induces a permutation on the family of all 1-factors of K_{2n} .

Definition 1.3. Suppose G is a 1-factorization of K_{2n}. Then FS(G) = {π: π ∈ S_{2n} and π[G] = G}. If π ∈ FS(G), then π induces a permutation ρ on the 1-factors of G.

$$Sym(G) = \{ρ: some π in FS(G) induces ρ\}$$

It turns out that for perfect 1-factorizations, FS(G) and Sym(G) are isomorphic groups. See [2] for information on this.

Definition 1.4. A 1-factorization F of K_{2n} is semi-regular iff for any F_i, F_j, F_k, F_m in F such that i ≠ j, k ≠ m, the cycle structures of $F_i \cup F_j$ and $F_k \cup F_m$ are identical. That is, if (i,j,n) is the number of cycles of length n in $F_i \cup F_j$, then for every positive integer n, (i,j,n) = (k,m,n).

2. SEMI-REGULAR 1-FACTORIZATIONS

We can now state the main result.

__Theorem 2.1.__ Suppose p is a prime and m is a positive integer such that $3 < p^m \equiv 3 \pmod 4$. If x is primitive in $GF[p^m]$, then

> (i) F(x) is a semi-regular 1-factorization of K_{2n}, $2n = p^m + 1$;

and   (ii) Sym(F(x)) is transitive and of order at least $p^m(p^m - 1)/2$.

__Proof.__ As before, $p^m = 2t + 1$, t odd. We make the following definition.

$$\Sigma S(x) = \{x^{2i} + x^{2i+1}: \{x^{2i}, x^{2i+1}\} \in S(x)\}.$$

Now $x^{2i} + x^{2i+1} = x^{2i}(1 + x)$. If $1 + x = x^{2j}$, $1 \leq j \leq t - 1$, then it is easy to show that $\Sigma S(x)$ is a multiplicative subgroup M of $GF[p^m]$. If, on the other hand, $1 + x = x^{2j+1}$, $1 \leq j \leq t - 1$, then $\Sigma S(x^{-1})$ is a multiplicative subgroup. We will assume the labeling is such that $\Sigma S(x) = M$. Note that $|M| = (p^m - 1)/2$.

The subgroup M has an important property. If $m \in M$, then $-m \notin M$. Supposition of the contrary leads to the equation $x^{2i} + x^{2j} = 0$. One may now follow the argument of [8, Lemma 1] to obtain a contradiction. Here is where use is made of the fact that t is odd.

As before, let $S*(x) = S(x) \cup \{\infty, 0\}$. It is clear that if $g \in GF[p^m]$ and

$$\tau_g(y) = \begin{cases} y + g, & y \in GF[p^m] \\ \infty, & y = \infty, \end{cases}$$

then $\tau_g[S*(x) + y] = S*(x) + (y + g)$. Thus Sym(F(x)) is transitive. Suppose $g \in GF[p^m] - \{0\}$ and

$$\sigma_g(y) = \begin{cases} gy, & y \in GF[p^m] \\ \infty, & y = \infty. \end{cases}$$

It is easy to show the following facts.

> (1) If $g \notin M$, then $\sigma_g[F(x)] = F(x^{-1})$

and   (2) If $g \in M$, then $\sigma_g[F(x)] = F(x)$,

> and $\sigma_g[S*(x)] = S*(x)$.

Thus, if $g \in M$, $\sigma_g \in FS(F(x))$. Each such σ_g induces an element of Sym(F(x)). We will abuse the notation slightly and say that if $g \in M$, then $\sigma_g \in \text{Sym}_0(F(x))$. It is obvious that different elements of M give rise to different members of $\text{Sym}_0(F(x))$.

We have, therefore, verified (ii) of the theorem.

The permutations known to be in $\mathrm{Sym}(F(x))$ make it easy to see that $F(x)$ is a semi-regular 1-factorization. Let m be a generator of the cyclic group M. We claim that the cycle structure of

$$S^*(x) \cup [S^*(x) + m]$$

is the same as the cycle structure of the union of any two distinct 1-factors of $F(x)$, say

$$[S^*(x) + g] \cup [S^*(x) + h] .$$

Since $g \neq h$, we know that either $(g - h) \in M$ or else $(h - g) \in M$. We may now suppose that $(g - h) \in M$. The permutation τ_{-h} shows that the cycle structure of

$$[S^*(x) + g] \cup [S^*(x) + h]$$

is identical to the cycle structure of

$$[S^*(x) + (g - h)] \cup S^*(x) .$$

Since $(g - h) \in M$, there is a positive integer k such that $m^k = (g - h)$. Let r be m^{k-1}. Then $r \in M$ and the permutation σ_r shows that the cycle structure of

$$[S^*(x) + (g - h)] \cup S^*(x)$$

is identical to the cycle structure of

$$[S^*(x) + m] \cup S^*(x) .$$

This completes the proof of the theorem.

We conclude by applying the above method to find some perfect 1-factorizations. First, we construct one on K_{28} . It may be verified that $x^3 + 2x^2 + 1$ is an irreducible polynomial over $GF[3]$, and that x is primitive in the resulting $GF[27]$. $F(x)$ is perfect. Note that by Theorem 2.1, this assertion may be checked by hand since all one has to do is show that $S^*(x) \cup [S^*(x) + g]$, $g \neq 0$, is a Hamiltonian circuit.

The method can be applied to twelve primes < 100. It turns out that in nine of these cases, the process described above yields perfect 1-factorizations. In the following table the first row gives the prime p and the second row a primitive root pr of $GF[p]$ that induces a perfect 1-factorization on K_{p+1}.

p	7	11	19	23	31	43	47	59	67	71	79	83
pr	−	8	2	5	−	−	31	50	61	7	3	8

185

REFERENCES

1. Anderson, B. A., "Finite Topologies and Hamiltonian Paths," *J. Combinatorial Theory* 14B (1973), 87-93.

2. Anderson, B. A., "Symmetry Groups of Perfect 1-Factorizations on Some K_{2n}," to appear.

3. Anderson, B. A., "A Perfectly Arranged Room Square," to appear, Proc. of the Fourth Southeastern Conference on Combinatorics, Graph Theory and Computing, Louisiana State University, Baton Rouge (1973).

4. Faber, V., Ehrenfucht, A., Mycielski, J., and Fajtlowicz, S., to appear in Proc. of Conference on Lattice Theory, Houstin (1973).

5. Gross, K., "A Multiplication Theorem for Strong Starters," to appear.

6. Kötzig, A., "Hamilton Graphs and Hamilton Circuits," *Theory of Graphs and its Applications*, (Proc. Sympos. Smolenice, 1963) 63-82.

7. Kötzig, A., "Combinatorial Structures and Their Applications, Gordon and Breach, New York, 1970, 215-221.

8. Mullin, R. C. and Nemeth, E., "An Existence Theorem for Room Squares," *Canad. Math. Bull.* 12 (1969), 493-497.

9. Wallis, W. D., "Duplication of Room Squares," *J. Austral. Math. Soc.* (14), 1972, Part I, 75-81.

CHROMATICALLY EQUIVALENT GRAPHS

Ruth A. Bari
George Washington University

ABSTRACT

Let G,H be graphs, and $P(G,\lambda)$, $P(H,\lambda)$ be the chromatic polynomials of G,H respectively. Then G is chromatically equivalent to H, (written $P \underset{c}{\sim} H$), if $P(G,\lambda) = P(H,\lambda)$.

In this paper, we first state some open questions relating to chromatic equivalence of graphs, and then give non-trivial examples of chromatically equivalent graphs and their chromatic polynomials.

CHROMATICALLY EQUIVALENT GRAPHS

1. INTRODUCTION

Let G be a graph, λ a positive integer. Then a λ-coloring of G is a mapping of the points of G into the integers $1, 2, \ldots, \lambda$, called colors, so that if two points of G are adjacent they are assigned different colors. We denote by $P(G,\lambda)$ the number of λ-colorings of G.

It can be shown that, for each graph G, $P(G,\lambda)$ can be expressed as a polynomial in λ with integral coefficients. We call this polynomial the chromatic polynomial (chromial) associated with G.

If λ is a positive integer, $P(G,\lambda) = 0$ if and only if G cannot be colored in λ colors. For example, if G is a triangulation of the sphere, $P(G,\lambda) = 0$ for $\lambda = 0, 1, 2$. If we could prove that $P(G,4) \neq 0$ for any planar triangulation G, or produce a planar triangulation for which $P(G,4) = 0$, we would solve the famous 4-color conjecture. The study of chromatic equivalence of graphs may help us to characterize chromials and their zeroes.

We present below, some of the unsolved problems concerning chromatic equivalence, and then give examples of chromatically equivalent graphs which are planar triangulations.

2. UNSOLVED PROBLEMS CONCERNING CHROMATIC EQUIVALENCE

(a) Necessary and Sufficient Conditions

Are there necessary and sufficient conditions that two graphs be chromatically equivalent?

It is well known that a necessary condition for chromatic equivalence of graphs is that they have the same number of points, and a sufficient condition for two p-point graphs to be chromatically equivalent is that they be trees.

L. A. Lee, my student, has partial results, which he will present in his doctoral thesis at the George Washington University, but the general question is still open.

(b) The Line Reconstruction Construction Conjecture
    and Chromatic Equivalence

Let G be a graph with at least 3 points and q lines,
$x_1, x_2, \ldots, x_q$.

The line reconstruction conjecture (LRC) first proposed
by Harary [4], states that if the q subgraphs G - x_i are
given, then the entire graph G can be reconstructed, uniquely
up to isomorphism, from these line-deleted subgraphs.

Following Harary [4] we associate a graph G of q lines
with a deck D of q numbered cards, with the ith card containing
the subgraph G_i = G - x_i. D is called the line deletion deck
of G.

F. Harary asked (verbal communication) whether we can
construct the chromatic polynomial of a graph G from its
line-deletion deck. When this question was presented by the
author to T. Brylawski, he proved [3] that if G is not a tree
or a cycle, we can indeed reconstruct $P(G,\lambda)$ from its line-
deletion deck.

Since it is well known that all trees with q lines have
the same chromatic polynomial, and that unicyclic graphs are
line-reconstructible [5], it is clear that the line-deletion
deck D of G determines $P(G,\lambda)$.

Thus we see that the following is equivalent to LRC:
There is a unique graph, chromatically equivalent to G,
which has the line-deletion deck D.

Is there a proof of the above form of LRC?

(c) Chromatically Equivalent Maps and Duals

Are there planar maps G,H which have planar duals G',H'
respectively, such that G $\underset{c}{\sim}$ H and G' $\underset{c}{\not\sim}$ H'?

This question was presented to me by T. Brylawski and
S. Beraha.

(d) Chromatic Equivalence Classes

Since chromatic equivalence is an equivalence relation, it partitions the set of graphs into chromatic equivalence classes.

1. Given a graph G, is there an algorithm which would enable us to generate all graphs chromatically equivalent to G?

2. Is there a canonical form for a representative of each chromatic equivalence class, so that we can count chromatic equivalence classes of p-point graphs rather than listing all p-point graphs?

This procedure could help in graphical enumeration problems.

(e) Chromatic Equivalence of Regular Major Maps

In studying the 4-color conjecture it is known that it is sufficient to restrict oneself to special maps, which we shall call regular major maps. These maps will all be planar duals of planar triangulations.

A map M is called a regular major map if every region has at least five sides, and there are no proper n-rings for n < 4. (See [1] for definitions)

M is called a 4-regular major map if it is a regular major map which has no 4-rings.

When the author expressed the opinion that there were no regular major maps which were chromatically equivalent, since there are no such maps with fewer than 20 regions, the following examples refuted this conjecture:

1. L. A. Lee found a regular major map which is not 4-regular, and which is chromatically equivalent to a map with a quadrilateral.

2. F. Bernhart, in a private communication, exhibited a pair of 4-regular major maps which are chromatically equivalent.

I am sure that there are no two regular major maps with at most 20 regions, and having proper 4-rings which are chromatically equivalent. What is the smallest number n of regions such that there are two regular major maps with n regions, both having 4-rings, which are chromatically equivalent?

Bernhart's example is much larger than 20 regions. In fact,it has more than 60 regions. What is the smallest number n of regions such that there are two 4-regular major maps with n regions which are chromatically equivalent?

3. CHROMATICALLY EQUIVALENT GRAPHS AND THEIR Q-CHROMIAL

Let G be a graph whose chromial is $P(G,\lambda)$. Then the Q-chromial $Q(G,u)$ of the graph G is given by

$$Q(G,u) = \frac{P(G,\lambda)}{\lambda(\lambda-1)(\lambda-2)(\lambda-3)}$$

with the chromial $P(G,\lambda)$ expanded in powers of $u = \lambda - 3$.

The Q-chromial, introduced by Birkhoff and Lewis [2], has degree $p - 4$ for a p-point graph, and smaller coefficients than does $P(G,\lambda)$. Thus, if G is a (p,q) graph with large p, it is more convenient to list $Q(G,u)$ rather than $P(G,\lambda)$.

Below is a list of twenty-five pairs and two triples of chromatically equivalent p-point graphs, with $11 \leq p \leq 17$.

Each of these graphs is a planar triangulation, drawn with an outer circle whose points are all to be connected to a single point, the point at infinity. The degree of the point at infinity is given by the circled number.

The label (n; p, q, r, ..., t) tells us that the graph has n points, of which p are of degree 4, q of degree 5, r of degree 6, etc. Below each set of chromically equivalent graphs is their common Q-chromial.

It is hoped that an examination of these non-trivial examples of chromatically equivalent graphs will help us to characterize chromatic equivalence.

191

p = 11

Chromatic Pair 11-1

⑥

⑦

(11; 4, 4, 3)

(11; 4, 5, 1, 1)

$u^7 + 0u^6 + 7u^5 - 4u^4 + 11u^3 + 0u^2 - 2u + 0$

Chromatic Pair 11-2

⑦

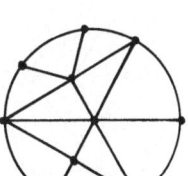

⑦

(11; 4, 5, 1, 1)

(11; 5, 3, 2, 1)

$u^7 + 0u^6 + 7u^5 - 3u^4 + 9u^3 - u^2 - 2u + 0$

Chromatic Pair 11-3

⑥

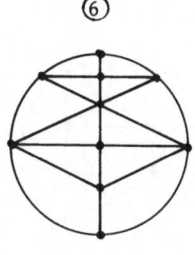

⑦

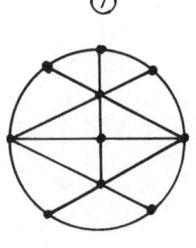

(11; 5, 2, 4)

(11; 6, 1, 3, 1)

$u^7 + 0u^6 + 7u^5 - u^4 + 12u^3 + 6u^2 + 3u + 1$

p = 12

Chromatic Pair 12-1

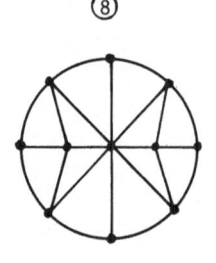

(12; 6, 4, 0, 0, 2)

(12; 6, 4, 0, 0, 2)

$u^8 + 0 + 8 + 1 + 9 - 2 + 3 + 2 + 0$

p = 13

Chromatic Triple 13-1

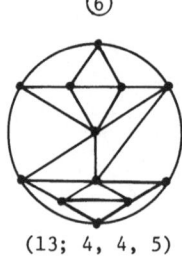

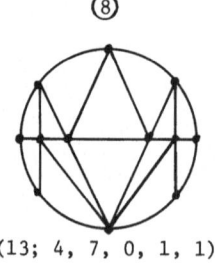

(13; 4, 4, 5)

(13; 4, 6, 2, 0, 1)

(13; 4, 7, 0, 1, 1)

$u^9 + 0 + 9 - 6 + 24 - 15 + 6 + 6 + 0 + 0$

Chromatic Pair 13-2

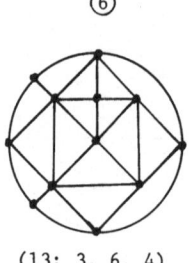

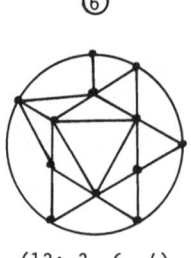

(13; 3, 6, 4)

(13; 3, 6, 4)

$u^9 + 0u^8 + 9u^7 - 13u^6 + 40u^5 - 43u^4 + 29u^3 + 0u^2 - 5u + 1$

p = 13

Chromatic Pair 13-3

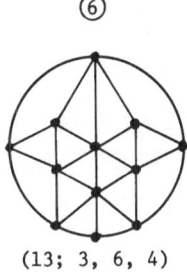

(13; 3, 6, 4)   (13; 3, 6, 4)

$$u^9 + 0u^8 + 9u^7 - 13u^6 + 39u^5 - 41u^4 + 31u^3 + 3u^2 - 4u + 1$$

p = 14

Chromatic Pair 14-1

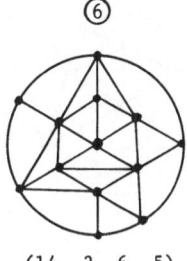

(14; 3, 6, 5)   (14, 3, 6, 5)

$$u^{10} + 0u^9 + 10u^8 - 15u^7 + 55u^6 - 79u^5 + 88u^4 - 32u^3 - 3u^2 + 8u - 1$$

Chromatic Pair 14-2

(14; 3, 7, 3, 1)   (14; 3, 7, 3, 1)

$$u^{10} + 0u^9 + 10u^8 - 15u^7 + 51u^6 - 72u^5 + 76u^4 - 32u^3 - 6u^2 + 6u - 1$$

p = 14

Chromatic Pair 14-3

⑨

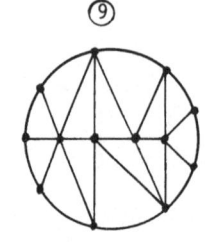

⑨

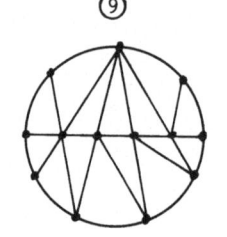

(14; 5, 5, 3, 0, 0, 1)

(14; 5, 6, 1, 1, 0, 1)

$$u^{10} + 0u^9 + 10u^8 - 5u^7 + 31u^6 - 14u^5 + 16u^4 - 2u^3 - 2u^2 + u + 0$$

Chromatic Pair 14-4

⑧

⑧

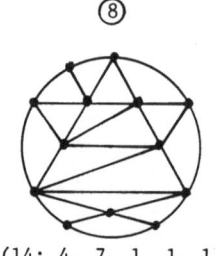

(14; 5, 6, 0, 2, 1)

(14; 4, 7, 1, 1, 1)

$$u^{10} + 0u^9 + 9u^8 - 9u^7 + 32u^6 - 19u^5 + 23u^4 + 7u^3 - 12u^2 - u + 1$$

p = 15

Chromatic Triple 15-1

⑦

⑧

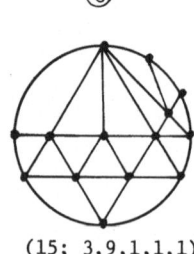

⑧

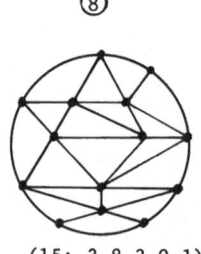

(15; 3,9,0,3)

(15; 3,9,1,1,1)

(15; 3,8,3,0,1)

$$u^{11} + 0u^{10} + 11u^9 - 14u^8 + 55u^7 - 77u^6 + 101u^5 - 58u^4 + 0u^3 + 11u^2 - 2u + 0$$

p = 15

Chromatic Pair 15-2

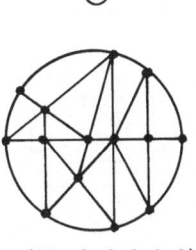

(15; 4,8,0,2,1)   (15; 3,9,1,1,1)

$u^{11} + 0u^{10} + 11u^9 - 13u^8 + 53u^7 - 64u^6 + 82u^5 - 50u^4 + 4u^3 + 10u^2 - 2u + 0$

Chromatic Pair 15-3

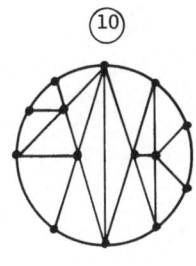

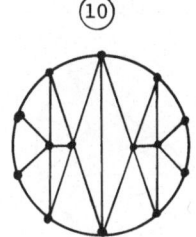

(15; 5,7,1,1,0,0,1)   (15; 4,8,2,0,0,0,1)

$u^{11} + 0u^{10} + 10u^9 - 4u^8 + 27u^7 - 18u^6 - 14u^5 + 6u^4 - 3u^3 + 2u^2 + u + 0$

Chromatic Pair 15-4

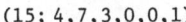

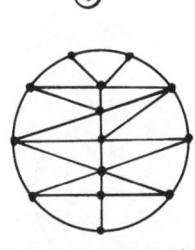

(15; 4,7,3,0,0,1)   (15; 4,7,3,0,0,1)

$u^{11} + 0u^{10} + 11u^9 - 8u^8 + 41u^7 - 37u^6 + 49u^5 - 9u^4 - 11u^3 + u^2 + 0u + 0$

p = 15

Chromatic Pair 15-5

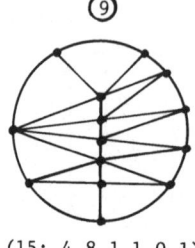

(15; 4,8,1,1,0,1)

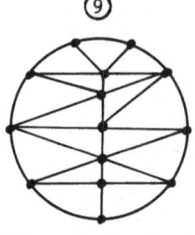

(15; 4,7,3,0,0,1)

$u^{11} + 0u^{10} + 11u^9 - 7u^8 + 40u^7 - 31u^6 + 42u^5 - 10u^4 - 10u^3 + u^2 + 0u + 0$

Chromatic Pair 15-6

(15; 5,7,0,2,0,1)

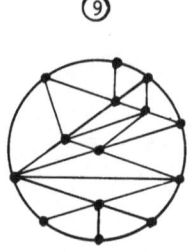

(15; 4,8,1,1,0,1)

$u^{11} + 0u^{10} + 10u^9 - 9u^8 + 39u^7 - 29u^6 + 41u^5 + u^4 - 15u^3 + 0u^2 + u + 0$

p = 16

Chromatic Pair 16-1

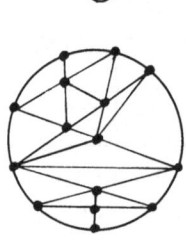
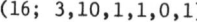

(16; 3,10,1,1,0,1)

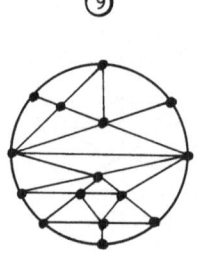

(16; 4,9,0,2,0,1)

$u^{12} + 0u^{11} + 11u^{10} - 11u^9 + 45u^8 - 53u^7 + 72u^6 - 31u^5 - 24u^4 + 13u^3 + 2u^2 - u + 0$

$p = 16$

<u>Chromatic Pair 16-2</u>

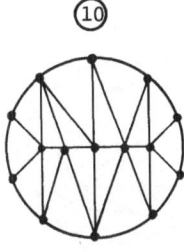

(16; 4,8,3,0,0,1)

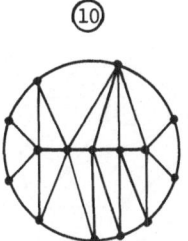

(16; 4,9,1,1,0,0,1)

$u^{12} + 0u^{11} + 12u^{10} - 8u^9 + 48u^8 - 44u^7 + 65u^6 - 35u^5 + 4u^4 + 9u^3 - u^2 + 0u + 0$

<u>Chromatic Pair 16-3</u>

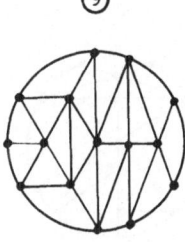

(16; 3,9,3,0,0,1)

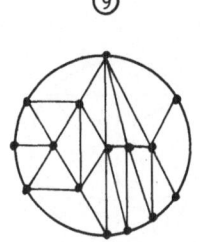

(16: 3,10,1,1,0,1)

$u^{12} + 0u^{11} + 12u^{10} - 14u^9 + 58u^8 - 83u^7 + 132u^6 - 89u^5 + 29u^4 + 18u^3 - 6u^2 + 0u + 0$

<u>Chromatic Pair 16-4</u>

(16; 4,9,0,2,0,1)

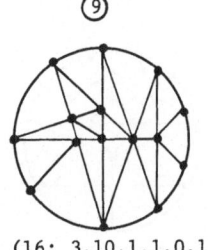

(16; 3,10,1,1,0,1)

$u^{12} + 0u^{11} + 12u^{10} - 14u^9 + 61u^8 - 82u^7 + 119u^6 - 97u^5 + 44u^4 + 0u^3 - 10u^2 + 2u + 0$

$$p = 16$$

Chromatic Pair 16-5

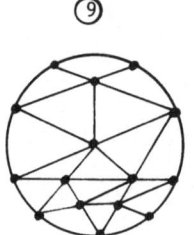

(16; 3,9,3,0,0,1)   (16; 3,9,3,0,0,1)

$$u^{12} + 0u^{11} + 12u^{10} - 16u^9 + 63u^8 - 99u^7 + 142u^6 - 136u^5 + 62u^4 + 9u^3 - 12u^2 + 2u + 0$$

Chromatic Pair 16-6

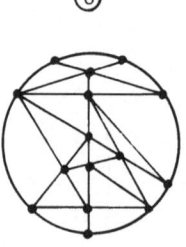

(16; 3,8,4,0,1)   (16; 3,9,2,1,1)

$$u^{12} + 0u^{11} + 12u^{10} - 16u^9 + 69u^8 - 114u^7 + 181u^6 - 162u^5 + 68u^4 + 15u^3 - 13u^2 + 2u + 0$$

Chromatic Pair 16-7

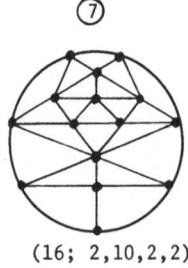

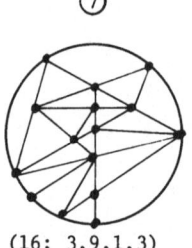

(16; 2,10,2,2)   (16; 3,9,1,3)

$$u^{12} + 0u^{11} + 12u^{10} - 16u^9 + 75u^8 - 124u^7 + 204u^6 - 193u^5 + 86u^4 + 8u^3 - 18u^2 + 3u + 0$$

p = 16

Chromatic Pair 16-8

⑦

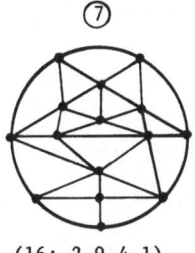

(16; 2,9,4,1)

⑦

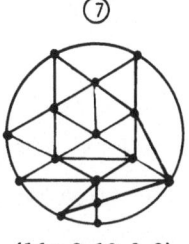

(16; 2,10,2,2)

$$u^{12} + 0u^{11} + 12u^{10} - 18u^9 + 80u^8 - 147u^7 + 262u^6 - 271u^5 + 151u^4 - 4u^3 - 30u^2 + 6u + 0$$

Chromatic Pair 16-9

⑦

(16; 2,9,4,1)

⑦

(16; 2,9,4,1)

$$u^{12} + 0u^{11}\ 12u^{10} - 20u^9 + 84u^8 - 182u^7 + 322u^6 - 382u^5 + 255u^4 - 51u^3 - 31u^2 + 16u - 2$$

p = 17

Chromatic Pair 17-1

⑧

(17; 3,10,1,2,1)

⑧

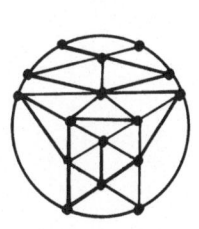

(17; 2,11,2,1,1)

$$u^{13} + 0u^{12} + 13u^{11} - 17u^{10} + 84u^9 - 147u^8 + 268u^7 - 303u^6 + 211u^5 - 65u^4 - 12u^3 + 18u^2 - 3u + 0$$

REFERENCES

1. Bari, R., "Absolute Reducibility of Maps of at Most 19 Regions," Doctoral Dissertation, John Hopkins University, 1966.

2. Birkhoff, G. D. and Lewis, D. C., "Chromatic Polynomials," Trans. Amer. Math. Soc., 60, No. 3 (1946), 355-451.

3. Brylawski, T. A., "Reconstructing Combinatorial Geometries," this volume p. 226.

4. Harary, F., Graph Theory, Addison-Wesley, Reading, Mass., 1969.

5. Manvel, B, "Reconstruction of Unicyclic Graphs," Proof Techniques in Graph Theory (F. Harary, ed.) Academic Press, New York, 1969, 103-107.

6. Tutte, W. T., "On Chromatic Polynomials and the Golden Ratio," J. Combinatorial Theory 9, 290-296.

7. Whitney, H., "The Coloring of Graphs" Ann. of Math. 33 (1932) 688-718.

ON COVERING THE POINTS OF A GRAPH
WITH POINT DISJOINT PATHS

F. T. Boesch, S. Chen, and J. A. M. McHugh
Bell Laboratories
Holmdel, New Jersey

ABSTRACT

The minimum number of point disjoint paths which cover all the points of a graph defines a covering number denoted by ζ. The relation of ζ to some other well-known graphical invariants is discussed, and ζ is evaluated for a variety of special classes of graphs. A simple algorithm is developed for determining ζ in the case of a tree, and it is shown that this tree algorithm can be generalized to yield ζ for any connected graph. Degree conditions are also derived which yield simple upper bounds for ζ.

ON COVERING THE POINTS OF A GRAPH
WITH POINT DISJOINT PATHS

1. INTRODUCTION

Different covering numbers for graphs (the terminology and notation used
here follow that of Harary [4]) arise naturally in network diagnostic schemes.
For example, consider the question of providing systematic procedures for testing
the integrity of the lines of a graph by examining appropriate line sequences.
One natural approach would be to cover all the lines of a graph with line-disjoint
trails. The smallest such number of line disjoint trails which cover all the lines
of a graph G is called the <u>trail-to-line covering number</u> denoted by $\eta(G)$, or simply
η. Clearly, η is related to the classical notion of eulerian transversability.
Indeed, for a connected graph G, η is just the max $(1, P_0)$, where P_0 is one-half
the number of points of odd degree (Harary [4]).

A more interesting parameter is obtained by considering point diagnostic
schemes for network integrity. Consequently, this paper investigates coverings of
the points of a graph with point-disjoint paths. The minimum number of point-
disjoint paths which cover all the points of G is denoted by $\zeta(G)$, or simply ζ,
and is called the <u>path-to-point covering number</u> of G. In order for ζ to be well-
defined, paths which consist of a single point must be admitted. Then, for example,
the star $K_{1,n}(n \geq 2)$ has $\zeta(K_{1,n}) = n - 1$.

Conceptually, the covering parameter most similar to η seems to be the path
number $\pi(G)$, the smallest number of line-disjoint paths which cover all the lines
of G ([5, 6, 7, 11]).

Although η and ζ are intuitively dual concepts, and indeed similar to other
covering parameters, the covering number ζ appears to have received much less
attention in the literature. However, ζ is clearly related to the study of hamiltonian graphs, and it has been introduced in this context by Barnette [1] to
examine the existence of hamiltonian paths in three-connected planar graphs. A
complete discussion of this problem is given by Klee in [8]. The relation of ζ
to hamiltonian graphs is also explored by Goodman and Hedetniemi [3].

The motivation for introducing $\pi(G)$ came from considering a graph theoretic model for information retrieval structures. The covering number ζ could likewise be introduced in an information retrieval context. For example, path-to-point covers are useful for certain problems in the theory of information structures. Thus consider a data structure organized as a 2-way linked list (Knuth [9]. It is often necessary to reclaim storage using garbage collection techniques as well as to search into the structure for desired data items. In such cases, one must have a systematic way for investigating all the items in the structure, and, for implementation on a real-time computer system, traversal must be very efficient. Clearly, maximum efficiency is achieved when the number of paths in the traversal is minimal.

2. COMPARISON WITH OTHER GRAPHICAL INVARIANTS

Trail-to-line and path-to-point covers are natural generalizations of eulerian trails and hamiltonian paths in the following sense:

> (i) $\eta(G) = 1$ iff G has an eulerian trail,

and

> (ii) $\zeta(G) = 1$ iff G has a hamiltonian path.

Note that although η and ζ are line and point analogs, they are not simply related to each other using line graphs. The example in Figure 1 shows that $\eta(G) \neq \zeta(L(G))$ and $\eta(L(G)) \neq \zeta(G)$.

Furthermore η and ζ are not related by a simple inequality, i.e., $K_{1,5}$ shows that $\zeta \not\leq \eta$ and K_4 shows that $\eta \not\leq \zeta$. On the other hand, ζ is related to the diameter and connectivity of a graph.

$\zeta(G) = 2$ $\zeta(L(G)) = 1$

$\eta(G) = 2$ $\eta(L(G)) = 1$

G   L(G)

Figure 1. A Graph and Its Line-Graph

Theorem 1.   Let G be a two-connected graph with connectivity $\kappa(G)$ and diameter $d(G)$. Then for $\kappa(G) \geq 2$,

(i) $$\zeta(G) \leq \kappa(G)[2-d(G)] + p - 3,$$

and for $\kappa(G) = 1$.

(ii) $$\zeta(G) \leq p - d(G).$$

Remark.   Notice that if G is merely an even cycle C_{2m} then the bound given by (i) of Theorem 1 is attained, and for the star $K_{1,n}$ the bound in (ii) of Theorem 1 is attained.

Proof.   (i) If v_1 and v_2 are two diametral points (endpoints of a longest geodesic), then by Menger's Theorem (Harary [4]), there are $\kappa \geq 2$ paths connecting v_1 and and v_2 which are point-disjoint except for v_1 and v_2.   Hence the points on all these paths can be covered by $\kappa(G) - 1$ point-disjoint paths.   Now as the diameter is $d(G)$, there must be $d(G) - 1$ or more points excluding v_1 and v_2 on each of these "Menger-paths."   Thus there remains at most $p - [\kappa(d(G) - 1) + 2]$ points of G, which can be covered by zero length paths.

(ii)   The path connecting two diametral points cover $d(G) + 1$ points, and zero length paths cover the remaining points of G.

Finally note that ζ can be defined equivalently as a point-disjoint covering of all the points of a graph with paths such that the total number of lines in these paths is a maximum.   However, the caterpillar example developed in Figures 2a, 2b, and 2c demonstrates that this equivalence does not imply that ζ can be obtained in the "greedy" sense of obtaining a longest path and then a next longest, etc.

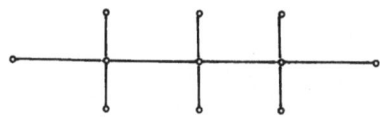

Figure 2a.   The Caterpillar

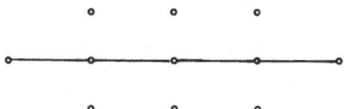

Figure 2b. A "Greedy" Seven-Path Covering

Figure 2c. A Minimum Covering

3. EVALUATING ζ FOR SPECIAL GRAPHS

It is obvious that the complete graph K_p has $\zeta(K_p) = 1$, and that $\zeta(K_{1,n}) = n - 1 \ (n \geq 2)$. A more interesting but still simple case is the complete bigraph $K_{n,m}$. Clearly, $\zeta(K_{n,m})$ depends on the magnitude of the difference between n and m. Rather than discuss the bigraph, consider the case of the complete n-partite graph.

Theorem 2. Let $K(p_1,p_2,\ldots,p_n) = (V,X)$ be the complete n-partite graph with parts $V_1,V_2,\ldots,V_n$ where $|V_k| = p_k$ and $p_1 \geq p_2 \geq \cdots \geq p_n$. Then

$$\zeta(K(p_1,p_2,\ldots,p_n)) = \max \left\{ 1, \ p_1 - \sum_{j=2}^{n} p_j \right\}$$

Proof. Two cases are considered. First it is shown that $\zeta = 1$ if

$$p_1 - \sum_{j=2}^{n} p_j \leq 1 .$$

To this end, note that the minimum degree δ of any point in $K(p_1,p_2,\ldots,p_n)$ is $\sum_{j=2}^{n} p_j$. Thus by Ore's condition for a Hamiltonian path, (Ore [10]), $\zeta = 1$ if

$$2\delta = 2 \sum_{j=2}^{n} p_j \geq p - 1 = \sum_{j=2}^{n} p_j + p_1 - 1 .$$

Suppose then that for the second case,

$$p_1 > \sum_{j=2}^{n} p_j + 1 .$$

If $\zeta = 1$, then since no two points of V_1 are adjacent, and since all points of V_1 must be covered by one path,

$$P_1 \leq \sum_{j=2}^{n} P_j + 1 .$$

Thus $\zeta \geq 2$. Now the endpoints of any 2 paths in a minimum cover cannot be adjacent; thus all the endpoints of the paths must be contained in the same part, say V_i. It is now demonstrated that the points of each path in a minimum cover must alternate between V_i and $\overline{V}_i$ (the complement of V_i). Hence assume to the contrary that one of these paths contains a subpath of the form

$$v_0 x_1 v_1 x_2 v_2 x_3 \cdots v_s$$

where $v_0, v_s \in V_i$ and $v_1, v_2 \notin V_i$; see Figure 3. Since $\zeta \geq 2$ there exists another path P' say $v_0' x_1' v_1' \cdots v_m'$ with $v_0', v_m' \in V_i$. The two paths could be combined into the following single path

$$v_0 x_1 v_1 x'' \left[v_0' x_1' v_1' \cdots v_m' \right] x''' v_2 \cdots v_s .$$

This, however, would contradict minimality.* Thus the points of each path in the minimal cover alternate between V_i and $\overline{V}_i$. Now this is impossible if $i \neq 1$ as the points of $V_1 \subseteq \overline{V}_i$ ($i \neq 1$) could not be covered by such paths. Hence suppose there are α paths of nonzero length which start and end in V_1 and β paths of zero length, each path being a point of V_1. Hence

$$P_1 = \sum_{j=2}^{n} P_j + \alpha + \beta$$

and

$$\zeta = P_1 - \sum_{j=2}^{n} P_j$$

It should be noted that if $\alpha > 1$, then two paths of non-zero length can be combined to increase β by one and decrease α by one. Hence, there is always a cover with $\alpha = 1$.

---

*Note that this contradiction is still valid when P' consists of a single point.

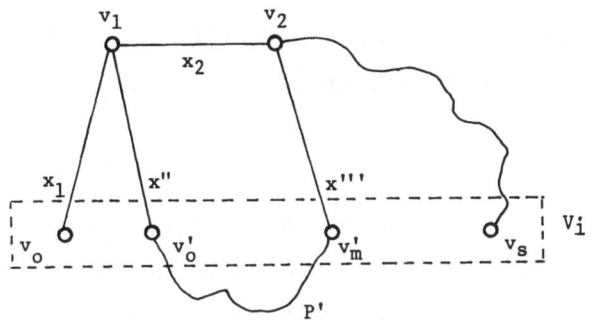

Figure 3. The Construction for Theorem 2

Moreover, observe that the bipartite case of Theorem 2 shows how to construct

non-hamiltonian graphs of arbitrarily large connectivity.

4. CALCULATING ζ FOR TREES AND CONNECTED GRAPHS

The value of ζ for an arbitrary tree does not appear to be simply related to

any other well-known tree invariant. However ζ can be calculated for a forest or

a tree using the algorithm presented in this section; this algorithm was discovered

independently by Goodman and Hedetniemi [3]. Recall that if V is the point set of

a graph then for $W \subseteq V$, the notation $\langle W \rangle$ means the induced subgraph on W.

Theorem 3.   (i)   If G has two or more endpoints $v_{a_1}, v_{a_2}, \ldots, v_{a_k}$

adjacent to a point v_0 , then

$$G' = \left\langle V - \{v_0, v_{a_1}, v_{a_2}, \ldots, v_{a_k}\} \right\rangle$$

satisfies

$$\zeta(G) = \zeta(G') + k - 1 ,$$

(ii)   If G has one endpoint w which is adjacent to a point v

of degree two, then

$$G'' = \left\langle V - \{w\} \right\rangle$$

satisfies

$$\zeta(G) = \zeta(G'')$$

Proof. Case (i) Clearly $\left\langle v_0, v_{a_1}, v_{a_2}, \ldots, v_{a_k} \right\rangle$ is a star $K_{1,k}$, and hence

$\zeta(G) \leq \zeta(G') + k - 1$ by covering G' and the star independently. It is now shown

that any minimum cover of G defines a cover of G' which has $\zeta(G) - (k-1)$ paths, and the proof of this case is then complete. Clearly, any minimum cover of G must include v_0 and v_0 cannot be on a path of zero length. Furthermore the path which covers v_0 cannot terminate at v_0. This path may terminate at some v_{a_i} or it may not. However, in either case there exists a cover of G' which has $\zeta(G') - (k-1)$ paths.

   Case (ii) Any minimum cover of G must include w, and w must be an endpoint of some path. Furthermore w cannot be covered by a path of zero length else v would be the endpoint of some path. Hence the path which covers w must include v and $\zeta(G'') \leq \zeta(G)$. On the other hand, any cover of G'' must have v as an endpoint of a path. Hence any minimum cover of G'' yields a cover of the same size in G, and thus $\zeta(G) \leq \zeta(G'')$.

   Suppose G has endpoints which satisfy either (i) or (ii) of Theorem 3. Then the determination of $\zeta(G)$ can be simplified to finding the ζ of a reduced graph. If the reduced graph has a known ζ, then $\zeta(G)$ is determined. Alternatively if the reduced graph again has endpoints of the desired type, then the reduction can be continued. The next lemma establishes that this reduction yields an algorithm for finding ζ of a forest.

Lemma. Any nontrivial forest has either two or more endpoints adjacent to the same point or it has at least one endpoint adjacent to a point of degree two.

Proof. Suppose not, then all endpoints are adjacent to different points, and these in turn are of degree greater than two. However this is impossible as removing the endpoints must produce another forest which in turn has endpoints.

   By observing that if G is a forest, then G' and G", as defined in Theorem 3, are also forests, it follows that this Lemma together with Theorem 3 provide an algorithm for calculating ζ for any forest. For certain special trees, this algorithm can be used to derive an explicit formula for ζ; in particular this is the case for regular acyclic data structures. To this end, let $T_{d_1, d_2, \ldots, d_{\ell-1}}$ denote a tree with one point v of degree d_1 such that for each $k(1 \leq k \leq \ell - 1)$ every point at distance k from v has degree $d_k + 1$; all points at distance ℓ from v are endpoints. Also let T_0 denote an isolated point. Thus Theorem 3 provides for the following recursion formula

$$\zeta(T_{d_1,d_2,\ldots,d_{\ell-1}}) = \zeta(T_{d_1,d_2,\ldots,d_{\ell-3}}) + (d_{\ell-1} - 1) \prod_{i=1}^{\ell-2} d_i \ .$$

The value of ζ can then be calculated by distinguishing odd and even ℓ and observing that $\zeta(T_{d_1}) = d_1 - 1$ and $\zeta(T_0) = 1$. Simple formulas are easily obtained for special cases of such trees by applying this recursion formula.

<u>Corollary 1.</u> Let T be a tree in which all of the nonendpoints have degree n and each endpoint is equidistant (say at distance ℓ) from some fixed point. Then

$$\zeta(T) = (n-1)^\ell$$

<u>Corollary 2.</u> Let T be a tree in which all of the nonendpoints have degree $(n+1)$ except for one point of degree n and each endpoint is equidistant (say at distance ℓ) from the degree n point. Then

$$\zeta(T) = \left[\frac{n^{\ell-1}+1}{n+1} \right]$$

where [x] denotes the largest integer $\leq$ x, as usual.

Another result is that the tree algorithm can be used to determine ζ of any graph

<u>Theorem 4.</u> Let G be a connected graph and let $T_1, T_2, \ldots, T_\ell$ denote all the spanning trees of G. Then

$$\zeta(G) = \min_{1 \leq k \leq \ell} \zeta(T_k)$$

<u>Proof.</u> Now if H is any spanning subgraph of G then certainly $\zeta(G) \leq \zeta(H)$. On the other hand, if there is a covering of the points of a graph, then the lines contained in the paths of this covering are contained in some spanning tree T of G. Hence $\zeta(T) \leq \zeta(G)$ for some spanning tree T.

Since there are many known algorithms for finding all the spanning trees of a connected graph, Theorem 4 actually provides an algorithm for determining ζ for any graph; see [3] for a discussion of the efficiency of this approach.

Before concluding this section, it is interesting to note that the homeomorphic reduction in Step (ii) of Theorem 3 is peculiar to the endpoints,viz., for the example of Figure 4, $\zeta(T) = 3$ and $\zeta(T_H) = 2$ where T_H is the homeomorphic reduction of T.

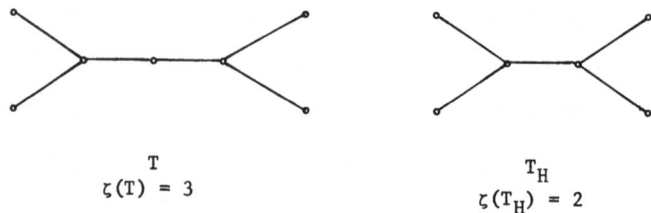

$$T$$
$$\zeta(T) = 3$$

$$T_H$$
$$\zeta(T_H) = 2$$

Figure 4. A Graph and Its Homeomorphic Reduction With Different Values of ζ

5. DEGREE CONDITIONS

Path-to-point covers generalize the concept of a hamiltonian path. Since there are a variety of degree conditions for hamiltonian graphs, it is only natural to seek degree conditions for ζ. In fact, an examination of Ore's work [10] on degree conditions for hamiltonian paths reveals a theorem on ζ. Ore introduced point-disjoint path coverings of points such that the paths contain a maximum number of lines. As previously noted, this is equivalent to the definition of ζ given here. Ore derived a basic lemma on ζ but only used it to derive a theorem for ζ, when ζ is equal to one. Since his final result on $\zeta = 1$ can be derived without ever introducing the covering concept, this particular lemma may not be well-known. The next theorem, gives a slightly more general version of this lemma; the proof being similar to that given by Ore.

Theorem 4. Let n be any number with $1 \leq n \leq p - 1$. If for each set N of $n + 1$ independent points of a graph G, $d_i + d_j \geq p - n$ for some $v_i, v_j \in N (i \neq j)$, then $\zeta(G) \leq n$.

The bound of this theorem is attainable in the following sense. For any $n \geq 1$, let G* be a complete bipartite graph with parts of size n and 2n. By Theorem 2, $\zeta(G*) = n$, and clearly $d_i \geq n$ for all of the 3n points of G*.

An immediate corollary of this theorem is a bound for ζ in terms of the minimum degree δ.

Corollary. If δ is the minimum degree of the points of G, then $\zeta(G) \le$ max
$\{1, p-2\delta\}$. Furthermore, this bound is attained by G.

Many interesting degree-condition questions remain unresolved. For example, conditions stronger than those of Theorem 4 are known for n = 1 (Chvátal [2]), but analogous conditions for n > 1 were only obtained several months after this conference by V. Chvátal. These two results are as follows.

Theorem. Let the non-decreasing sequence $\Pi(G) = (d_1, d_2, \ldots, d_p)$ be the degree sequence of a graph G, and let n be a given number ($1 \le n \le p$). Then $\zeta(G) \le n$, if

$$d_i + n \le i < \frac{p+n}{2} \quad \text{implies} \quad d_{p+n-1} \ge p-i \ .$$

Furthermore if $\Pi(G)$ fails to satisfy this condition at i = k, then the graph

$$\hat{G} = K_{k-n} + (\overline{K}_k \cup K_{p-2k+n})$$

with non-decreasing degree sequence $(\hat{d}_1, \hat{d}_2, \ldots, \hat{d}_p)$ has $\zeta(\hat{G}) = n+1$, and $\hat{d}_i \ge d_i$ for all i.

Corollary. (The line extremal result). For any $k, n \ge 1$ and $p \ge 2$, let

$$f(k) = k(k-n) + \frac{(p-k)\ (p-k-1)}{2}$$

and

$$q_0 = 1 + \max \left\{ f(n),\ f\left(\left\lceil \frac{p+n-1}{2} \right\rceil\right) \right\}.$$

Then all (p,q) graphs with $q \ge q_0$ lines have $\zeta \le n$. Furthermore there exists a graph G_0 with $q_0 - 1$ lines such that $\zeta(G_0) = n+1$.

Similarly, though the line extremal result can be derived when n = 1 by using Theorem 4 (Ore [10]), the same type of result for general n is still unavailable. In particular, direct application of Theorem 4 does not yield a sharp (extremal) result when $n \ge 2$. Instead, it only supplies a bound q_0 such that for all graphs G with more than q_0 lines, $\zeta(G) \le n$. However, this bound is not sharp since there are cases where all graphs with $q_0 - 1$ lines also have $\zeta \le n$. Furthermore, the obvious generalization $(K_{p-n} \cup \overline{K}_n)$ of the extremal graph for n = 1 does not yield the general extremal graph. For example, $K_4 \cup \overline{K}_3$ (with 36 lines) has a ζ of 3, while $K_4 + \overline{K}_8$ (with 38 lines) has a ζ of 4.

REFERENCES

1. Barnette, D. Trees in Polyhedral Graphs, <u>Canad. J. Math.</u> 18 (1966), 731-736.

2. Chvatal, V., On Hamilton's Ideals, <u>J. Combinatorial Theory</u> 12B (1972), 163-168.

3. Goodman, S. and Hedetniemi, S., On the Hamiltonian Completion Problem, this volume p. 262.

4. Harary, F., <u>Graph Theory</u>, Addison-Wesley, Reading, Mass. (1969).

5. Harary, F., Covering and Packing in Graphs, I, <u>Ann. New York Acad. Sci.</u>, 175, (1970), 198-205.

6. Harary, F. and Hsiao, D., A Formal System for Information Retrieval from Files, <u>CACM</u> 13, (1970), 67-73.

7. Harary, F. and Schwenk, Evolution of the Path Number of a Graph: Covering and Packing in Graphs II, <u>Graph Theory and Computing</u>, (R. C. Read, ed.) Academic Press, New York, 1972, 39-45.

8. Klee, V. Long Paths and Circuits on Polytopes, Chapter 17, <u>Convex Polytopes</u> (by B. Grünbaum), Wiley-Interscience, New York, 1967.

9. Knuth, D. <u>The Art of Computer Programming</u>, Vol. I, Addison-Wesley, Reading, Mass., 1968.

10. Ore, O., Arc Coverings of Graphs, <u>Ann. Mat. Pura Appl.</u> 55 (1961) 315-322.

11. Stanton, R., Cowan, D. D., and James, L. O., Some Results on Path Numbers, <u>Proceedings of the Louisiana Conference on Combinatorics, Graph Theory and Computing</u>, Baton Rouge, 1970, 112-135.

A USEFUL FAMILY OF BICUBIC GRAPHS

T. G. Boreham, I. Z. Bouwer
University of New Brunswick

R. W. Frucht
Universidad Tecnica Federico Santa Maria
Santiago, Chile

ABSTRACT

Let the vertices of a 2n-gon be labelled, in an order of traversal: 0, 1', 1, 2', 2, ..., (n-1)', n-1, 0'. Let $G(n,m)$ denote the bicubic graph derived from this 2n-gon by adjunction of the chords $(i,(i+m)')$, $i = 0, 1, 2, ..., n-1$, the addition being taken modulo n. Restricting ourselves to the case when n is prime, we determine the isomorphism classes of the graphs $G(n,m)$, and the corresponding automorphism groups. Various applications are discussed.

A USEFUL FAMILY OF BICUBIC GRAPHS

1. INTRODUCTION

Let n, m be any natural numbers, with $1 < m \leq n-1$. The bicubic (= bipartite trivalent) graph $G(n,m)$ is defined as follows: It has $2n$ vertices, which we shall label:

$$0, 1, 2, \ldots, n-1, 0', 1', 2', \ldots, (n-1)'$$

and the $3n$ edges:

$$[i,i'], \quad [i, (i+1)'], \quad [i, (i+m)'],$$
$$i = 0, 1, 2, \ldots, (n-1)$$

where the addition is taken modulo n. The graph $G(n,m)$ may be viewed as derived from the $(2n)$-gon with vertices (in an order of traversal):

$$0', 0, 1', 1, 2', 2, \ldots, (n-1)', n-1,$$

by the adjunction of the chords

$$[i, (i+m)'], \qquad i = 0, 1, 2, \ldots, n-1.$$

In the notation of Frucht [6], $G(n,m)$ may be represented by the symbol

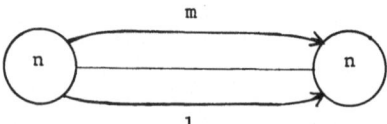

This family of graphs was introduced by Coxeter [2, p. 426], as a generalization of the well-known 6-cage or Heawood graph, which is $G(7,3)$. In [3, p. 107], Coxeter and Moser give a detailed study of a certain subfamily of these graphs, namely the graphs associated with certain hexagonal maps $\{6,3\}_{b,c}$ on the torus, and they determine the groups of these maps. While the graphs $G(n,m)$ are in general only vertex-transitive, this subfamily corresponds to graphs that are also edge-transitive, and hence symmetrical. In particular, in 1967, C. C. Sims pointed out [letter to R. M. Foster] that when b and c are coprime, $b + c > 3$, $bc(b - c) \neq 0$, then the graphs $\{6,3\}_{b,c}$ of Coxeter's family are 1-regular (in the sense of Tutte [9]; 1-unitransitive, in the terminology of Harary [7]) and not 2-regular (as reported by Coxeter [2], p. 427). This provided an infinite family of 1-regular trivalent graphs. Earlier, such graphs were thought to be rather exceptional; the first known one (with 432 vertices) was described by Frucht [5], and only 4 with fewer than 400 vertices were listed in Foster's census [4]. Recently, Miller [8] found all 1-regular trivalent graphs with girth 6. In terms of the graphs G(n,m), the family described by Sims and Miller corresponds to the case where m is the least positive solution of the congruence

$$m(m-1) \equiv -1 \pmod{n}, \quad n > 7 .$$

The smallest member is G(13,4) (= $\{6,3\}_{3,1}$) with 26 vertices, as found by Sims.

However, it seems that the symmetry properties of the class of graphs G(n,m) as a whole have not yet been systematically studied, although these properties are of use in relation to various problems, of which we mention the following.

(1)   In [10], M. Watkins raised the question as to which finite groups have regular representations as automorphism groups of graphs. He showed that, in particular, the dihedral groups of order at least 12 admit such representations on 5-valent graphs. We see that dihedral groups of order at least 18 admit such representations on trivalent graphs, these being G(n,3), where 2n is the order of the dihedral group.

(2)   Bouwer and Frucht, in [1], left open the following problem for primes p $\geq$ 11:

How many different graphs are there with the least possible number of edges (namely 4p) such that their automorphism group is isomorphic to the cyclic group of order p?

The graphs G(p,m) are an essential tool in the construction of these edge-minimal graphs, and the question can be answered as soon as one can determine the number of non-isomorphic graphs G(p,m) for each prime p $\geq$ 11.

(3) Let $F(n,m)$ be the following cubic form in n variables,
v_0, v_1, v_2, ..., v_{n-1}:

$$F(n,m) = \sum_{i=0}^{n-1} v_i \, v_{i+1} \, v_{i+m} \, ,$$

where addition of the subscripts is taken modulo n. If
we let the group of permutations of the variables that
leave the value of F fixed be $C(n,m)$, we can see that
this is a subgroup of index 2 in the automorphism group
of $G(n,m)$. Thus, determining the groups of the graphs
$G(n,m)$ will enable us to find $C(n,m)$.

Our main results may be briefly summarized as follows:
We have found the isomorphism classes and automorphism
groups of the graphs $G(p,m)$ for any given prime p, with
$2 \leq m \leq \frac{p+1}{2}$. One isomorphism class is always formed by
$G(p,2)$ and $G(p, \frac{p+1}{2})$. If $p \equiv 5 \pmod 6$ then the graphs
$G(p,m)$ with $3 \leq m \leq \frac{p-1}{2}$ are mutually isomorphic in triples, so that there are $\frac{p-5}{6}$ different isomorphism
classes of this kind. If $p \equiv 1 \pmod 6$ then the graphs
$G(p,m)$ with $3 \leq m \leq \frac{p-1}{2}$ are mutually isomorphic in
triples with one exception. Thus in this case, the
number of isomorphism classes of the graphs $G(p,m)$ with
$3 \leq m \leq \frac{p-1}{2}$ is $\frac{p-1}{6}$. The exception just mentioned
corresponds for $p = 7$ to the 6-cage with a group of order
336 [2, p. 425], and for $p = 13, 19, 31, \ldots$ to the one-
regular graphs found by Sims whose automorphism group is
known [3, p. 107] to be of order 6p. In the remaining
cases the groups are isomorphic to dihedral groups, and
of order 4p if $m = 2$ or $m = \frac{p+1}{2}$, and of order 2p if
$3 \leq m \leq \frac{p-1}{2}$. Since 2p is also the number of vertices
of $G(p,m)$ and since these graphs are vertex-transitive,

we have as a corollary the existence of regular representations of the dihedral group D_p on trivalent graphs for any prime $p \geq 11$. Avoiding the exceptional one-regular graphs by choosing m = 3, we can, in particular, exhibit G(p,3) as yielding the desired representation of D_p. It has already been mentioned above that also if n is not a prime, then G(n,3) still furnishes a regular representation of D_n on a trivalent graph, whenever $n \geq 9$. In general, however, the results obtained in this paper for prime n are not valid for composite n. This latter case will be considered in a subsequent paper.

2. PETRIE PATHS: DEFINITION

Lemma 2.1. Let n, m be any natural numbers such that $n \geq 3$ and $2 \leq m \leq \frac{n+1}{2}$. Then

(i)   G(n,m) is of girth 6 iff $2 < m < \frac{n+1}{2}$.

(ii)   Each edge of G(n,m) belongs to exactly two hexagons iff $3 < m < \frac{n-1}{2}$ and $m \neq [\frac{n+2}{3}]$.

Proof.

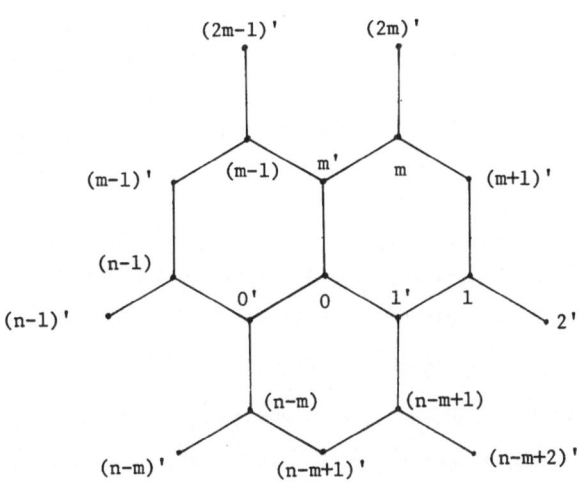

Figure 1

Consider any vertex of G(n,m), say 0. In Figure 1 we have the subgraph of G(n,m) containing all vertices at a distance ≤ 3 from the vertex 0. The vertices at a distance ≤ 2 from 0 are easily seen to be all distinct iff $3 \leq m \leq \frac{n-1}{2}$. Thus

for these values of m, the graph $G(n,m)$ does not contain quadrilaterals. Similarly,

the nine vertices at a distance 3 from the vertex 0 are all distinct if $3 < m < \frac{n-1}{2}$,

$m \neq [\frac{n+2}{3}]$ (square brackets denoting integral part). In this case, we note that

each edge incident with 0 belongs to exactly two otherwise disjoint hexagons. By

transitivity, any edge of $G(n,m)$ belongs to exactly two hexagons, which are other-

wise disjoint.

We now restrict ourselves to the latter class of graphs, i.e., $G(n,m)$ with

$3 < m < \frac{n-1}{2}$, $m \neq [\frac{n+2}{3}]$. In this class of graphs, any two incident edges belong

to exactly one hexagon, so we can define a Petrie path as follows:

Definition 2.2. Let e_1, e_2 be any two adjacent edges of $G(n,m)$. Then a Petrie

path $P(e_1,e_2)$ is defined as a walk containing e_1, e_2 as consecutive edges,

such that no three consecutive edges belong to the same hexagon.

It is readily seen from the above definition that given any two edges in a

Petrie path, it is uniquely determined. We remark that Dr. W. Harries showed

[private communication] that given any graph G, there exist a surface S and an

embedding of G in S such that there is only one Petrie path of the embedding, which

therefore covers each edge of the graph twice. We will show that our Petrie path(s)

do not do this.

Lemma 2.3. Each vertex in a Petrie path is of degree 2 in the subgraph defined by

the path.

Proof. Consider a Petrie path containing in the natural way the following sequence

of edges and vertices ... e_1, v_1, e_2, v_2, e_3, v_3, e_4,.... . Now assume that v_1 is of

degree three in the path, and without loss of generality that edges e_2, e_3 ... e_{d+1}

with $e_{d+1} \neq e_1$, form a simple circuit of length d leading back to v_1. Let this be

the shortest such circuit. Now there is an automorphism of the graph which will

take e_1, v_1, e_2 to e_3, v_3, e_4 respectively. Thus, since a Petrie path is completely

determined by any two consecutive edges, the path defined by e_3 and e_4 is the same

as that defined by e_1 and e_2 . But now we have that we can begin at e_1 and continue

with e_2, e_3 ... e_{d+1} to arrive back at v_1 without repeating edges. It is clear

that there can be no walk within the Petrie path beginning e_3, e_4 and arriving back at v_3 in d steps without repeating edges. Thus we have a contradiction and there can be no vertices of degree three in the Petrie path. Thus every vertex must be of degree two in the subgraph defined by the path.

Lemma 2.4. If p is prime, each Petrie path in $G(p,m)$ is a Hamiltonian circuit.

Proof. If it were not a Hamiltonian circuit, it would decompose the vertex set into a number of sets of the same cardinality, by transitivity. The cardinality of the sets would have to divide 2p, and would therefore be either 2 or p. Sets of cardinality 2 are not possible since the graph has no multiple edges; sets of cardinality p are not possible since the graph is bipartite, and p is odd.

Henceforth we restrict ourselves to those graphs $G(p,m)$ with p prime and $p \geq 7$.

3. PETRIE PATHS: DESCRIPTION

Since a Petrie path is determined by any given choice of two consecutive edges, there are exactly three Petrie paths passing through a given vertex. Thus there are three Petrie paths in all. One of these is the outer polygon, 0' 0 1' 1 2' 2 ... (p-1)' (p-1). The following argument now shows that a given graph $G(p,m)$ is isomorphic to at most two graphs $G(p, m_1)$, $G(p, m_2)$.

Any isomorphism must take Petrie paths to Petrie paths. Thus interchanging the outer polygon with one of the other Petrie paths may result in a different value for m. The only maps preserving the outer polygon are rotations and reflections about a vertex or the mid-point of an edge. These either leave the value of m unchanged or change it to $(p + 1 - m)$. Thus there are at most two values m_1, m_2 in the range $3 < m_1, m_2 < (\frac{p-1}{2})$, $m_1, m_2 \neq [\frac{p+2}{3}]$ such that $G(p, m_1) \cong G(p, m_2) \cong G(p,m)$.

We now find the three Petrie paths explicitly. We consider the paths as they pass through the vertex 0.

1. The path defined by the edges 0' 0, and 0 1'. This Petrie path is the outer polygon.

2. The path defined by the edges 1' 0, 0 m'. For this Petrie path we let $u_0' = 1'$, $u_0 = 0$, $u_1' = m'$. Then $u_1 = (m-1)$ and in general $u_n' = (n(m-1))'$, $u_n = (n(m-1) + 1)'$, both of these values being taken modulo p. We define an isomorphism of the graph $G(p,m)$ with the

graph $G(p,k)$ by $u_n' \to n$. Then k satisfies the equation

$$u_k' = 0', \quad \text{or}$$

(1) $\qquad k(m-1) \equiv -1 \bmod p$.

So k is the smallest positive integer staisfying this equation.

3. The path defined by the edges $0'$ 0, 0 m'. For this Petrie path we let $v_0' = 0'$, $v_0 = 0$, $v_1 = m'$. Then $v_1 = m$ and in general $v_n' = (nm)'$, $v_n = nm$, both modulo p. We can define an isomorphism of $G(p,m)$ with the graph $G(p,s)$ by $v_n' \to n'$, $v_n \to n$. Then s satisfies

$$v_s' = 1', \quad \text{or}$$

(2) $\qquad sm \equiv 1 \bmod p$.

Again, s is the smallest positive integer satisfying (2).

__Lemma 3.1.__ If $3 \le m \le \frac{p-1}{2}$, then k and s satisfy $3 \le k, s \le p-2$, $k, s \ne \frac{p+1}{2}$.

__Proof.__ It is clear that $k, s > 2$. It is easy to show that $k = p-1$ or $s = \frac{p+1}{2}$ implies $m=2$, and $k = \frac{p+1}{2}$ or $s=p-1$ implies $m = p-1$, both giving rise to a contradiction.

We recall that any graph $G(p,q)$ with $\frac{p+3}{2} \le q \le p-2$ is isomorphic to the graph $G(p, p+1-q)$, and $3 \le p+1-q \le \frac{p-1}{2}$. We therefore introduce the following notation.

__Notation.__ For any $x \in [3, \frac{p-1}{2}] \cup [\frac{p+3}{2}, p-1]$ we let $\underline{x}$ denote that member of x and $p+1-x$ lying in the interval $[3, \frac{p-1}{2}]$.

4. ISOMORPHISM CLASSES

__Theorem 4.1.__ If p is a given prime, then the isomorphism classes of the graphs $G(p,m)$, $3 \le m \le \frac{p-1}{2}$, can be characterized by the triples $\{m, \underline{k}, \underline{s}\}$ and at most one singleton $\{m = \underline{k} = \underline{s}\}$, these triples and possible singleton forming a partition of the set of integers $\{3, 4, \ldots \frac{p-1}{2}\}$. A singleton occurs iff $p \equiv 1 \pmod 6$.

__Proof.__ We proceed by proving a series of lemmas.

Lemma 4.2. With m, k <u>and</u> s <u>defined as above, we have that either</u>:

    (1) $\underline{m} = \underline{k} = \underline{s}$, <u>or</u>

    (2) $\underline{m}$, $\underline{k}$ and $\underline{s}$ <u>are all different</u>.

We consider every other possibility (i.e., exactly two of $\underline{m}$, $\underline{k}$, $\underline{s}$ equal) and show that they cannot occur.

    (a)   Let $\underline{k} = \underline{s}$, $\underline{k} \neq \underline{m}$.

        (i)   If $k = s$, we have $k(m-1) \equiv -1 \bmod p$

                 and $km \equiv 1 \bmod p$.

         Adding, we get $k(2m-1) \equiv 0 \bmod p$, which is a contradiction

         since both k and $(2m-1)$ are less than p.

      (ii)   If $k = p+1-s$, we have $k(m-1) \equiv -1 \bmod p$,

               and $(p+1-k)m \equiv 1 \bmod p$,

                so that $m-k \equiv 0 \bmod p$,

         which implies $m = k$, i.e. $\underline{m} = \underline{k} = \underline{s}$.

    (b)   Let $\underline{k} = \underline{m}$, $\underline{k} \neq \underline{s}$.

        (i) $k = m$ iff $m(m-1) \equiv -1 \bmod p$

                  iff $(p+1-m)m \equiv 1 \bmod p$

                  iff $s = p + 1 - m$ i.e. $\underline{k} = \underline{m} = \underline{s}$.

      (ii) $k = p+1-m$ so $(p+1-m)(m-1) \equiv -1 \bmod p$

                so $m(2-m) \equiv 0 \bmod p$

                so $m = 2$ or $p \mid m(2-m)$ which is a contradiction.

    (c)   Let $\underline{s} = \underline{m}$, $\underline{k} \neq \underline{s}$.

        (i) $s = m$

              so $m^2 \equiv 1 \bmod p$ by (2)

              so $m^2 - 1 \equiv 0 \bmod p$

              so $p \mid (m+1)(m-1)$, which cannot be.

      (ii)   The case $s = p+1-m$ has already been covered under (b) (i).

         Thus any graph $G(p,m)$ satisfying $\underline{m} = \underline{k} = \underline{s}$ has $m = k = p+1-s$.

Lemma 4.3.   <u>For any</u> p <u>there is at most one graph</u> $G(p,m)$ <u>with</u> $\underline{m} = \underline{k} = \underline{s}$, <u>with</u> $3 \leq m \leq p$.

Proof. When $m = k = p+1-s$, the congruences $k(m-1) \equiv -1 \bmod p$ and $sm \equiv 1 \bmod p$ reduce to

$$m^2 - m + 1 \equiv 0 \bmod p .$$

If there were m_1 also satisfying $m_1^2 - m_1 + 1 \equiv 0 \bmod p$, we would have, subtracting

$$(m-m_1)(m+m_1-1) \equiv 0 \bmod p,$$

hence $m = m_1$ or

$$m + m_1 \equiv 1 \bmod p, \text{ so } m_1 = p + 1 - m \geq \frac{p+3}{2} .$$

We note that we can also define the paths algebraically to satisfy the same equations. This leads to $G(p,3) \cong G(p, [\frac{p+2}{3}]) \cong G(p, \frac{p-1}{2})$. We note that 3, $[\frac{p+2}{3}]$, $\frac{p-1}{2}$, are always distinct unless $p = 7$, in which case they are all equal.

For any m, $3 \leq m \leq \frac{p-1}{2}$, we consider the triple $(\underline{m}, \underline{s}, \underline{k})$ with s and k defined as above. The same three entries $\underline{m}$, $\underline{s}$ and $\underline{k}$ are found regardless of which one we start with.

For a triple $(\underline{m}', \underline{s}', \underline{k}')$,

if $m' = \underline{s}$, then $s' = \underline{m}$ and $k' = p + 1 - \underline{k}$;

if $m' = \underline{k}$, then $s' = p + 1 - \underline{m}$ and $k' = p + 1 - \underline{s}$.

Thus we have shown that the integers in the interval $[3, \frac{p-1}{2}]$ are partitioned into triples $(\underline{m}, \underline{s}, \underline{k})$ with at most one singleton $(\underline{m} = \underline{s} = \underline{k})$, and this partitioning characterizes the isomorphism classes of the graphs $G(p,m)$ for given p and $3 \leq m \leq \frac{p-1}{2}$.

Lemma 4.4. There will be a graph $G(p,m)$ satisfying $m = \underline{s} = \underline{k}$ iff $p \equiv 1 \bmod 6$.

Proof. p must be either 1 mod 6 or 5 mod 6.

(i) $p = 1 \bmod 6$. Let $p = 6n + 1$, then the number of integers in the interval $[3, \frac{p-1}{2}]$ is $3n - 2 = 3(n-1) + 1$. There must accordingly be a singleton graph $G(p,m)$ with $\underline{m} = \underline{s} = \underline{k}$, and $(n-1)$ is isomorphic triples $(\underline{m}, \underline{k}, \underline{s})$.

(ii) $p = 5 \bmod 6$. Let $p = 6n + 5$, then the number of integers in the interval $[3, \frac{p-1}{2}]$ is $3n$. There must be n isomorphic triples $(\underline{m}, \underline{k}, \underline{s})$ and there can be no graph $G(p,m)$ with $m = \underline{k} = \underline{s}$.

5. AUTOMORPHISM GROUPS

We now consider the groups of these graphs. We divide the graphs into three types for given p, and $3 \leq m \leq \frac{p-1}{2}$,

(i) Those graphs characterized by a triple $\{\underline{m}, \underline{k}, \underline{s}\}$ not containing the integer 3, with $\underline{m}, \underline{k}, \underline{s}$ distinct.

(ii) Those graphs characterized by a triple $\{\underline{m}, \underline{k}, \underline{s}\}$ containing the integer 3, with $\underline{m}, \underline{k}, \underline{s}$ distinct.

(iii) Those graphs characterized by a singleton $\{\underline{m} = \underline{k} = \underline{s}\}$.

For graphs of type (i), the Petrie paths are defined by topological properties. Therefore they have to be preserved. If a map interchanges two Petrie paths, the value of m changes, and it follows that the map cannot be an automorphism.

It follows that automorphisms of the graph must leave each Petrie path fixed globally. The only maps which leave the outer polygon fixed globally are rotations through any multiple of $\frac{\pi}{p}$ and reflections about a vertex or the mid-point of an edge. The only maps which are also automorphisms are rotations about any multiple of $\frac{2\pi}{p}$ and reflections about the mid-point of an edge. The others are isomorphisms with G(p, p+1-m). Thus the group of the graph is abstractly isomorphic to the dihedral group of order 2p.

The graphs of type (ii) have the same automorphisms as specified above, but we cannot use the Petrie path argument to show that these are the only automorphisms of the graph. Instead we consider the stabilizer of a vertex of G(p,3), 0 say. We see that $d(1', 2') = d(2', 3') = d(2', 0') = 2$.

It can be seen from the diagram below that 2' is the only vertex at a distance 2 from all the neighbors of 0. Also, from the diagram, assuming p > 9, we see that there are three more vertices at a distance two from both 0' and 1', namely 3', (p-1)' and (p-2)' whereas there are only two more (namely 4', 0') at a distance two from both 1' and 3', and only one (1') at a distance two from both 0' and 3'. It follows that the stabilizer of a vertex is the identity, and the maximum order of the group of the graph is the number of vertices, 2p. A similar argument holds for the case p = 9. The group of this type of graph is therefore also isomorphic to the dihedral group of order 2p, p ≥ 9.

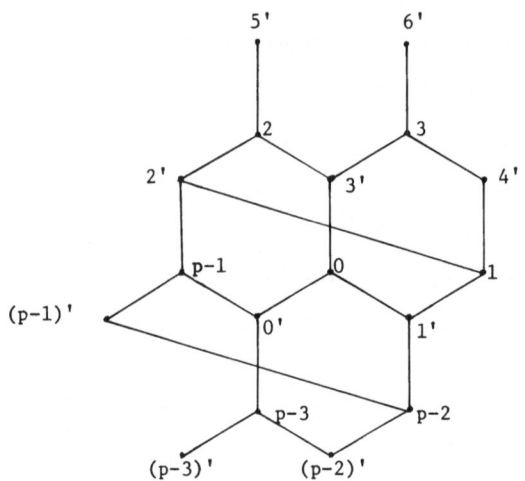

Figure 2

(All the vertices in this
diagram are distinct if
p > 9.)

For graphs G(p,m) of type (iii) (i.e. where m = k̲ = s̲), p > 7, we have as
before that the group of automorphisms preserving the edge set of any Petrie path
is isomorphic to the dihedral group D_p. Now consider any 2-path, 0' 0 1' say. It
uniquely determines a Petrie path. No automorphism can reverse this 2-path, for
this would reverse the Petrie path, and the graph would be mapped to G(p, p+1-m).
Thus, for any two 2-paths there can be at most one automorphism transforming the
edge set of one to the edge set of the other. Since we know that the three Petrie
paths are mutually interchangeable, they must be so with respect to the cyclic
group C_3. The group of this type of graph is described in [3, p. 108].

REFERENCES

1. Bouwer, I. Z., and Frucht, R., "Line Minimal Graphs with Cyclic Groups," <u>A Survey of Combinatorial Theory</u>, (Ed. J. N. Srivastava), North-Holland, New York, 1973, 53-67.

2. Coxeter, H. S. M., "Self-dual Configurations and Regular Graphs," <u>Bull. Amer. Math. Soc</u>. 56 (1950), 413-455.

3. Coxeter, H. S. M. and Moser, W. O. J., "Generators and Relations for Discrete Groups," 3rd Edition, Springer-Verlag, New York, 1972, Section 8.4, 107-109.

4. Foster, R. M., "A Census of Trivalent Symmetrical Graphs I," presented at the conference on Graph Theory and Combin. Analysis, Waterloo, Ontario, 1966.

5. Frucht, R., "A One-regular Graph of Degree Three," <u>Canad. J. Math</u>. 4 (1952), 240-247.

6. Frucht, R., "How to Describe a Graph," <u>Ann. N. Y. Ac. Sci</u>. 175, Part I (1970), 159-167.

7. Harary, F., "Graph Theory," Addison-Wesley, Reading, Mass., 1969.

8. Miller, R. C., "The Trivalent Symmetric Graphs of Girth at Most 6", <u>J. Combinatorial Theory</u>, Series B, 10 (1971), 163-182.

9. Tutte, W. T., "A Family of Cubical Graphs," <u>Proc. Cambridge Philos. Soc</u>. 43 (1947), 459-474.

10. Watkins, M. E., "On the Action of Non-abelian Groups on Graphs," <u>J. Combinatorial Theory</u> 11 (1971), 95-104.

RECONSTRUCTING COMBINATORIAL GEOMETRIES

Thomas H. Brylawski
University of North Carolina

ABSTRACT

Ulam-type reconstruction problems are given for combinatorial geometries. In particular the subgeometry generating function is evaluated from the subgeometry generating functions of single element deletions. This result is applied to graphs to give the chromatic polynomial (for vertices or regions) and a number of other invariants which can be computed from the deck of edge deletions.

RECONSTRUCTING COMBINATORIAL GEOMETRIES

1. INTRODUCTION

In [5] it was hoped that readers would appreciate the link between graphs and matroids (combinatorial geometries) since theorems (and we may add problems) for the former can be generalized to the latter and since general matroid considerations often clarify aspects of graph theory. In this spirit we mention some of the ways Ulam's reconstruction problem [3,4,8] for graphs may be rephrased for geometries. In particular, the vertex problem (§3) has a number of generalized conjectures for geometries (although the most obvious one is false) leading to one which is equivalent to the Ulam conjecture when restricted to any geometry represented by a four-connected graph.

However, it is the edge problem (§4) that most easily generalizes to geometries. The phrasing of this problem in a geometrical setting is not only natural but allows a geometric computation which, when applied to graphical geometries, gives an easy way to compute the chromatic polynomial (and hence the chromatic number and Möbius function), complexity, number of acyclic orientations, number of forests, number of spanning subsets of edges and number of blocks of a graph (or its planar dual when relevant) given its "deck" of edge deletions. We first define some relevant notions and properties of geometries (§2) adding in brackets the parallel notion for graphs when the geometry is graphical. The reader desiring a fuller treatment of these notions can find it in [1], [2], or [5]. We would like to thank Dr. J. Longyear for first suggesting this problem in a remark in "The Graph Theory Newsletter," and Professor R. Bari for her helpful discussions relating the Ulam conjecture to the chromatic theory of graphs -- an idea originating with Professor F. Harary.

In fact it was Harary who first asked if the edge reconstruction problem could be solved if one knew the chromatic polynomial of the reconstructed graph.

We hope our computation of the chromatic polynomial will lead to new reconstruction techniques which use this information.

2. BASIC CONCEPTS

A finite _pregeometry_ or _matroid_, is a finite set of _points_ with a closure operator $J_G(\cdot)$ satisfying the _exchange property_: For any points $p, q \in G$ and any subset $P \subseteq G$, if $p \in \overline{P \cup \{q\}}$ but $p \notin \overline{P}$, then $q \in \overline{P \cup \{p\}}$. It is easy to see that the edges of a finite graph (more generally multigraph) satisfy this property where closure means circuit completion ($p \in \overline{A}$ if $p \in A$ or p is in a circuit in $A \cup p$). For such pregeometries each geometric concept has a graphical interpretation which we will give in brackets. Since edges are under consideration we will henceforth assume that no subgraphs have any isolated vertices.

A _geometry_ [Michigan graph] is a pregeometry in which the empty set and each point [edge] are closed [no loops or multiple edges]. The lattice, L, of closed sets or _flats_ [subgraphs whose components are induced subgraphs] is called a _geometric lattice_ [lattice of contractions]. Flats covered by 1 in L are called _hyperplanes_ [maximal subgraphs which have one less vertex or one more connected component than G]. A _bond_ [cutset] B of G is the set complement of a hyperplane. In L each lattice element x has a well-defined _rank_, $r(x)$, equal to the length of any maximal chain from the 0 element [the set of loops] to x. $r(A)$, the _rank_ of a subset of points [subgraph] $A \subseteq G$ is defined as $r(\overline{A})$. [This is given for a subgraph A by the number of vertices of A, $|V(A)|$, minus its number of connected components, $k(A)$].

For any subset A, the _cardinality_ of A, $|A|$, denotes the number of points [edges] it contains. The _corank_ of A, $c(A)$, is the non-negative integer $r(G) - r(A)$, and the _nullity_ of A, $n(A)$, is the non-negative integer $|A| - r(A)$. A subset of points $A \subseteq G$ represents a _spanning set_ for G if $\overline{A} = G$ [$V(A) = V(G)$ and $k(A) = k(G)$]. A subset of points is independent [a forest] if $r(A) = |A|$ [it contains no circuit]. Otherwise, $r(A) < |A|$ and A is _dependent_. An independent spanning set is called a _basis_ [spanning forest, or, in the case of a connected graph, spanning tree]. A _circuit_ [circuit] is a minimal dependent set. Two pregeometries are isomorphic [2-isomorphic] denoted $G \cong H$ if there is a $1 - 1$ correspondence, f, between the points of G and H and the circuits of G and H such that for any point p and circuit C,

$p \in C$ iff $f(p) \in f(C)$. [This is weaker than graphical isomorphism: if G and H are two isomorphic graphical pregeometries then the corresponding graph G can be made graphically isomorphic to H by a sequence of operations consisting of separating and/or reconnecting G at a cutpoint or by separating G into subgraphs G_1 and G_2 at any two vertex cut set and reconnecting at the same vertices in the opposite order. Hence for three-connected graphs, graphical and geometric isomorphism are equivalent.]

The (Whitney) dual [planar dual] of G, G*, is the unique pregeometry on the same point set with a set of bases consisting of base complements of G. [It is a celebrated theorem of Tutte [7] that a graphical pregeometry has a graphical dual if and only if it is planar in which case the dual is the usual graph-theoretic one.]

We say G is the direct sum of two pregeometries $G_1 \oplus G_2$ if the points of G and circuits of G, C(G), are the disjoint unions $G_1 \uplus G_2$ and $C(G_1) \uplus C(G_2)$ respectively; G is then said to be a direct sum factor [block] of G, and G is termed separable with flats G_1 and G_2 as separators. If no such nontrivial direct sum decomposition [block decomposition] exists, any two distinct points of G are contained in a circuit and G is termed connected [two-connected]. A one-point direct sum factor, p, is either contained in no circuits and termed an isthmus [isthmus or bridge], I, or is itself a circuit and termed a loop [loop], L. A point is a non-factor if it is neither a loop nor an isthmus. A Boolean algebra [forest] B_n is a geometry which is the direct sum of n isthmuses.

For $p \in G$ we define two derived pregeometries on the point set $\{G\} - \{p\}$: the deletion G - p [subgraph G - p] and the contraction, G/p [contraction and removal of the edge p]. If $A \subseteq \{G\} - \{p\}$, and $\bar{A}$ denotes its closure in G, then the closure of A in G - p is defined as $\bar{A} - \{p\}$; while its closure in G/p is defined as $\overline{A \cup \{p\}} - \{p\}$. [Graphically this amounts to identifying to a new vertex $\bar{v}$ the vertices v and v' incident with p while connecting to $\bar{v}$ any edge formerly incident with v or v', and deleting the loop from the resulting multigraph.] An arbitrary sequence of contractions and deletions leads to a minor [homomorphic subgraph] of G. We then have $G* - p = (G/p)*$ and $G*/p = (G - p)*$.

A geometry is binary if it can be represented as a set of vectors in a vector space over the field with two elements with closure corresponding to linear closure.

A theorem of Whitney [7] states that a pregeometry is binary if and only if it has no minor consisting of a rank-two geometry with four rank-one flats (a four-atom line). Graphical geometries are binary.

An _invariant_ is a function f defined on the class of all pregeometries such that $f(G) = f(H)$ if $G \simeq H$.

Examples of invariants used in this paper include $C(G)$, the _complexity_ or number of bases of G; the _subgeometry generating function_ $S(G) = \sum_{i,j} a_{ij} u^i v^j$ where a_{ij} is the number of subgeometries or subsets [subgraphs] of G with corank i and nullity j; and $\mu(G)$, the _Möbius function_ which is defined as $\mu(0,1)$ evaluated on the geometric lattice L, where for $x \leq y$, $\mu(x,y)$ is given by the recursion: $\mu(x,x) = 1$, $\mu(x,y) = - \sum_{x \leq z < y} \mu(x,z)$. Two other invariants evaluated on L are the _characteristic polynomial_ $\chi(G,\lambda) = \chi(G) = \sum_{x \in L} \mu(0,x) \lambda^{r(1)-r(x)}$ if G is loopless and 0 otherwise, and the _Crapo invariant_ $\beta(G)$ [1]. [The chromatic polynomial $\bar{\chi}(G)$ of a graph is given by the formula $\bar{\chi}(G) = \lambda^{k(G)} \chi(G)$ (see, e.g. [1]).]

3. THE HYPERPLANE [VERTEX] RECONSTRUCTION PROBLEM

Definition 3.1. A pregeometry G is _hyperplane reconstructible_ if there is no non-isomorphic pregeometry with the same set of hyperplanes.

Conjecture 3.2. A binary (in particular graphic) pregeometry is hyperplane reconstructible.

This is related to the Ulam conjecture since for a graph G and vertex v, the subgraph $G - v$ of G with all edges incident with v removed is a hyperplane (the complement of a vertex cutset). The two differences with the Ulam conjecture when (3.2) is applied to graphical pregeometries is that in (3.2) all hyperplanes are given; and that we only know those hyperplanes (and can only reconstruct G) up to 2-isomorphism. However, we may modify (3.2) so that it will agree more closely with the Ulam conjecture.

Conjecture 3.3. A binary (or graphic) pregeometry of known cardinality and rank
is reconstructible from its connected hyperplanes.

We note that for a graph any hyperplane which does not come from the deletion
of a vertex bond is separable and while the converse is false in general it is true
for three-connected graphs (in which case all vertex hyperplanes are two-connected
and hence geometrically connected). If in fact the graph is four-connected then
the connected hyperplanes are three-connected and their geometric structure strictly
determines their graphical structure. Hence in this case we have a precise reformu-
lation of the Ulam conjecture:

Conjecture 3.4 (Ulam). A geometry which is represented by a four-connected graph
is reconstructible from its connected hyperplanes.

The following counterexample shows that (3.2) is false in general for geometries.

Example 3.5. The geometries given by the affine diagrams and

have isomorphic hyperplane structures (two three points lines and

nine two points lines) but are not isomorphic.

Other related problems can be posed. For example, what geometric invariants
can be recovered from the set of hyperplanes. Even the number of points seems
to be a nontrivial proposition (although the number of edges of a graph is easily
recoverable from its set of vertex deletions). Another useful invariant it would
be nice to recover is the characteristic polynomial $\chi(G) = \lambda^n + a_{n-1}\lambda^{n-1} + \ldots + a_1\lambda + a_0$.
However, even the Möbius function, a_0, seems difficult; the only coefficient
easily recoverable being $a_1 = \sum \mu(H)$ the sum being taken over all hyperplanes. A
class of geometries for which the characteristic polynomial (in fact the subgeometry
generating function) is recoverable is the class of Hartmanis partitions (cf. 7.9
of [1]).

A much more optimistic situation exists for edge reconstructions.

4. THE DELETION [EDGE] RECONSTRUCTION PROBLEM

Definition 4.1. A pregeometry G is deletion reconstructible if no nonisomorphic
pregeometry has the same multiset or deck $\mathcal{D}$ of deletions $\{G - p_i | p_i \in G\}$.

Conjecture 4.2. A pregeometry which is not a circuit or Boolean algebra is deletion reconstructible.

We excepted circuits and Boolean algebras since B_n and an n-point circuit have isomorphic decks (n Boolean algebras of cardinality n - 1). However, these exceptions should give graph theorists no pause as trees are graphically reconstructible from edge deletions [3] as are circuits for n > 3.

Proposition 4.3. A pregeometry G is an (n-point) circuit C_n or an (n-element) Boolean algebra B_n iff all (n) deletions are Boolean algebras (of rank n - 1).

Proof. If all deletions of G are Boolean algebras, the nullity of G, n(G), is at most 1 since 0 = n(G-p) > n(G) - 1. If the nullity of G is one, then $G = C_k \oplus B_{n-k}$. But if 0 < k < n, C_k would appear as a subgeometry of at least one of the {G - p}.

We will henceforth assume in what follows that G is never a Boolean algebra.

Proposition 4.4. G ($\neq B_n$) has an isthmus iff r(G-p) < r(G-p') for some p ≠ p' in which case $G \cong I \oplus (G-p)$ where I represents an isthmus.

Proof: A point p is an isthmus of G iff r(G) = r(G-p) + 1. Since not all points are isthmuses, for some (any) nonisthmus p', r(G) = r(G-p'). But an isthmus is a direct sum factor so if p is an isthmus of G, $G \cong I \oplus (G-p)$.

In light of (4.3) and (4.4) we will now consider only those pregeometries (including circuits) without isthmuses so that r(G) = r(G-p) and hence n(G) = n(G-p) + 1 for all p∈G . Not only are geometries with isthmuses reconstructible but suitable modifications (involving changing the corank for subsets of isthmus deletions) will make the following theorem (4.6) applicable to such geometries.

Theorem 4.5. The subgeometry generating function S(G) is the unique polynomial function defined on all pregeometries with the following properties:

$$S(G) = S(G-p) + S(G/p) \text{ for all nonfactors p}$$

$$S(G \oplus H) = S(G)S(H)$$

$$S(I) = u + 1 \text{ for an isthmus I}$$

$$S(L) = v + 1 \text{ for a loop L.}$$

In addition, $S(G)$ gives the following information about G:

1. $S(1,1)$ counts the number of subsets [subgraphs] $= 2^{|G|}$.

2. $S(0,0)$ counts the number of bases [spanning forests], $C(G)$.

3. $S(1,0)$ counts the number of independent sets [forests].

4. $S(0,1)$ counts the number of spanning sets.

5. $S(0,-1) = |\mu(0,1)|$, the absolute value of the Möbius function of G.

6. $\partial S/\partial u\big|_{u=-1, v=-1} = \beta(G)$, the Crapo invariant which is 0 iff G is separable (or a loop) and is 1 iff G is a (connected) series-parallel network.

7. $(-1)^{r(G)}S(-\lambda,-1) = \chi(G,\lambda)$, the characteristic polynomial of G.

8. $(-1)^{n(G)}S(-1,-\lambda) = \chi(G^*,\lambda)$.

9. In particular if G is graphic: $(-1)^{r(G)}\lambda^{k(G)}S(-\lambda,-1)$ counts the number of ways to properly (vertex) color G with λ colors (where $k(G)$ is the number of connected components of G).

10. $(-1)^{n(G)}\lambda S(-1,-\lambda)$ counts the number of ways to properly color the regions of G (a planar map) with λ colors.

11. $S(1,-1)$ counts the number of ways to acyclically orient the edges of G.

<u>Proof</u>: These along with other properties of S can be inferred from [1] where they are all proved in terms of the Tutte polynomial $t(G) = S(z-1,x-1)$. Statement 11 of Theorem 4.5 can be found in [6].

<u>Theorem 4.6</u>: If G has no isthmuses the subgeometry generating function of G can be reconstructed from the subgeometry generating functions of all its deletions by the formula:

$$S(G) = v^N + \sum_{\mathcal{D}} \sum_{i,j} \frac{a^p_{i,j}}{N+i-j} u^i v^j$$

$$= v^N + \int_0^1 t^{N-1} \sum_{\mathcal{D}} S(G-p; ut, vt^{-1})dt$$

where $\mathcal{D}$ is the deck of deletions of G; $a^p_{i,j}$ is the number of subsets of $G - p$ with corank i and nullity j; and N is the nullity of G $(N = n(G-p) + 1)$.

<u>Proof</u>: The integral representation for $S(G)$ is easily seen to be equivalent to the summation representation and is due to Professor L. Geissinger.

If A is a subgeometry of G, it appears in exactly those G - p with p ∉ A.
Further the geometric structure of both A and G - p is given by internal calcu-
lations in the supremum subsemilattice of L(G) generated by A and G - p respectively. Hence if p ∉ A, the rank and cardinality of any subset A is the same
when viewed in G or G - p. But since r(G) = r(G-p) for all p, the corank and
nullity of A is also the same when computed in either G or G - p.

If the corank of A is i and its nullity is j, A contributes $u^i v^j$ to S(G)
and contributes $\frac{u^i v^j}{N+i-j}$ exactly $|G| - |A| = r(G) + n(G) - r(A) - n(A) = N + i - j$
times to the right-hand expression. Note that N + i - j is always positive unless
A = G in which case i = 0 and j = N.

Hence we see that all the invariants in (4.5) can be computed from the deck.

<u>Example 4.7.</u> Let G be the geometry represented in real affine space by the diagram

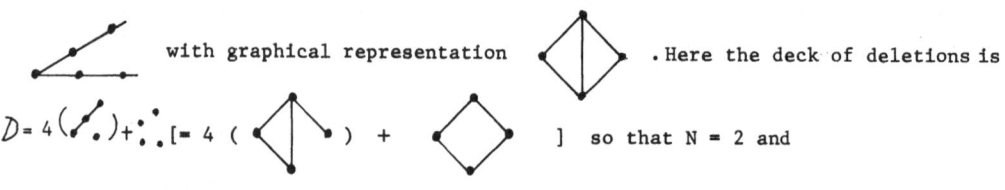

with graphical representation   . Here the deck of deletions is

$\mathcal{D} = 4$ (   ) + ⋮ [= 4 (   ) +   ] so that N = 2 and

$$\sum_{\mathcal{D}} S(G-p) = 4(u^3 + 4u^2 + 6u + uv + 3 + v) + u^3 + 4u^2 + 6u + 4 + v$$
$$= 5u^3 + 20u^2 + 30u + 4uv + 16 + 5v.$$

Therefore, $S(G) = u^3 + 5u^2 + 10u + 2uv + 8 + 5v + v^2$.

Hence, for example, C(G) = 8, there are 24 independent sets, and
$\chi(G) = \lambda^3 - 5\lambda^2 + 8\lambda - 4$ so that $\bar{\chi}(G) = \lambda^4 - 5\lambda^3 + 8\lambda^2 - 4\lambda$. The graph has
18 acyclic orientations, and as a planar map, the regions have chromatic poly-
nomial $\lambda^3 - 3\lambda^2 + 2\lambda$.

<u>Conjecture 4.8.</u> If G and H are binary (or graphic) and S(G) = S(H), then G ≅ H.

Note that (4.8) implies (4.2) since S(G) is recoverable from $\mathcal{D}$.

Finally, if we rephrase each conjecture and theorem to G* and dualize back,
we note that everything holds for pregeometries which are not all loops if we
replace deletion by (point) contraction where in (4.6) we must interchange the
roles of corank and nullity as well as A and G - A so that

$$S(G) = u^r + \sum_{\mathcal{D}} \sum_{i,j} \frac{a_{i,j}^{(p)}}{r-i+j} u^i v^j$$

where r is the rank of G and $a_{i,j}^{(p)}$ is the number of subgeometries of G/p with corank i and nullity j.

REFERENCES

1. Brylawski, T. H., A Decomposition for Combinatorial Geometries, Trans. Amer. Math. Soc. 171 (1972), 235-282.

2. Crapo, H. H., and Rota, G. -C., Combinatorial Geometries, M.I.T. Press, Cambridge, Mass., 1970.

3. Greenwell, D. L., and Hemminger, R. L., Reconstructing Graphs, The Many Facets of Graph Theory, Springer-Verlag, Berlin (1969), 91-114.

4. Harary, F., A Seminar on Graph Theory, Holt, Rinehart, and Winston, New York, New York, 1967.

5. Harary, F., and Welsh, D. J. A., Matroids Versus Graphs, The Many Facets of Graph Theory, Springer-Verlag, Berlin (1969), 155-170.

6. Stanley, R., Acyclic Orientation of Graphs, (to appear).

7. Tutte, W. T., Lectures on Matroids, J. Res. Nat. Bur. Stand. 69B(1965), 1-48.

8. Ulam, S. M., A Collection of Mathematical Problems, Wiley, New York, (1960), 29.

9. Whitney, H., 2-isomorphic Graphs, Amer. J. Math. 55 (1933), 245-254.

ON TRIANGULAR AND CYCLIC RAMSEY NUMBERS WITH k COLORS

Fan Chung
University of Pennsylvania

ABSTRACT

Define $r(G;k)$ to be the smallest integer with the following property: For any $n \geq r(G;k)$, color the edges of K_n in k colors, then there exists a mono-chromatic graph isomorphic to G. In this paper, we discussed the bounds for $r(K_3;k)$ and $r(C_4;k)$.

ON TRIANGULAR AND CYCLIC RAMSEY NUMBERS WITH k COLORS

Let G be a finite graph and k be a positive integer. Define $r(G;k)$ to be the smallest integer with the following property: For any $n \geq r(G;k)$, color the edges of K_n in k colors; then there exists a monochromatic graph isomorphic to G. The existence of $r(G;k)$ is assured by Ramsey's theorem [1,2].

In the case of $G = K_3$ and $k = 2$, $r(K_3;2) = 6$. This is one of the most interesting fundamental problems that appeared in Putnam Mathematics Competition [3] in 1953. The problem can be stated as follows: Color the edges of K_6 in red or blue; then either a red triangle or a blue triangle exists.

In 1955 Greenwood and Gleason [4] proved that $r(K_3;3) = 17$ and $r(K_3;4) > 41$. The value of $r(K_3;4)$ is still unknown. Whitehead and Taylor [5] proved that $r(K_3;4) > 49$ in 1971. In 1972, G. J. Porter (unpublished) and the author [6] proved independently that $r(K_3;4) > 50$ and a lower bound for $r(K_3;k)$ was obtained. A simpler proof will now be presented.

<u>Theorem 1</u>. Let $f(k) = r(K_3;k) - 1$ and let $t = 0.103 \ldots$ be the only positive root of $x^3 + 6x^2 + 9x - 1 = 0$ and $C = 50t^2 = 0.5454 \ldots.$ Then $f(k+1) \geq 3f(k) + f(k-2)$ for $k > 3$ and $f(k) \geq (3+t)^k C$.

We need the following lemma.

<u>Lemma 1</u>: The edges of K_n can be colored in k colors without any monochromatic triangle if and only if its adjacency matrix A_n is the sum of k symmetric binary matrices $M_1, M_2, \ldots, M_k$ where the componentwise product of M_i and M_i^2 is zero for $i = 1,2,\ldots,k$, i.e., $A_n = M_1 + M_2 + \ldots + M_k$ and $(M_1 * M_i^2)_{jm} = (M_i)_{jm}(M_i^2)_{jm} = 0$ for $i = 1,2,\ldots,k$.

<u>Proof</u>: If the edges of K_n are colored in k colors without monochromatic triangles, then define

$$(M_i)_{jm} = \begin{cases} 1 & \text{if the edge (jm) has color i} \\ 0 & \text{otherwise} \end{cases}$$

Obviously $(M_i^2)_{jm}$ is the number of paths of length 2 joining points j and m. But $(M_i)_{jm}$ should be zero when (M_i^2) is non-zero. Hence $A_n = M_1 + M_2 + \ldots + M_k$ and $M_i * M_i^2 = 0$ for all i.

Conversely, given k symmetric binary matrices $M_1, M_2, \ldots, M_k$ with their sum A_n and $M_i * M_i^2 = 0$ for $i = 1, 2, \ldots, k$, we have a k-coloring of K_n without any monochromatic triangles.

<u>Proof of theorem</u>. The edges of $K_{f(k)}$ can be colored in k colors withoug monochromatic triangles. By Lemma 1 there exist $M_1, M_2, \ldots, M_k$ and $N_1, N_2, \ldots, N_{k-2}$ such that

$$A_{f(k)} = M_1 + M_2 + \ldots + M_k \quad \text{and} \quad M_i * M_i^2 = 0 \quad \text{for} \quad i = 1, 2, \ldots, k$$

$$A_{f(k-2)} = N_1 + N_2 + \ldots + N_{k-2} \quad \text{and} \quad N_j * N_j^2 = 0 \quad \text{for} \quad j = 1, 2, \ldots, k-2$$

and let J be the $f(k-2) \times f(k)$ matrix with all entries 1.

Let $L_1, L_2, \ldots, L_{k+1}$ be square matrices of order $(3f(k) + f(k-2))$; then symmetric matrices are defined as follows:

$$L_1 = \begin{bmatrix} 0 & & & \\ M_2 & M_2 & & \\ M_1 & I & M_1 & \\ J & 0 & 0 & 0 \end{bmatrix}$$

$$L_2 = \begin{bmatrix} M_1 & & & \\ M_1 & 0 & & \\ I & M_2 & M_2 & \\ 0 & J & 0 & 0 \end{bmatrix}$$

$$L_3 = \begin{bmatrix} M_2 & & & \\ I & M_1 & & \\ M_2 & M_1 & 0 & \\ 0 & 0 & J & 0 \end{bmatrix}$$

$$L_i = \begin{bmatrix} M_i & & & \\ M_i & M_i & & \\ M_i & M_i & M_i & \\ 0 & 0 & 0 & N_{i-3} \end{bmatrix} \quad \text{for } i = 4, 5, \ldots, k+1.$$

It is clear that $L_1 + L_2 + \ldots + L_{k+1} = A_{3f(k) + f(k-2)}$ and $L_i * L_i^2 = 0$ for $i = 1, 2, \ldots, k+1$.

Since the complete graph $K_{3f(k)} + f(k-2)$ can be colored in k+1 colors without any monochromatic triangle,

$$f(k+1) \geq 3f(k) + f(k-2) \quad \text{for} \quad k \geq 3 .$$

From the above inequality we can get $f(k) \geq (3+t)^k C$ where $t = 0.103...$ is the only positive root of $x^3 + 6x^2 + 9x - 1 = 0$ and $C = 50t^2 = 0.5454...$

The classical upper bound [4] for $r(K_3;k)$ is $[k!e] + 1$. Whitehead [5] proved $r(K_3;4) \leq 65$ $[4!e] + 1$. Combining these, we get the next inequality.

Theorem 2. $r(K_3;k) \leq [k!(e-1/24)] + 1.$

From Theorems 1 and 2, we know that the limit of k'th root of $f(k)$ will be between $3+t$ and ∞ if it exists.

Lemma 2. $f(jk) \geq (f(k))^j .$

Proof. Let $f(k) = n$ so that the edges of K_n can be k-colored without any monochromatic triangles.

Define K_{nj} with vertices the vectors $(i_1, i_2, ..., i_j)$, $i_s = 1, 2, ..., n$. Let c_s, $s = 1, ..., j, m=1, ..., k$, be the jk colors available.

The edge joining $(i_1, i_2, ..., i_j)$ and $(i'_1, i'_2, ..., i'_j)$ is colored in the color c_{jm} if and only if $i_1 = i'_1, ..., i_{j-1} = i'_{j-1}, i_j \neq i'_j$ and the edge joining i_j and i'_j has color m.

It is clear that this gives a coloring of edges of K_{nj} without any monochromatic triangle in kj colors.

Therefore
$$f(jk) \geq (f(k))^j .$$

Theorem 3. $\lim_{k \to \infty} (f(k))^{1/k}$ exists.

Proof: Let $x = \lim \sup (f(k))^{1/k}$

There exists an integer m such that $f(m)^{1/m} > x-\varepsilon$

For any $n \geq m/\varepsilon$, $f(n)^{1/n} \geq f(m[n/m])^{1/n}$
$$\geq f(m)^{[n/m]/n}$$
$$\geq (x-\varepsilon)^{(1-\varepsilon)}$$

Hence $\lim \inf f(k)^{1/k} = \lim \sup f(k)^{1/k} \geq 3.103...$

<u>Theorem 4.</u> Let $r(K_m;k)$ be the classical Ramsey number $N(\underbrace{m,m,\ldots,m}_{k\ \text{times}};2)$. Then $\lim r(K_m;k)^{1/k}$ exists for any m and is greater than m-1 .

<u>Proof:</u> By a similar method we can prove $r(K_m;kj)^{1/kj} \geq r(K_m;k)^{1/k}$ and the limit exists.

Let $\xi_m = \lim r(K_m;k)^{1/k}$. Then $\xi_3 = 3.103\ldots$ It is not known that ξ_3 is finite or infinite. It was shown in [7] that $\xi_4 \geq \sqrt{17}$, $\xi_5 \geq \sqrt{37}$, $\xi_6 \geq \sqrt{101}$, $\xi_7 \geq \sqrt{109}$, $\xi_8 \geq \sqrt{281}$, $\xi_9 \geq \sqrt{373}$ and ξ_m is strictly increasing.

Some upper and lower bounds for $r(C_4,k)$ have been obtained.

<u>Lemma 3.</u> The edges of K_n can be colored in k colors without any monochromatic triangle if and only if the matrix A_n is the sum of k symmetric binary matrices $M_1,M_2,\ldots,M_k$ where $(M_i^2)_{jm} \leq 1$ for $j \neq m$ i = 1,2,\ldots,k.

The proof of Lemma 3 is clear.

<u>Lemma 4.</u> Let M be an n x n symmetric binary matrix and $(M^2)_{jm} \leq 1$ for $j \neq m$. Then

$$S = \sum_{i,j=1}^{n} M_{ij} \leq n\sqrt{n-3/4} + n/2 .$$

<u>Proof:</u> $\sum_{j=1}^{n} M_{ij}M_{jk} \leq 1 \ (i \neq k)$

Sum over k = 1,\ldots,n , $k \neq j$, to get

$$\sum_{j=1}^{n} M_{ij} \sum_{\substack{k=1 \\ k \neq 0}}^{n} M_{jk} \leq n-1 ,$$

or

$$\sum_{j=1}^{n} M_{ij}(r(i)-M_{ji}) \leq n-1,$$

where r(i) is the i'th column sum or row sum.

Then sum over j, to get $\sum_{j=1}^{n} r(i)^2 - \sum_{i,j=1}^{n} M_{ij} \leq n(n-1)$,

$$\sum_{i=1}^{n} r(i)^2 \leq n(n-1) + S .$$

For any positive numbers r(1), r(2), \ldots, r(n),

$$\sum_{i=1}^{n} r(i)^2 \geq (\sum_{i=1}^{n} r(i))^{2/n} .$$

So $\qquad s^2/n \le n(n-1) + S$

and $\qquad S \le n/2 + n\sqrt{n-3/4}$.

The equality holds when all the $r(i)$ are equal.

Theorem 5. $k^2 + k + 1 \ge r(C_4;k)$.

Proof. Let $r(C_4;k) - 1 = n$. By Lemma 3 we know that $\quad A_n = \sum\limits_{i=1}^{n} M_i \quad$ and

$(M_i)_{jm} \le 1$ for $j \ne m$, $i = 1,2,\ldots,$ k. There is some M_i with the property that

$$\sum\limits_{j,m=1}^{n} (M_i)_{jm} \ge n(n-1)/k .$$

By Lemma 4, we have $n/2 + n\sqrt{n-3/4} \ge n(n-1)/k$.

Then $k^2 + k + 1 \ge n.$

The equality holds when the row sums of M_i are all equal to $k+1$. In this case M_i is the adjacency matrix of a projective plane. But there does not exist [8] an adjacency matrix of a projective plane of trace 0.

Hence $\qquad k^2 + k + 1 > n$

and $\qquad k^2 + k + 1 \ge r(C_4;k).$

Theorem 6. $r(C_4;k) \ge k^2/16$ for infinitely many k's.

The proof is established by an explicit construction.

After the conference the author proved that $r(C_4;(1+\epsilon)k \ge k^2$ for any small ϵ and large k and that $r(C_4;k)$ is asymptotically equal to k^2.

REFERENCES

1. Ramsey, F. P., "On a Problem of Formal Logic", _Proc. London Math. Soc._, 30 (1930) 264-286.

2. Ryser, H. J., _Combinatorial Mathematics_, Wiley, New York, 1963.

3. Bush, L. E., "The William Lowell Putnam Mathematical Competition", _Amer. Math. Monthly_, 60 (1953) 539-542.

4. Greenwood, R. E., and Gleason, A. M., "Combinatorial Relations and Chromatic Graphs", _Canad. Math._ 7 (1955) 1-7.

5. Whitehead, E. G., Jr., "Ramsey Numbers of the Form N(3,...,3;2)", Doctoral Dissertation, University of Southern California, 1971.

6. Chung, F. R. K., "On the Ramsey Numbers N(3,3,...,3;2)", _Discrete Math._, to appear.

7. Burling, J. P., and Reyner, S. W., "Some Lower Bounds of the Ramsey Numbers n(k,k)", _Combinatorial Theory_ 13B (1972) 168-169.

8. Wilf, H. S., "The Friendship Theorem", Combinatorial Mathematics and Its Applications, (D. J. A. Welsh, ed.), Academic Press, New York, 1971, 307-309.

THE MINIMALITY OF THE MYCIELSKI GRAPH

V. Chvátal
Centre de Recherches Mathématiques
Université de Montréal
Montréal 101, Canada

ABSTRACT

It is proved that the smallest 4-chromatic triangle - free graph has eleven vertices and is unique.

THE MINIMALITY OF THE MYCIELSKI GRAPH

Mycielski [6] constructed a sequence $G_1, G_2, \ldots, G_k$ of graphs such that no G_k contains a triangle and each G_k is k-chromatic. (The existence of graphs with any prescribed girth and chromatic number has been established by Erdös [3]; an explicit construction of such graphs has been given by Lovász [5].) I remark that Mycielski's graph G_4 is the unique triangle-free 4-chromatic graph with at most eleven vertices. The purpose of this note is to give a complete proof of this state, which is a stronger statement than that in Harary [4], since uniqueness is shown here.

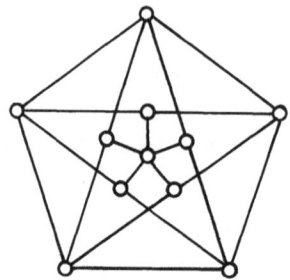

Figure 1

Let $G = (V, E)$ be a triangle-free 4-chromatic graph with at most eleven vertices. Let u be a vertex of largest degree in G and let A be the set of neighbours of u. By Brooks' theorem [1], we have $|A| \geq 4$ (otherwise G is 3-colorable). Besides, A is an independent set (otherwise G contains a triangle). Next, set $B = V - \{u\} - A$, so that $|B| = |V| - 1 - |A| \leq 6$. The subgraph of G induced by B is triangle-free and 3-chromatic (otherwise $B \cup \{u\}$ can be colored by two colors and A by a third one). Therefore B contains a 5-cycle (so that $|A| = |V| - 1 - |B| \leq 5$). Let its vertices (in the natural cyclic order) be u_1, u_2, u_3, u_4, u_5.

Case 1. $|A| = 5$. Firstly, let us assume that given any i, there is always a $v \in A$ which is adjacent to both u_i and u_{i+2} (with addition modulo five). Then G must contain the graph G_4. Since the addition of any new line to G_4 creates a triangle, we conclude that $G = G_4$.

Now, we may assume that no v ε A is adjacent to both u_1 and u_3. (Note that each v ε A is adjacent to at most two u_i's (otherwise G contains a triangle). But then G can be colored in three colors as follows: u_1 by 1, u_2 by 3, u_3 by 2, u_4 by 1, u_5 by 2, each v ε A by 1 or 2 and finally u by 3.

Case 2. $|A| = 4$. The set $V - \{u\} - A - \{u_1,u_2,u_3,u_4,u_5\}$ includes at most one vertex; if there is such a vertex, denote it by w. First of all, we shall show that there is an i such that

      (i)   w is not adjacent to u_i,

      (ii)   no v ε A is adjacent to both u_{i-1} and u_{i+1}.

Since $|A| = 4$, there is at least one i satisfying (ii); without loss of generality, we may assume i = 2. If w is not adjacent to u_2, we are done. Otherwise both i = 1 and i = 3 satisfy (i). If neither i = 1 nor i = 3 satisfies (ii) then there are v', v" ε A such that v' is adjacent to both u_2, u_5 and v" is adjacent to both u_2 and u_5, and v" is adjacent to both u_2 and u_4. To avoid the triangle v'$u_4 u_5$, we must have v' $\neq$ v". But then u_2 is adjacent to five distinct vertices (namely, $u_1, u_3, w, v', v"$), contradicting maximality of $|A|$.

Now, we can assume that i = 2 satisfies both (i), (ii). Color w by 3 and the rest of G as in Case 1. The proof is finished.

Now, let f(k) be the smallest number of vertices in a k-chromatic triangle-free graph. If $k \geq 4$, Brooks' Theorem implies $f(k) \geq 1 + k + f(k-1)$ and so

$$f(k) \geq \binom{k+2}{2} - 4 \qquad (k \geq 4).$$

On the other hand, a theorem due to Erdös [4] yields

$$f(k) < c(k \log k)^2$$

for some absolute constant c.

REFERENCES

1. Brooks, R. L., On Colouring the Nodes of a Network, <u>Proc. Cambridge Phil. Soc.</u> 37 (1941), 194-197.

2. Chvátal, V., The Smallest Triangle-free 4-chromatic 4-regular Graph, J. Com <u>J. Combinatorial Theory</u> 9 (1970), 93-94.

3. Erdös, P., Graph Theory and Probability, <u>Canad. J. Math</u>. 11 (1959), 34-38.

4. Harary, F., <u>Graph Theory</u>, Addison-Wesley, Reading, Mass., 1969, p. 149, Exercise 12.19.

5. Lovász, L., On the Chromatic Number of Finite Set-systems, <u>Acta. Math. Acad. Sci. Hungar</u>. 19 (1968), 59-67.

6. Mycielski, J., Sur le coloriage des graphes, <u>Colloq. Math</u>. 3 (1955), 161-162.

ON THE RAMSEY NUMBER OF THE FIVE-SPOKED WHEEL

V. Chvátal
Université de Montréal

A. Schwenk
University of Michigan
and Oxford University

ABSTRACT

Let $r(W_6)$ be the smallest n such that the five-spoked wheel $W_6 \subset G$ or else $W_6 \subset \bar{G}$ for every graph G with n vertices. Recently, Erdös asked whether whether $r(W_6) \geq 18$; we prove that $17 \leq r(W_6) \leq 20$. In so doing, we establish that $r(C_5, W_6) = 13$. We conjecture that $r(W_6) = 20$.

ON THE RAMSEY NUMBER OF THE FIVE-SPOKED WHEEL

1. INTRODUCTION

Let F_1, F_2 be finite graphs without isolated vertices (for graph-theoretical definitions and notation, see [7]). By Ramsey's theorem [8], there is a positive integer n such that $F_1 \subset G$ or $F_2 \subset \overline{G}$ holds for every graph G with n vertices. As in [2], we denote the smallest such n by $r(F_1, F_2)$; when $F_1 \cong F_2$, we shorten this to $r(F_1)$. In 1947, Erdös [5] proved that

$$r(K_n) > 2^{n/2}$$

for every n > 2. Chvátal and Harary [3] noticed that Erdös's technique yields

$$(1) \qquad r(F) > \left(\frac{s}{2}\right)^{1/p} 2^{q/p}$$

where s is the number of automorphisms of F, q is the number of edges and p is the number of vertices of F. Now, let F be n-chromatic. Then F contains a vertex-critical n-chromatic graph F_0 and each vertex of F_0 has degree at least $n-1$ [4]. Writing q_0 for the number of edges and p_0 for the number of vertices of F_0, we have $q_0/p_0 \geq (n-1)/2$ and so (1) yields

$$r(F) \geq r(F_0) > c2^{n/2}$$

where c is a positive constant independent of n. Recently, Erdös conjectured that

$$(2) \qquad r(F) \geq r(K_n)$$

whenever F is n-chromatic. Evidently, (2) holds whenever $n \leq 3$. (In fact, the inequality is strict.) Greenwood and Gleason [6] found that $r(K_4) = 18$, so the conjecture is that $r(F) \geq 18$ for every 4-chromatic graph. If F has more than 6 points, the coloring $G = 3K_6$ proves that $r(F) > 18$, for $F \not\subset G$, and if $F \subset \overline{G}$, then F is 3 colorable. Similarly, if $K_4 \subset F$, we clearly have $r(F) \geq r(K_4)$. Thus, it only remains to test the conjecture on 4-chromatic graphs with at most 6 points such that $K_4 \not\subset F$. An exhaustive search reveals that there is only one graph: the 5-spoked wheel W_6 (see figure 1). Thus, for 4-chromatic graphs, Erdös' conjecture will be verified if $r(W_6) \geq 18$. We are going to show that $17 \leq r(W_6) \leq 20$.

W_6:

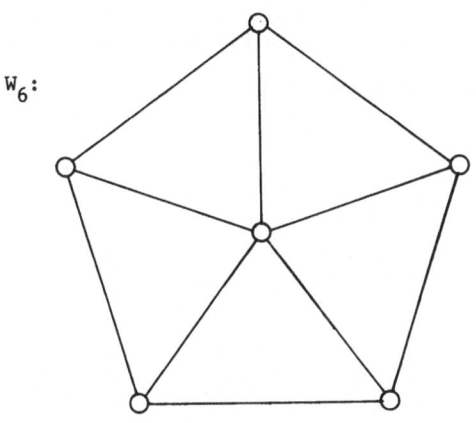

<u>Figure 1</u>

2. THE LOWER BOUND

We define the neighborhood subgraph N(G;v) as the subgraph of graph G induced by the neighbors of point v. Now let G be the graph with vertex-set $\{1, 2,...,8\}$ × {1,2} where two <u>distinct</u> vertices (i, j), (r, s) are adjacent if and only if $|i-r| \varepsilon \{0, 1, 4, 7\}$. We shall show that $W_6 \not\subset G$ and $W_6 \not\subset \overline{G}$. Firstly, assume that $W_6 \subset G$. Since the antomorphism group of G is transitive, we may assume that (1, 1) is the hub of $W_6 \subset G$. That is, N(G; (1,1)) must contain C_5. The neighborhood subgraph N(G, (1,1)) is depicted in Figure 2; obviously it contains no C_5. Hence, $W_6 \not\subset G$.

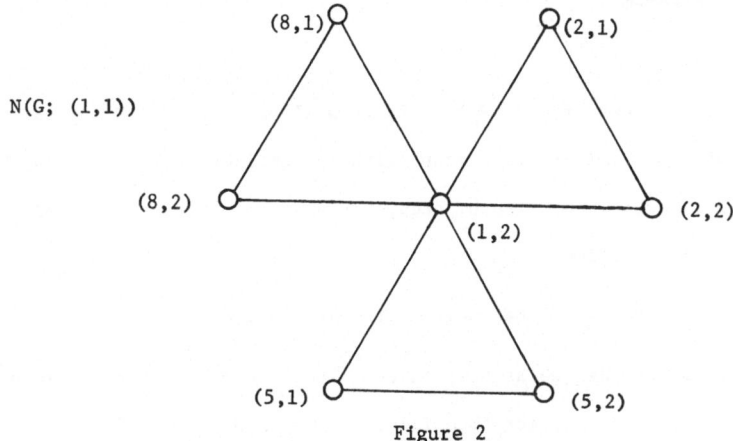

N(G; (1,1))

<u>Figure 2</u>

Secondly, assume $W_6 \subset \overline{G}$. Again, it follows that $N(\overline{G}; (1,1))$ must contain C_5. But this subgraph is bipartite as can be seen from Figure 3. Hence $W_6 \not\subset \overline{G}$, and so $r(W_6) \geq 17$.

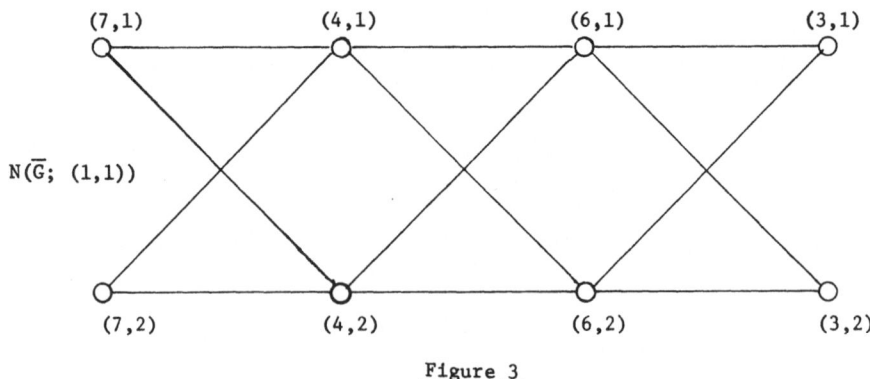

$N(\overline{G}; (1,1))$

Figure 3

3. THE UPPER BOUND

In order to prove 20 is an upper bound, we must derive several preliminary results. First, we evaluate $r(C_5 , W_6)$. Next we prove two extremal results that if G has enough lines but no C_5, then G must be a bigraph. Theorems 4 and 5 give conditions which force $W_6 \subset \overline{G}$. Finally, we quote a result from [9] before presenting the main theorem.

Theorem 1. $r(C_5 , W_6) = 13$.

Proof. The graph $3K_4$ shows $r(C_5 , W_6) > 12$; we will show $r(C_5 , W_6) \leq 13$. Throughout the proof, G will denote a graph with 13 vertices such that $C_5 \not\subset G$ and $W_6 \not\subset \overline{G}$. We will establish, step by step, various properties of G and eventually arrive at a contradiction.

Proposition 1. Each vertex of G has degree at least four.

Proof. Assume that u has degree at most three. Then there are nine vertices of G that are not adjacent to u; let them induce a subgraph H of G. Since $r(C_5) = 9$ (see [1]), we have either $C_5 \subset H \subset G$ or else $C_5 \subset \overline{H}$, so that $W_6 \subset \overline{G}$.

Proposition 2. $K_4 \not\subset G$.

Proof. Assume $K_4 \subset G$. By Proposition 1 and the hypothesis $C_5 \not\subset G$, there must be a set S of 4 distinct points, each adjacent to just one point of K_4 . Now two points of S cannot be adjacent, so each S-point must be joined to at least 3 of the remaining points. But then one of these points is adjacent to two S-points, forming $C_5 \subset G$. Thus $K_4 \not\subset G$.

Proposition 3. $K_4 - x \not\subset G$.

Proof. Assume $K_4 - x \subset G$ with points v_1 , v_2 , v_3 , v_4 where v_3 and v_4 are not adjacent. The set S of points adjacent to $K_4 - x$ must contain four distinct points, u_i and w_i adjacent to v_i for $i = 3$ and 4. If there are also two distinct points u_i adjacent to v_i for $i = 1$ and 2, u_1 and u_2 cannot be adjacent to any previously labeled points (besides v_1 and v_2) without forming C_5 . Thus each is joined to two of the remaining 3 points, and so one of these last points w_0 is adjacent to both. But then $w_0 u_1 v_1 v_2 u_2$ is a C_5 . On the other hand, if there is one point u_1 adjacent to v_1 and v_2 , then u_1 must be adjacent to at least two of the remaining points, say w_1 and w_2 , both of which must be joined to each other as well as to the last two points a_1 and a_2 . Since the degree of u_3 is at least 4, the lines $u_3 a_1$ and $u_3 a_2$ are forced, but this forms the cycle $u_3 a_1 w_1 w_2 a_2$.

Proposition 4. $K_3 \not\subset G$.

Proof. Let $v_1 v_2 v_3$ be a triangle in G . Then by Proposition 3 there must be six distinct points u_i , w_i for $i = 1, 2, 3$ where u_i and w_i are adjacent to v_i . Label the last 4 points a_1 , a_2 , a_3 , a_4 . Suppose u_1 is adjacent to w_1 . Then by Proposition 3, no a-point may be adjacent to both u_1 and w_1 . Thus we may assume without loss of generality that a_1 and a_2 are joined to u_1 and a_3 and a_4 are joined to w_1 . But now no line from u_2 , w_2 , u_3 , w_3 to an a-point is possible, which forces the a-points to comprise K_4 . Thus there is no line $u_i w_i$. Consequently, each u_i and w_j is joined to 3 a-points, totaling 18 such lines. Thus, some a_k is joined to 5 u- or w-points, forming C_5 .

Proposition 5.   G is a bigraph.

Proof. If not, G has a shortest odd cycle C_{2n+1} with $n \geq 3$. By Proposition 1, each point of this cycle must be joined to at least 2 of the remaining 12-2n points. But since 12-2n < 2n+1 for $n \geq 3$, this implies one of these remaining points is joined to 3 points on the cycle, forcing a shorter odd cycle.   Thus G must have no odd cycles.

But finally, if G is a bigraph with 13 points, we have $W_6 \subset K_7 \subset \overline{G}$, and so no such graph G can exist. This completes the proof of Theorem 1.

Theorem 2.   If $C_5 \not\subset H$ and H has 9 points and at least 16 lines, then H is a bigraph.

Proof. We shall demonstrate that H must be a bigraph by assuming it is not and then deriving properties of H until a contradiction is reached.

Proposition 6. $2K_4 \not\subset H$.

Proof. If $2K_4 \subset H$, the ninth point can have degree at most 2 since $C_5 \not\subset H$. Consequently, there must be at least 2 lines joining the two copies of K_4 , but such a pair of lines forces $C_5 \subset H$.

Proposition 7. $K_4 \cup K_4 - x \not\subset H$.

Proof. We routinely find it impossible to add 5 lines joining K_4 , $K_4 - x$, and K_1 without producing C_5 .

Proposition 8. Let F be the graph $K_{1,3} + x$ formed by adding a line to the 4-point star. $K_4 \cup F \not\subset H$.

Proof. The ninth point cannot be joined to F by 3 lines without violating Proposition 7. Furthermore, it can have only 1 line joining it to K_4 since $C_5 \not\subset H$. Thus there are at least 3 lines joining K_4 to F . To avoid forming C_5 , this must be comprised of one point of the K_4 adjacent to the three points in the triangle in F . But again, adding 3 more lines to the ninth point forces a C_5 .

Proposition 9. $K_4 \cup K_3 \not\subseteq H$.

Proof. There can be at most 3 lines joining K_4 to K_3 producing no C_5 . Thus, there remain 4 lines incident with the last two points. By Proposition 8, none of these 4 can be incident with K_3 , and so one of the last two points is adjacent to two points of K_4 forcing $C_5 \subseteq H$.

Proposition 10. $K_4 \not\subseteq H$.

Proof. Let S be the subgraph $H - K_4$. Since $K_3 \not\subseteq S$, $C_5 \not\subseteq S$, and S has 5 points, we conclude that S is a bigraph. If S has 6 lines, then $S = K_{2,3}$ and each choice of 4 lines joining S to K_4 produces C_5 . Similarly, if S has 5 lines, then $S = K_{2,3} - x$ and any 5 lines joining S to K_4 must produce C_5 . Finally, if S has less than 5 lines, the 6 lines joining S to K_4 must include two incident with a single point of S , but this forms a C_5 .

Proposition 11. $K_3 \subseteq H$.

Proof. By assumption, H is not a bigraph, so if $K_3 \not\subseteq H$, and $C_5 \not\subseteq H$, then H must contain an induced C_7 . But now H has at least 8 lines joining C_7 to the last two points, and so one of these points is adjacent to at least 4 points of C_7 , forcing $K_3 \subseteq H$.

Proposition 12. $K_4 - x \subseteq H$.

Proof. Let $S = H - K_3$. We assume $K_4 - x \not\subseteq S$. By inspection, S has at most 9 lines. If S has 9 lines, $S = K_{3,3}$ and cannot be joined to K_3 by 4 lines without producing C_5 . If S has 8 lines, S must be $K_{2,4}$ or $K_{3,3} - x$, and in either case cannot be joined to K_3 by 5 lines. If S has 7 lines, either some point of S is joined to two points of K_3 forming $K_4 - x$, or else S has 6 points each joined to exactly 1 point of K_3 . But notice that two adjacent points cannot be adjacent to different points of K_3 without forming C_5 . Thus, any point with degree 2 or more in S produces a $K_4 - x$ containing one point of K_3 and 3 points of S . Finally, if S has less than 7 lines, one point of S must be adjacent to two points of K_3 , completing the proof.

Proposition 13. $K_2 + \overline{K}_3 \subset H$.

Proof. Let $S = H-(K_4 - x)$. By Proposition 10, $K_4 \not\subset S$, and we may assume $K_2 + \overline{K}_3 \not\subset S$. Then we find that S has at most 6 lines. No point of S may be joined by two lines to $K_4 - x$ without forming either C_5 or $K_2 + \overline{K}_3$. Thus, we conclude that S has exactly 6 lines, and each point of S is joined to one point in $K_4 - x$. As in the preceding proof, adjacent points in S must be joined to the same point in $K_4 - x$. Since S has a point of degree 3 or more, H contains $K_2 + \overline{K}_3$.

Proposition 14. $K_2 + \overline{K}_4 \subset H$.

Proof. Let $S = H - K_2 + \overline{K}_3$. If S has 5 lines, we find it impossible to join S and $K_2 + \overline{K}_3$ without forming C_5 . On the other hand, if S has fewer lines, one point of S must be joined to two points of $K_2 + \overline{K}_3$ forming $K_2 + \overline{K}_4$.

But now it is easily seen to be impossible to join $K_2 + \overline{K}_4$ to the other three points of H without forming C_5 . This contradiction implies that our assumption that H is not a bigraph is wrong, proving Theorem 2.

Theorem 3. If G has 10 points and at least 19 lines, $C_5 \not\subset G$, and $W_6 \not\subset \overline{G}$, then G is a bigraph.

Proof. Assume G is not a bigraph and remove a point of minimum degree to obtain the subgraph $H = G-v$ with at least 16 lines. By the previous theorem, H must be a bigraph. G must contain a triangle, for otherwise $C_7 \subset G$ and there are 10 lines joining the remaining 3 points to this cycle. Thus one point is joined to 4 points of C_7 forcing a triangle. Of course this triangle must contain v, and so we conclude that v is joined to at least 2 adjacent points in H , say u_1 and w_1 . The line $u_1 w_1$ must not lie in $C_4 \subset H$, for otherwise $C_5 \subset G$. There are 7 bigraphs with 9 points and 17 lines, but only $K_2 \cdot K_{4,4}$ (see [7, p. 23]) has a line (namely the endline) lying in no C_4 . Thus G might be $K_3 \cdot K_{4,4}$, but we observe that $\overline{G} \supset K_6 - x \supset W_6$. On the other hand, if H has only 16 lines, the minimum possible number, then v must be joined to 3 points in H , say u_1 and u_2 in the same set, and w_1 adjacent to u_1 . With some effort, we find the 18 possible bigraphs with 9 points and 16 lines, but only 4 of them (formed by adding an endline in either of 2 ways to either $K_{3,5}$ or $K_{4,4} - x$) have a line lying in no C_4. In all four cases, the line vu_2 forces an occurrence of C_5. Thus our assumption was incorrect, and so G must be a bigraph.

Theorem 4. If $p \geq 11$, $C_5 \not\subset G$, and $K_5 \subset \overline{G}$, then $W_6 \subset \overline{G}$.

Proof. Let S be the subgraph induced by the points not contained in the induced $\overline{K}_5 \subset G$, and assume $W_6 \not\subset \overline{G}$. Now each point of S is joined to at least 3 points of $\overline{K}_5$ to prevent forming $W_6 \subset \overline{G}$. Moreover, if S has less than 3 lines, $W_6 \subset \overline{S} \subset \overline{G}$, so we may assume S has at least 3 lines. Let $v_1 v_2$ be one line of S. If any point of S has 4 lines joining it to $\overline{K}_5$, $v_1 v_2$ lies in C_5. On the other hand, if every point of S is adjacent to exactly 3 points of $\overline{K}_5$, and $C_5 \not\subset G$, then G minus the lines of S (except for $v_1 v_2$) must be the graph formed by identifying a point of $K_2 + \overline{K}_3$ with a point from $K_{3,p-7}$ as shown in Figure 4. But adding the second of the three lines known to be in S must form C_5. This contradiction implies $W_6 \subset \overline{G}$.

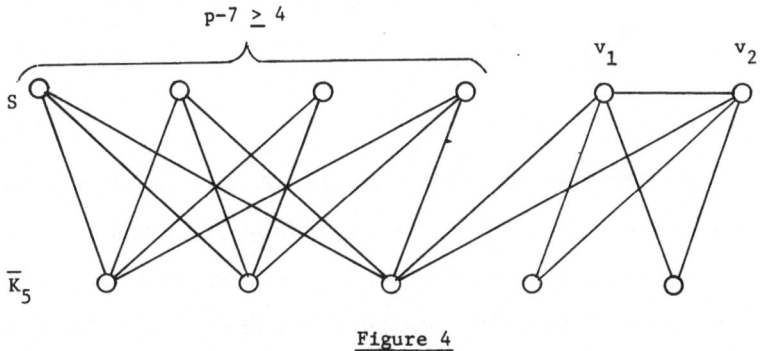

Figure 4

Theorem 5. If $C_5 \not\subset G$ and G has 12 points and at least 26 lines (or 11 points and at least 22 lines), then $W_6 \subset \overline{G}$.

Proof. Let $H = G - v_0$ be the subgraph obtained by removing a point of v_0 of minimum degree from G. Then H has 11 points and at least 22 lines, so we proceed with one proof for both the 11 and 12 point versions. We may assume H is not a bigraph since otherwise $W_6 \subset K_6 \subset \overline{H}$.

Case 1. If H is not regular, then H has a subgraph $H' = H-v_1$ with at least 19 lines. Now by Theorem 3, H' is a bigraph, and so $K_5 \subset \overline{H}' \subset \overline{H}$. Thus by Theorem 4, $W_6 \subset \overline{H}$.

Case 2. If H is regular of degree 4, we proceed to specify properties of H until we force a contradiction.

Proposition 15. $K_4 \not\subset H$.

Proof. If $K_4 \subset H$, let S be the set of exactly 4 points each adjacent to a point of K_4 , and let R be the last three points. Any line within S would create a C_5 , so we must have 12 lines joining S to R . But now any two of these lines which are incident with the same R-point lie in a C_5 .

Proposition 16. $K_4 - x \not\subset H$.

Proof. If $K_4 - x \subset H$, then S must have 5 or 6 points. If S has 5 points, we notice that the point adjacent to 2 points of $K_4 - x$ must also be joined to both of the last two points of H . But now any additional line joining one of these points to S forms a C_5 . Similarly, if S has 6 points the S-point adjacent to one of the points with degree 3 in $K_4 - x$ cannot have degree 4 without forming C_5 .

Proposition 17. Let $F = K_{1,3} + x$. $F \not\subset H$.

Proof. If $F \subset H$, then there are 8 lines joining F to S . Since S has at most 7 points, one point of S is adjacent to two points of F . This must occur without forming C_5 or $K_4 - x$, and so we see that $K_3 \cdot K_3 \subset H$. But now the graph $R = H - K_3 \cdot K_3$ has 6 points and 8 lines joining it to $K_3 \cdot K_3$. Two of these lines incident with the same point of R force an occurrence of either C_5 or $K_4 - x$.

Proposition 18. $K_3 \not\subset H$.

Proof. This is now trivial since H is regular of degree 4. If $K_3 \subset H$, then $F \subset H$.

But now we have shown that H is not a bigraph and H has no odd cycle of length 3 or 5. Hence H has an induced C_7 , and so each of the remaining points is joined to at most 2 points on C_7 . Thus, H has at most $7 + 4 \cdot 2 + 4 = 19$

lines, a contradiction. That is, Case 2 cannot occur, and Theorem 5 has been proved.

Let $\mu(G)$ denote the sum of the number of triangles in G plus the number of triangles in $\bar{G}$. In other words, μ is the number of monochromatic triangles in the two-coloring of K_p represented by G. We also write $\mu(v)$ for the number of monochromatic triangles containing the point v. Note that $3\mu(G) = \sum \mu(v_i)$. Recall that $N(G;v)$ is the graph induced by the points adjacent to v in G. Then clearly $\mu(v)$ is the sum of the number of lines in $N(G;v)$ plus $N(\bar{G},v)$. The next result is proved in [9].

__Theorem 6.__ If d_1, d_2, $\ldots$, d_p denote the degrees of the points in G, then
$$\mu(G) = \binom{p}{3} - \frac{1}{2} \sum d_i(p - 1 - d_i) .$$

Define $\nu(v)$ the net number of monochromatic triangles containing v to be
$$\nu(v_i) = \mu(v_i) - 3[(p - 1)(p - 2) / 6 - d_i(p - 1 - d_i) /2] .$$

This definition is chosen so that

(3) $\qquad \sum \nu(v_i) = \sum \mu(v_i) - 3\mu(G) = 0$

Thus, $\nu(v)$ measures the difference between the number of monochromatic triangles containing v and the number of monochromatic triangles forced to be in G by the degree of v.

At last we have assembled the tools needed to prove the upper bound.

__Theorem 7.__ $17 \le r(W_6) \le 20$.

__Proof.__ The lower bound was demonstrated in the previous section. We now prove the upper bound. Assume G represents one color class of a two coloring of K_{20} containing no monochromatic W_6. We proceed to derive properties of G until we obtain a contradiction.

__Proposition 19.__ If v has degree 9 or 10, $N(G;v)$ and $N(\bar{G};v)$ cannot both be bigraphs.

__Proof.__ If both are bigraphs, then $K_5 \subset \bar{N}(\bar{G};v)$ and $\bar{K}_5 \subset N(G;v)$ as an induced subgraph. If any point of $\bar{K}_5$ is joined to 3 points of K_5, we have $W_6 \subset K_6 - P_3 \subset G$, so there are at most 10 lines joining $\overline{K}_5$ and K_5 . Thus, some point of K_5 is joined to at most 2 points of $\overline{K}_5$ forcing $W_6 \subset K_6 - P_3 \subset \overline{G}$. This contradiction proves the proposition.

Proposition 20. G has at least 7 points of degree 9 or 10.

Proof. We first establish upper bounds on the value of $\nu(v)$ depending upon d, the degree of v. This degree cannot be less than 7 (or greater than 12) without Theorem 1 forcing a monochromatic W_6 . If d = 7 or 12, we note that $\nu(v) = \mu(v) - 45$. But as mentioned above, μ is just the sum of the number of lines in $N(G;v)$ and $N(\overline{G};v)$. It can be shown that, for $p \geq 6$, the maximum number of lines in a graph not containing C_5 is the same as in Túran's Theorem for K_3 (see [7, p. 17]). Applying this and Theorem 5, $\mu(v) \leq 12 + 25 = 37$, so that $\nu(v) \leq -8$. Similarly, if v has degree 8 or 11, we compute that $\nu(v) \leq 16 + 21 - 39 = -2$. Finally, if d = 9 or 10, we recall Theorems 2 and 3 as well as Proposition 19 to conclude that $\mu(v) \leq$ $\mu(v) \leq$ max {15 + 25, 20 + 18}. Thus $\nu(v) \leq 40 - 36 = +4$. By Equation (3) and the upper bounds just computed, we observe that at least one third of the points must have degree 9 or 10, completing the proof.

Let v_0 be a point of degree 9 or 10 whose ν is maximal. By redefining G to be $\overline{G}$ if necessary, we may assume the degree is 10.

Proposition 21. $N(G; v_0)$ cannot be a bigraph.

Proof. Assume $N(G; v_0)$ is a bigraph. Then $N(\overline{G}; v_0)$ is not a bigraph, and we proceed to estimate the total number of lines in G.

Let $N(G; v_0)$ have 25-a lines. Note that this forces $\nu(v_0) \leq 4-a$. Of course, there are 10 lines joining v_0 to N. Finally, just as in the proof of Proposition 19, we observe that each point in $N(\overline{G}; v_0)$ must be joined to at least 6 points in $N(G; v_0)$ to avoid forming $W_6 \subset K_6 - P_3 \subset \overline{G}$. Thus, there are at least 54 lines joining $N(G; v_0)$ and $N(\overline{G}; v_0)$, and so G has at least 21 + (25-a) + 10 + 54 = 110-a lines. It is easy to see that for a given number of lines, our upper bound for $\sum \nu(v_i)$ is maximized by taking every point to have degree as close to the average as possible. (Otherwise adding 1 to the degree of a point of minimum degree and subtracting 1 from a point of a maximum degree is guaranteed to increase the bound for $\sum \nu(v_i)$). Thus, if a = 0, the average degree is 11 and so $\sum \nu(v_i) \leq 20(-2) = -40$.

Since this sum must equal 0, $a \geq 1$. If $a = 1$, recall that now $\nu(v) \leq 3$ for points of degree 9 and 10, and so since the average degree is at least 10.9, we see that

$$\sum \nu(v_i) \leq 18(-2) + 2(3) = -30 < 0$$

Thus $a \geq 2$. If $a = 2$, then $\overline{d} \geq 10.8$ and

$$\sum \nu(v_i) \leq 16(-2) + 4(2) = -24 < 0 \ .$$

Thus $a \geq 3$. If $a = 3$, then $\overline{d} \geq 10.7$, so that

$$\sum \nu(v_i) \leq 14(-2) + 6(1) = -22 < 0 \ ,$$

and so $a \geq 4$. But if $a = 4$, then $\overline{d} \geq 10.6$ and

$$\sum \nu(v_i) \leq 12(-2) + 0 = -24 < 0 \ .$$

But a can be no larger, for otherwise $\nu(v) < 0$ for all v, and so $\sum \nu(v_i) < 0$. This contradiction completes the proof of Proposition 21.

Now if $N(\overline{G}; v_0)$ is not a bigraph, then by Theorems 2 and 3, $\nu(v_0) \leq 15+18-36 = -3$, violating the choice of v_0. Thus $N(\overline{G}; v_0)$ must be a bigraph and we proceed to estimate the number of lines in G.

<u>Proposition 22.</u> Let $N(\overline{G}; v_0)$ have 20-a lines. Then G has at most 86 + 2a lines.

<u>Proof.</u> $\overline{N}(\overline{G}; v_0)$ has 36-(-20-a) = 16 + a lines, $N(G; v_0)$ has at most 18 lines, and there are 10 lines joining v_0 to $N(G; v_0)$. Thus we will be through if we can show that there are at most 42+a lines joining $N(G; v_0)$ and $N(\overline{G}; v_0)$. Assume G has 43+a such lines. If a point in $N(G; v_0)$ is joined to 3 points in $K_5 \subset \overline{N}(\overline{G}; v_0)$, we have $W_6 \subset K_6 - P_3 \subset G$, so each point in $N(G; v_0)$ is joined to at most 6 points in $N(\overline{G}; v_0)$. If some point is joined to 6 points, then there must be a second point joined to 5 points, again forcing $W_6 \subset G$. Thus, we may assume no point in $N(G; v_0)$ is joined by more than 5 lines to $N(\overline{G}; v_0)$, and there are 3+a points, $v_1, v_2 \ldots, v_{3+a}$ joined to exactly 5 points. Each of these points must be joined by 2 lines to K_5 and by 3 to K_4. Furthermore, if any two v_i are joined to a different subset of 3 points in K_4, we have $W_6 \subset K_6 - P_4 \subset G$. Label the omitted point v_{4+a}. We see that any line joining v_i to v_j for $1 \leq i < j \leq 4+a$, forms a wheel, and so we have an induced $\overline{K}_{4+a}$. Now a = 0 or 1 for otherwise $W_6 \subset K_{4+a} \subset \overline{G}$. If a = 1, the four

points $v_1, \ldots, v_4$ must each be joined by at most 2 lines to $K_5 \subset \overline{N}(\overline{G}; v_0)$. Thus, some point w in K_5 is joined to only 1 point of these 4. This forms a W_6 in $\overline{G}$ on the points $v_1, v_2, v_3, v_4, v_5, w$. Thus a = 0, and we note that $N(\overline{G}; v_0) = K_{4,5}$. Now v_1, v_2, and v_3 are joined to $K_5 \subset \overline{N}(\overline{G}; v_0)$ by at most 6 lines, and there are no lines joining v_4 to K_5. Thus K_5 has two points w_1 and w_2 joined by at most 2 lines to v_1, v_2, v_3, forcing $W_6 \subset \overline{G}$ with point set $\{v_1, v_2, v_3, v_4, w_1, w_2\}$. This final contradiction completes the proof of this proposition.

Proposition 23. If v_1 has degree 9, then $\nu(v_1) < \nu(v_0)$.

Proof. Since v_0 was chosen to have maximal ν, assume $\nu(v_1) = \nu(v_0) \geq 0$. Now if $N(\overline{G}; v_1)$ is a bigraph, we could replace G by $\overline{G}$ and v_0 by v_1 violating Proposition 21. Thus $N(\overline{G}; v_1)$ is not a bigraph. Then $\nu(v_1) \geq 0$ requires that $N(G; v_1)$ is a bigraph. Furthermore, the 2 sets of points in $N(G; v_1)$ must have sizes 4 and 5, for otherwise $W_6 \subset K_6 \subset \overline{N} \subset \overline{G}$. But can such a point exist? If $v_1 \in N(\overline{G}; v_0)$, then $v_1 \subset K_4$, so that $K_3 \subset N(G; v_1)$. Thus, v_1 must lie in $N(G; v_0)$. Let v_1 be joined to n other points in $N(G; v_0)$ and to 8-n points in $N(\overline{G}; v_0)$. Observe that these n points must induce $\overline{K}_n$ for otherwise 2 adjacent points plus v_0 form a triangle in $N(G; v_1)$. Now $n \leq 5$ for otherwise $K_6 \subset \overline{G}$. If n = 5, v_1 is joined to 3 points in $\overline{N}(\overline{G}; v_0)$. Two of these must be adjacent, so this pair contributes one point to each set of vertices in the bigraph $N(G; v_1)$. But since we already had $\overline{K}_5 \subset N(G; v_1) \cap N(G; v_0)$ this again provides $K_6 \subset \overline{G}$. If n = 4, then v_1 is also joined to 4 points in $\overline{N}(\overline{G}; v_0)$, which either forces $K_3 \subset N(G; v_1)$ or else contributes 2 points to each set again forming $K_6 \subset \overline{G}$. Finally, if $n \leq 3$, the 5 lines joining v_1 to $\overline{N}(\overline{G}; v_0)$ force a triangle in $N(G; v_1)$. Thus, we have shown that no point of degree 9 can have ν equal to $\nu(v_0)$.

We now conclude the proof. If a = 0, G has 86 lines, and the maximum estimate for $\sum \nu(v_i)$ occurs when we choose 12 points of degree 9 and 8 of degree 8 to obtain

$$\sum \nu(v_i) \leq 8(-2) + 12(1) = -4 < 0.$$

If a = 1, then G has 88 lines, $\nu(v_0) \leq 1$, and points of degree 9 have $\nu(v_1) \leq 0$. Then $\sum \nu(v_i) \leq 4(-2) + 16(0) = -8 < 0$. If a = 2, then G has 90 lines, $\nu(v_0) \leq 0$, and points of degree 9 have $\nu(v_1) \leq -1$. Then $\sum \nu(v_i) \leq 20(-1) = -20 < 0$. The value of a can be no larger without forcing $\nu(v_0)$ to be negative.

Thus the graph G cannot exist, and so $r(W_6) \leq 20$.

Acknowledgement. The authors are grateful to Stefan Burr for his many helpful comments.

REFERENCES

1. Chartrand, G. and Schuster, S. On the Existence of Specified Cycles in Complementary Graphs. Bull. Amer. Math. Soc., 77 (1971) 995-998.

2. Chvátal, V. and Harary, F. Generalized Ramsey Theory for Graphs. Bull. Amer. Math. Soc. 78 (1972) 423-426.

3. Chvátal, V. and Harary, F., Generalized Ramsey Theory for Graphs, I. Diagonal Numbers. Periodica Math. Hungar. 3 (1973) 115-124.

4. Dirac, G. A. A Property of 4-chromatic Graphs and Some Remarks on Critical Graphs. J. London Math. Soc. 27 (1952) 85-92.

5. Erdös, P. Some Remarks on the Theory of Graphs. Bull. Amer. Math. Soc. 53 (1947) 292-294.

6. Greenwood, R. E. and Gleason, A. M. Combinatorial Relations and Chromatic Graphs. Canad. J. Math. 7 (1955) 1-7.

7. Harary, F. Graph Theory. Addison-Wesley, Reading, Mass., 1969.

8. Ramsey, F. P. On a Problem of Formal Logic. Proc. London Math. Soc. 30 (1930) 264-286.

9. Schwenk, A. J. Acquaintance Graph Party Problem. Amer. Math. Monthly 79 (1972) 1113-1117.

ON THE HAMILTONIAN COMPLETION PROBLEM

S. Goodman and S. Hedetniemi
University of Virginia

ABSTRACT

We define the Hamiltonian completion number of a graph G, denoted hc(G),
to be the minimum number of lines that need to be added to G in order to make it
Hamiltonian. The Hamiltonian completion problem asks for hc(G) and a specific
Hamiltonian cycle containing hc(G) new lines. We derive an efficient algorithm
for finding hc(T) for any tree T, and show that if S is the set of spanning
trees of an arbitrary connected graph G, then

$$hc(G) = \min_{T_i \epsilon S} \; hc(T_i) \; .$$

A number of other general results are presented including an efficient heuristic
procedure which can be used on arbitrary graphs.

ON THE HAMILTONIAN COMPLETION PROBLEM

1. INTRODUCTION

The Hamiltonian completion number of a graph G, denoted hc(G), is the minimum number of lines which need to be added to G to make it Hamiltonian. By definition, $0 \leq hc(G) \leq p$, where p is the number of points in G.

The desirability of completing a graph in the above sense arises naturally in problems involving routing and the periodic traversal and updating of data structures.

The Hamiltonian completion problem is a special case of the traveling salesman problem in which each line in G is assigned a weight of 0 and each line in K_p not in G is given a weight of 1. This relationship can be used to show that the problem is a member of Karp's polynomial complete class of difficult computational problems [1]. This approach to the completion problem is being pursued elsewhere.

An equivalent formulation of the completion problem for non-Hamiltonian graphs involves the notion of island decomposition. An island decomposition of a graph G is an acyclic spanning subgraph of G in which every point has degree ≤ 2. An island decomposition is essentially a collection of point disjoint paths (islands) that cover the points of G. An isolated point can be a member of such a set. Let i(G) denote the minimum number of islands and $\ell(G)$ the maximum number of lines in any island decomposition of G. Then it is not difficult to show the next statement.

Lemma 1. For any graph G having p points

    i)   i(G) + ℓ(G) = p, and

   ii) either G is Hamiltonian or i(G) = hc(G).

Lemma 1 shows that we can now Hamilton thread the i(G) islands into a Hamiltonian cycle with the addition of i(G) new lines. Note that this approach cannot distinguish between a Hamiltonian cycle and a Hamiltonian path in G. In both cases we get i(G) = 1, although hc(G) = 0 if G has a Hamiltonian cycle. In what follows we assume G is connected (the generalization is immediate).

As far as we can tell, the Hamiltonian completion problem has never been formally studied as such. The superficially related problem of a minimum line cover of a graph by line disjoint paths has attracted some recent attention (cf. [2]); and the problem of finding a point disjoint path cover for the point set of a graph has been independently studied in a different context by Boesch, Chen and McHugh [3] in a paper first presented at this conference.

2. THE HAMILTONIAN COMPLETION NUMBER OF A TREE

This is one of the few problems involving paths and cycles where the solution for trees is not straightforward. However, we can construct an efficient algorithm for finding hc(T) for any tree T with p points. A very similar algorithm has been independently developed by Boesch, Chen and McHugh [3] which is presented in these proceedings. The proofs are different. Our proof is based on the following three lemmas.

<u>Lemma 2</u>. Let T be a tree and let v be an endpoint of T which is adjacent to a point u of degree 2 in T. Then hc(T) = hc(t-v).

<u>Proof</u>. Consider tne graph T-v. Assume that T-v is threaded with hc(T-v) threads. Since u is an endpoint of T-v, at least one thread (and possibly two) will be incident with point u. Let such a thread be uw. By re-directing this thread uw to vw in T we will produce a Hamilton threading of T with hc(T-v) threads. Hence hc(T) ≤ hc(T-v).

Conversely, let T be threaded with hc(T) new lines. Since v is an endpoint of T, at least one thread, and possibly two, must be incident with v.

<u>Case 1</u>. One thread, say wv is incident with v. In this case the correspond - ing Hamiltonian walk must at some point proceed from w to v to u. Thus by re-directing wv to wu we will obtain a Hamilton threading of T-v with hc(T) threads. Hence hc(T-v) ≤ hc(T).

<u>Case 2</u>. Two threads, say w_1v, vw_2, are incident with v. In this case, by adding a thread (if necessary) between w_1 and w_2 we will obtain a Hamilton threading of T-v with no more than hc(T) - 1 threads. Thus hc(T-v) < hc(T).

Lemma 3. Let u and v be endpoints of a tree T both of which are adjacent to a point w of degree $\geq$ 3. Then

$$hc(T-\{u,v,w\}) = hc(T) - 1 .$$

Proof. In everything that follows, we adopt the convention that $hc(K_1) = hc(K_2) = 1$. This handles the cases where $T-\{u,v,w\}$ is either K_1 or K_2.

We first show that $hc(T) \leq hc(T-\{u,v,w\}) + 1$. Assume that $T-\{u,v,w\}$ has been Hamilton threaded with $hc(T-\{u,v,w\})$ threads. Since this threading uses at least one thread, call it w_1w_2, we can eliminate it and add two new threads w_1u and vw_2 and produce a threading of T with $hc(T-\{u,v,w\}) + 1$ threads, i.e.,

$$hc(T) \leq hc(T-\{u,v,w\}) + 1 .$$

In order to establish the reverse inequality (and therefore equality), we will show that any threading of T with $hc(T)$ threads can be modified to produce a threading of $T-\{u,v,w\}$ with at most $hc(T) - 1$ threads.

Any threading of T with $hc(T)$ threads must place at least one thread adjacent to point u and one thread adjacent to point v. There are 11 possible threadings involving points u and v; these are illustrated in Figure 1. The heavy lines denote lines of T, the lighter lines denote threads in a completion solution for T, and the dotted lines denote threads in a candidate completion solution for $T-\{u,v,w\}$. For example, in Figure 1(1) an optimum completion solution for T requires the addition of four new lines through u,v,w while that solution can be modified for $T-\{u,v,w\}$ so as to include only three new lines. By inspection one can see that in all cases, $hc(T-\{u,v,w\}) \leq hc(T) - 1$. Note that a certain number of point coalescences are possible, e.g. Figure 1(12) shows a coalesence of Figure 1 (1). However, such cases do not change the thread counts adversely.

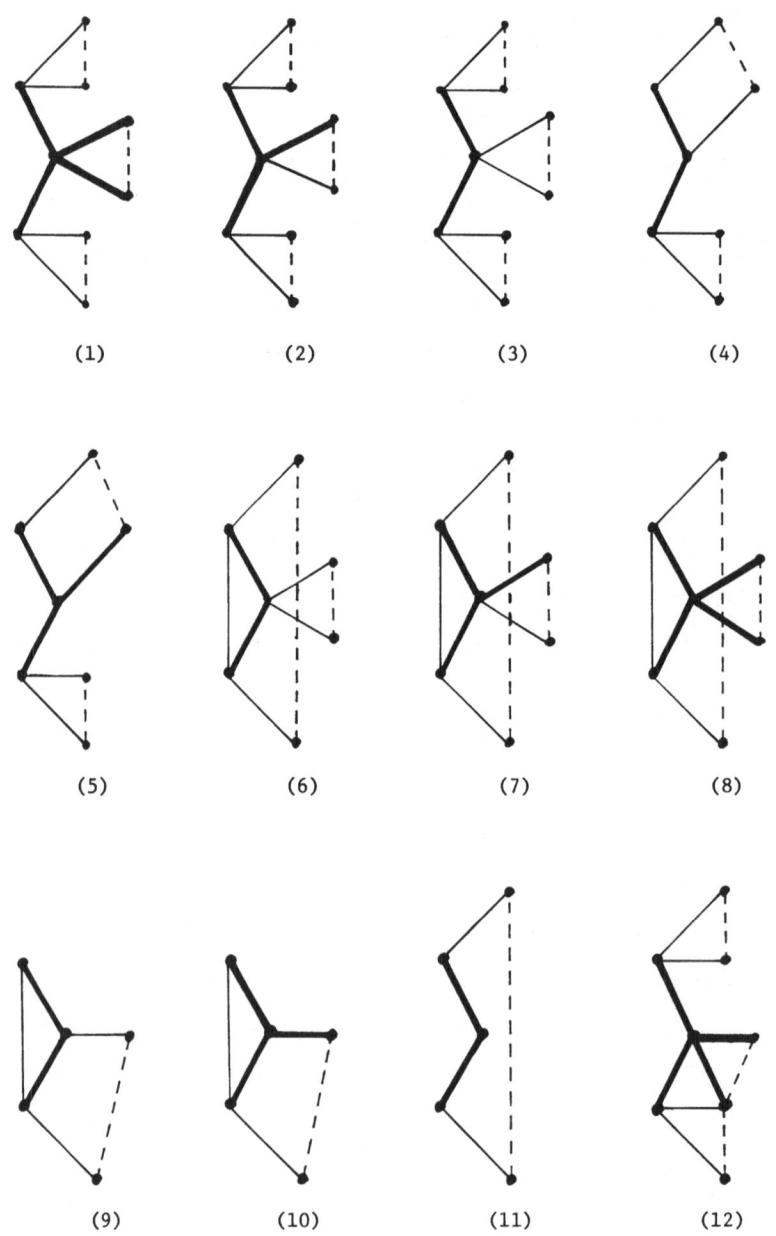

$$(1) \qquad (2) \qquad (3) \qquad (4)$$

$$(5) \qquad (6) \qquad (7) \qquad (8)$$

$$(9) \qquad (10) \qquad (11) \qquad (12)$$

Figure 1

We define:

 Operation A. Remove from T any endpoint u of T which is adjacent to a point v of degree 2 in T.

Lemma 4. Let T be a tree and let T* denote the tree which results from T by repeatedly applying Operation A above until it can no longer be applied. Then T* either equals K_2 or contains two endpoints u and v which are both adjacent to a point w of degree $\geq$ 3 in T*.

Proof. Clearly if T consists only of a single path, i.e., has no points of degree $\geq$ 3, then T* = K_2. On the other hand, if T contains at least one point of degree $\geq$ 3, then T* will contain at least one point of degree $\geq$ 3. In this case, let u_1 be any endpoint of T*. By definition of T*, u_1 must be adjacent to a point u_2 of degree $\geq$ 3 in T*. If u_2 is adjacent to another endpoint then we are done. If u_2 is not adjacent to a second endpoint, let $u_3 \neq u_1$ be a point adjacent to u_2.

If u_3 has degree 2, let u_4 be the other point adjacent to u_3. By definition, u_4 cannot be of degree 1. If u_4 has degree 2 then we continue on to the other point u_5 which is adjacent to u_4. Continuing in this way we ultimately must reach a point say u_i, having degree $\geq$ 3. If this point u_i is adjacent to two endpoints then we are done. Otherwise, we continue on to one of the non-endpoints, say u_j, adjacent to u_i, considering as before the other points adjacent to u_j.

Continuing in this fashion down some path in T we ultimately must reach a point v of degree $\geq$ 3 which is adjacent to two endpoints of T. This follows from the finiteness of T*, the fact that T* has no cycles and the fact that T* has no endpoints which are adjacent to points of degree 2.

We now introduce a second operation to be used in our algorithm:

Operation B. Remove from tree T points u,v,w,where u and v are endpoints adjacent to w,and w has degree $\geq$ 3 in T.

Our algorithm for determining hc(T) for any tree T can be given as follows.

Algorithm A.

Step 0. Set hc(T) = 0. Set G = T.

Step 1. Apply Operation A to G, repeatedly, until it can no longer be applied.

Step 2. Apply Operation B to G, increasing hc(T) by one.

Step 3. Remove from G all isolated points and all isolated K_2's, increasing hc(T) by one for every such subgraph.

Step 4. Repeat Steps 1, 2, and 3 until they can no longer be applied and stop.

By Lemma 2, Step 1 does not alter the value of hc(T). Lemma 3 shows that the subgraph removed from G in Step 2 is an island in a completion solution for T and we count it as such. The removal of such an island may force the isolation of single points and these must also be counted as islands in the computation of hc(T). Lemma 4 is necessary to justify the statement that Steps 1, 2, and 3 can be repeated until T is totally decomposed into islands.

The above algorithm can be modified very easily to keep track of the islands produced in a Hamilton threading of T. We illustrate this in Figure 2.

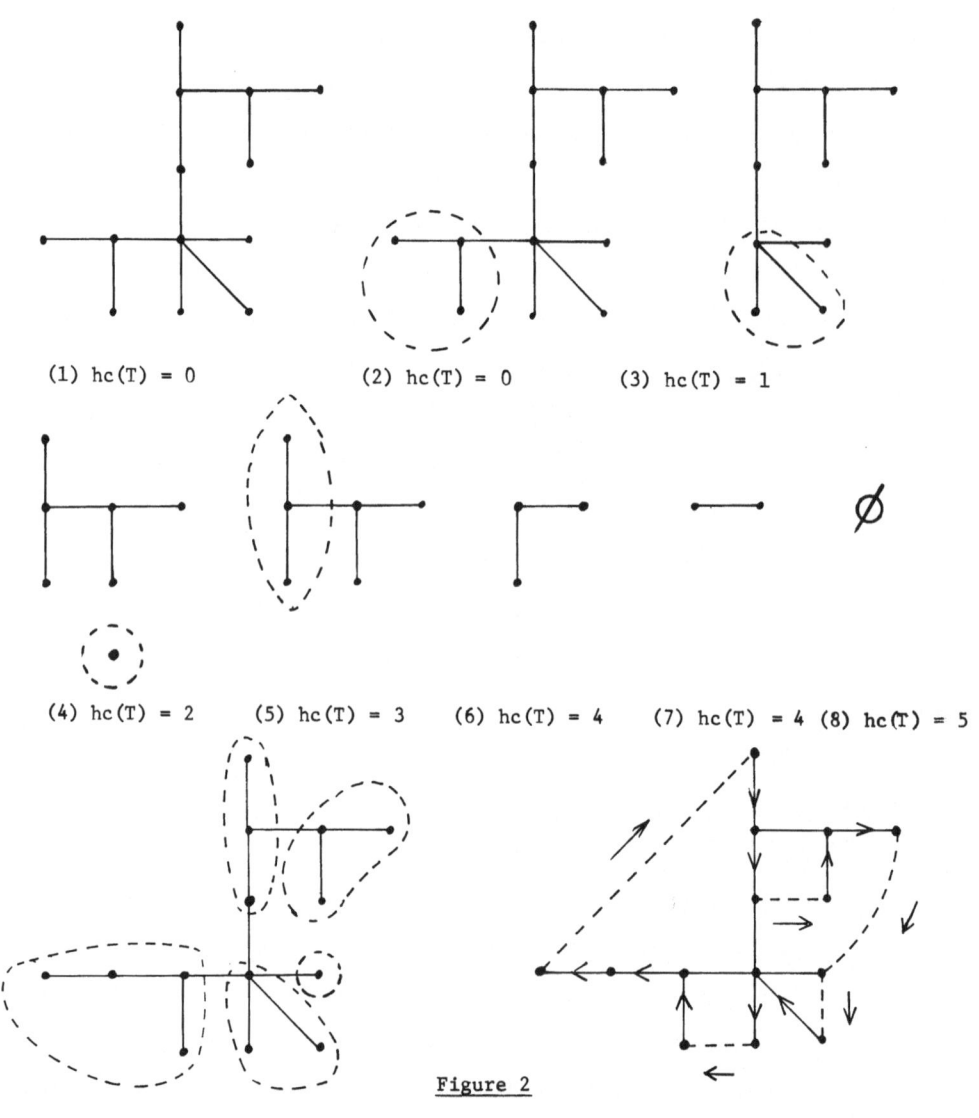

(1) hc(T) = 0   (2) hc(T) = 0   (3) hc(T) = 1

(4) hc(T) = 2   (5) hc(T) = 3   (6) hc(T) = 4   (7) hc(T) = 4   (8) hc(T) = 5

Figure 2

The last two figures above show the five islands of T produced by Algorithm A with a corresponding Hamilton threading of T.

To demonstrate that Algorithm A is bounded by $O(p^2)$ operations we proceed as follows. The repeated application of Operation A in Step 1 is $O(p)$. This involves checking at most p points for being endpoints and applying Operation A at most $O(p)$ times. Step 2 is also $O(p)$. It essentially involves checking every current endpoint, dropping its label into an urn labelled with its adjacent point and then checking the urns to see if any have 2 or more labels in it. Step 3 is clearly bounded by $O(p)$. Thus the number of operations in one iteration of Steps 1-3 can be expressed as a linear function of p. Since each such iteration reduces G by at least one point we never have to iterate more than $O(p)$ times. Therefore Algorithm A is bounded by $O(p^2)$ operations. We suspect a better bound is possible.

We conclude this section by presenting an easily computable and fairly tight upper bound for hc(T) for any tree T. The bound also holds for arbitrary graphs but is not as tight. A simple proof using induction can be given but is omitted.

<u>Proposition 5</u>. For any graph G having p points, q lines and e endpoints,

$$hc(G) \leq \sum_{d_i > 2} (d_i - 2) + 1 = 2q - 2p + e + 1 \, ,$$

where d_i is the degree of point v_i in G.

<u>Corollary 5a</u>. For any tree T, $[\frac{e}{2}] \leq hc(T) \leq e - 1$, where $[\frac{e}{2}]$ denotes the least integer greater than or equal to e/2.

3. THE HAMILTONIAN COMPLETION NUMBER OF ARBITRARY GRAPHS

Since the Hamiltonian completion problem is equivalent to a symmetric traveling salesman problem, it is quite possible that there is no polynomial time algorithm for finding hc(G) for an arbitrary graph G. However, some combinatorial search algorithms are much better than others, and to this end we present some theory which can greatly reduce the combinatorics involved in computing hc(G).

<u>Theorem 6</u>. For any connected graph G, let S denote the set of all spanning trees of G. Then either G is Hamiltonian or $hc(G) = \min_{T_i \in S} hc(T_i)$.

<u>Proof</u>. Clearly $hc(G) \leq \min_{T_i \in S} hc(T_i)$.

In order to establish the reverse inequality we only need to show that if G is
Hamilton threaded with hc(G) threads, then there exists some spanning tree T_j of G
for which the same hc(G) threads produce a Hamiltonian cycle through the points of
T_j. But this is easy, for let W be a Hamiltonian cycle through the Hamilton threaded
graph G. Color every line of G which is on this cycle, say blue, and color every
thread red. Next, remove from G any uncolored line, say e, whose removal does not
disconnect the graph G, i.e., $G_1 = G - e$ is a connected graph. Continue in this
fashion, i.e., remove another uncolored line e_2 such that $G_1 - e_2 = G_2$ is a connected
graph. By repeating this process until no more lines can be removed, without producing a disconnected graph, we will obtain a spanning tree T_j of G which contains
every blue line of W; hence $hc(T_j) \le hc(G)$, completing the proof.

It is possible to implement Theorem 6 using a search algorithm which systematically enumerates all the spanning trees of G.

Some further information relating spanning trees and completion numbers can be
obtained by introducing the tree graph of a graph. Let the points of the tree graph
of G, T(G), represent the spanning trees of G. Two points in T(G) are adjacent if
and only if the corresponding spanning trees can be transformed into each other by
removing exactly one line and replacing it with another.

Theorem 7. Let T' and T" be two spanning trees which are adjacent in T(G).
Then $|hc(T') - hc(T")| \le 1$.

Proof. If T' and T" are adjacent in T(G) then T" can be obtained by removing
one line, say e', from T' and adding one new line, say e", to T'-e'. Let $hc(T') = k$
and let P_1, P_2, ..., P_k, be the islands of T' produced by Algorithm A. Furthermore,
assume that $k = hc(T') < hc(T")$. Consider what happens if we remove a line e' from
T' and add a new line e", thereby obtaining the tree T".

Case 1. The line e' is not on any island $P_1, P_2, ..., P_k$, i.e. e' is a bridge between two islands. It follows in this case that by removing e' and adding a new line
e" that $hc(T") \le hc(T')$, which contradicts our hypothesis that $hc(T') < hc(T")$.

Case 2. The line e' is on one of the islands $P_1, P_2, ..., P_k$, say P_1. Then by
removing e' we will produce k+1 islands, $P_{11}, P_{12}, ..., P_k$. Adding a new line e" to
T'-e' cannot increase further the number of islands in T", Thus $hc(T") \le hc(T') + 1$.

<u>Corollary 7a.</u> Let G be a connected unicyclic graph, let $e_1, e_2, \ldots, e_k$ be the lines on the only cycle in G, and let $T_i = G - e_i$, $i = 1, \ldots, k$, be the spanning trees of G. Then for $1 \leq i, j \leq k$,

$$\left| hc(T_i) - hc(T_j) \right| \leq 1.$$

<u>Proof.</u> This follows since in this case the tree graph of G is the complete graph on k points, $T(G) = K_k$, that is every pair of spanning trees T_i and T_j are adjacent in $T(G)$.

If a graph G is unicyclic, Corollary 7a tells us that if we find any two spanning trees T_i and T_j such that $hc(T_i) < hc(T_j)$, then in fact $hc(G) = hc(T_i)$. We suspect that there is a way of finding $hc(G)$ for unicyclic graphs G by just inspecting the lines on the cycle.

We conclude with an efficient heuristic algorithm for computing $hc(G)$ for an arbitrary graph G. By a current labeling of a line uv in a graph H we mean the assignment of weight $d(uv) = d(u) + d(v) - 2$ to uv, where $d(u)$ and $d(v)$ are the degrees of points u and v in H.   The algorithm finds a "good" (but not necessarily optimum) spanning tree of G.

<u>Algorithm B.</u>

Step 0. Let $H = G$.

Step 1. Currently label all the lines of H.

Step 2. Remove the line with the largest current label which does not disconnect H (in the case of ties, choose arbitrarily).

Step 3. Repeat Steps 1 and 2 until H is a tree. Then compute $hc(H)$ as an approximation to $hc(G)$.

Figure 3 illustrates Algorithm B; Figure 4 gives an example where Algorithm B does not necessarily find $hc(G)$.

The authors have made considerable, but unsuccessful, efforts to ascertain some of the properties of optimum spanning trees. Nothing seems to be true <u>all</u> the time. The situation is complicated by the fact that it is very hard to take even a modestly sized multi-cycled block, work out $hc(G)$ in some ad hoc fashion, and have confidence in the result. We can only recommend Algorithm B as a good heuristic because 1) it is not easy to construct examples where it does not find hc(G), and 2) such examples seem to require a fairly delicate positioning of their pieces.

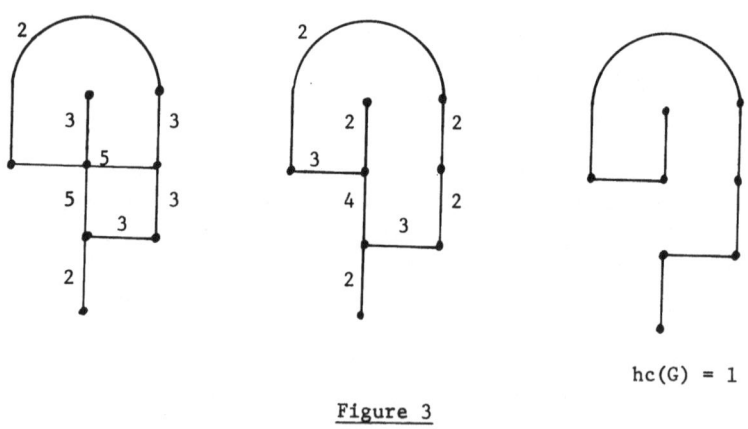

hc(G) = 1

Figure 3

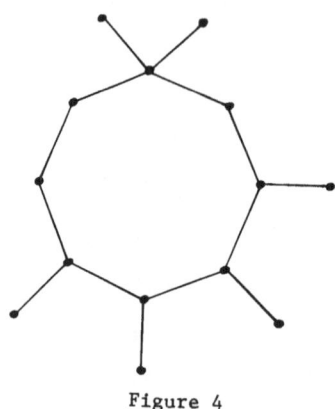

Figure 4

We have developed a fairly extensive theory which can be used to significantly reduce the combinatorics of finding hc(G) for several important classes of graphs. These results will be presented in a forthcoming paper.

REFERENCES

1. Karp, R. M., "Reducibility Among Combinatorial Problems," Complexity of Computer Computations (R. Miller and J. Thatcher, Eds.)

2. Harary, F., and Schwenk, A., "Evolution of the Path Number of a Graph, Covering and Packing in Graphs, II," Graph Theory and Computing (R. C. Read, ed.) Academic Press, New York, 1972, 39-45.

3. Boesch, F. T., Chen, S., and McHugh, N.A.M., "On Covering the Points of a Graph with Point Disjoint Paths," this volume p. 201.

COLORING SEVEN-CIRCUITS*

Dick Wick Hall
State University of New York at Binghamton

ABSTRACT

Recent work of Beraha, Hall, and Tutte has made it clear that it is interesting to express the constrained chromials of planar graphs containing an open n-
circuit in terms of free chromials. This problem was solved for n = 4 and n = 5,
and a partial solution given for n = 6 by Birkhoff and Lewis. The solution for
the case n = 6 was completed by Hall and Lewis. There are 162 constrained
chromials for the seven circuit. One of these has all 7 colors different, the
remaining 161 divide themselves into 23 groups of seven each. The sum of the
constrained chromials in each of these 23 groups is called a constrained rochromial.
The sum of the corresponding 7 free chromials in each group is called a free
rochromial. Equations are obtained expressing each of the constrained rochromials
in terms of free rochromials. A new Beraha number appears exactly where it should.

_____
*This paper was partially supported by grants from the SUNY Research Foundation

COLORING SEVEN-CIRCUITS

The chromial of a graph G is written P[G,n] and gives the number of ways in which the vertices of G may be colored using some or all of n given colors, with the restriction that no two adjacent vertices receive the same color. If the graph G contains a loop, then P[G,n] is identically zero. It is well known that P[G,n] is a polynomial in n of degree k, where k is the number of vertices of the graph G. Since P[G,n] is a polynomial, it has proven useful to consider properties of P[G,n] where n is not restricted to being an integer. Much work has been done along these lines by Tutte [6,7]. All graphs considered are assumed to be planar and connected.

Perhaps the largest non-trivial chromial thus far published is that for the dual of the truncated icosahedron [3]. This is a polynomial of degree 32 which has 8 real zeros. The four real zeros which are not integers are approximately

$$n_1 = 2.61803399$$
$$n_2 = 3.2469919$$
$$n_3 = 3.41539930$$
$$n_4 = 3.52004593$$

S. Beraha has pointed out that the sequence B_n defined by

$$B_n = 2 + 2\cos(2\pi/n)$$

appears to be important in the study of chromials, and this sequence has come to be known as the sequence of Beraha numbers. It should be noted that B_5, B_7, B_8, B_9 are quite close to n_1, n_2, n_3, n_4, respectively. In general, Beraha notes that there is an apparent tendency for the chromials of triangulations to have zeros close to Beraha numbers. Early in 1973 Tutte proved that all the Beraha numbers are important in the theory of chromials [8].

In an investigation of the zeros of chromials of planar triangulations, Tutte found that each of these chromials had a zero near

$$\frac{3 + \sqrt{5}}{2} = \tau + 1$$

$$= 2.6180339887,$$

where τ is the golden ratio. Accordingly, $\tau + 1$ became known as the golden root. Almost immediately Tutte pointed out the spectacular agreement between the golden root and n_1. He was later to prove that the golden root is never a zero of the chromial of a loopless planar graph.

In a monumentally important paper [1] published in 1946, Birkhoff and Lewis presented an exhaustive study of chromials. The portion of that paper which we are presently interested in is their introduction of the concept of a constrained chromial which we now explore.

To introduce this concept consider the four labeled and colored graphs shown in Figure 1

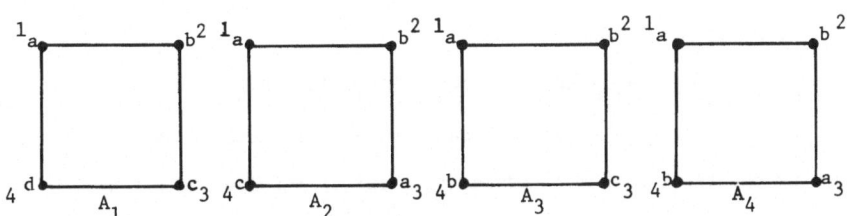

Figure 1.

and the four labeled but uncolored graphs shown in Figure 2.

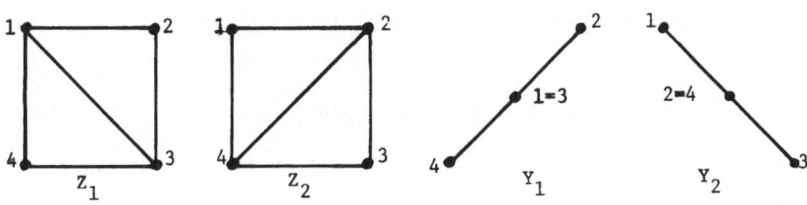

Figure 2.

The statement that these graphs are labeled means that each vertex has been assigned a numeral. The statement that the graphs are colored means that we have placed constraints on the manner in which the four vertices are colored. (We shall see quite soon that the formulas we derive work for any graphs containing the ones illustrated above.) Clearly, Z_1 and Z_2 are isomorphic, but graphs containing Z_1 and Z_2 need not be isomorphic.

The letters associated with the vertices in Figure 1 mean precisely the following: in A_1 the four colors assigned to the vertices are all different; in A_2 and A_4, vertices 1 and 3 receive the same color; however, in A_2 vertices 2 and 4 receive different colors, while in A_4 they receive the same color. No other interpretation should be given to the letters in these drawings.

As is customary, we use capital letters to denote either a graph, or its chromial, or its constrained chromial. We see, for example, that the number of ways of coloring A_1 (since it requires 4 different colors) is $n(n-1)(n-2)(n-3)$. Accordingly,

$$(1) \qquad A_1 = n(n-1)(n-2)(n-3)$$

$$(2) \qquad A_2 = A_3 = n(n-1)(n-2)$$

$$(3) \qquad A_4 = n(n-1)$$

It follows from (1), (2), and (3) that

$$(4) \qquad A_1 + (n-2)(n-3)A_4 = (n-3)(A_2 + A_3) \ .$$

This last equation is true in general and may be easily proven using induction on N where N is the total number of edges and vertices outside the circuit.

Inasmuch as this type of induction is fundamental in proving many theorems on chromials, we briefly describe how it is accomplished. To this end let G be any graph containing an edge A. Denote by G'_A the graph obtained by deleting the edge A, and by G''_A the graph obtained by contracting the edge A to a single point. It is a simple exercise to establish the _fundamental_ _reduction_ _formula_ for chromials, namely,

$$(5) \qquad P[G,n] = P[G'_A,n] - P[G''_A,n]$$

This formula enables us to make the inductive step in many of our proofs. When dealing with theorems involving circuits it is essential that we do not destroy the circuit by either deleting or contracting the edge A. Thus we must choose an edge A which has at least one of its endpoints not on the circuit. With this restriction it is clear that (5) holds also for constrained chromials where no constraints are placed on any vertex not in the circuit under consideraton.. The upshot of this is that in checking the "particular cases" to start our induction we must verify that our theorem is true for all planar connected graphs having no vertices outside the circuit. Clearly, although our graphs may contain loops, no generality is lost in removing all but one of the edges joining any pair of vertices. This follows since the only purpose (chromatically speaking) of an edge in a graph is to insure that the end points of this edge are colored differently.

Returning now to the graphs shown in Figures 1 and 2, and using, as always, a capital letter to denote either the graph or its chromial (or constrained chromial), we see at once from the definitions of these graphs that

(6)
$$Z_1 = A_1 + A_3$$

(7)
$$Z_2 = A_1 + A_2$$

(8)
$$Y_1 = A_2 + A_4$$

(9)
$$Y_2 = A_3 + A_4$$

The fundamental problem with which we are concerned is to solve the five equations (4), (6), (7), (8), (9), which are linearly dependent, for the constrained chromials A_1, A_2, A_3, A_4 in terms of the free chromials Z_1, Z_2, Y_1, Y_2, and to see what can be done about this problem for higher circuits. This is the Birkhoff-Lewis approach. In [1] they obtain

(10)
$$f_1 A_4 = -Z_1 + (n-2)Y_2 + (n-3)Y_1$$

and similar equations for A_1, A_2, and A_3, where, in each case

(11)
$$f_1 = n^2 - 3n + 1$$

is the coefficient of the left hand member. Solving a more complicated system of equations with polynomial coefficients, they solve the same problem for the 5-circuit, and, again, f_1 is the coefficient of the left member of every equation they obtain.

The striking fact to be noted here is that the larger zero of f_1 is precisely $B_5 = \tau + 1$.

In their solution of this problem for the 6-circuit [2], Hall and Lewis find that the coefficient of the left member of each equation is either $6f_1f_2$ or $6(n-2)f_1f_2$, where

$$f_2 = n^3 - 5n^2 + 6n - 1.$$

The largest zero of this cubic is called the silver root and has the value

$$B_7 = 3.24698...,$$

which is quite near n_2.

We are now faced with solving a similar problem for the 7-circuit, and of devising methods which will work for even larger circuits. My student, Robert Rector, in his Maryland dissertation [5], tabulated 1505 linear relations between constrained chromials for the 7-circuit. He then selected from these relations a maximal subset of 126 which he showed were linearly independent. He also showed that the entire collection of 1505 relations could be generated from this collection of 126. Finally, he obtained a set of 36 equations in 36 specific constrained chromials and 36 free chromials of graphs associated with the 7-circuit. The solution of this system of equations (which has not yet been obtained) would then complete the solution of the problem for the 7-circuit.

Clearly, if there is to be any hope of solving this problem for larger circuits, it is time for a new approach!

We work with a planar connected graph G containing a labeled open 7-circuit, and containing exactly 7 vertices. The statement that <u>the 7-circuit is open</u> means that no edge or vertex of G lies inside the 7-circuit. Each edge which lies outside the 7-circuit must have both of its end points on the 7-circuit. We choose as labels for our seven vertices the integers 1,2,3,4,5,6,7, and, in describing our graphs we work modulo 7. We name the graph by giving the endpoints of the edges which lie outside the graph. Figure 3 shows the graphs (00) [i.e., no outside edges], (1315), and (131447). Graphs of this type will be called <u>elementary graphs</u>.

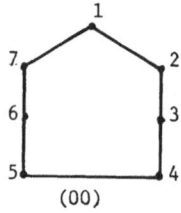
(00)

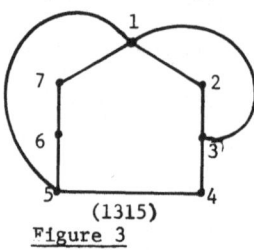
(1315)

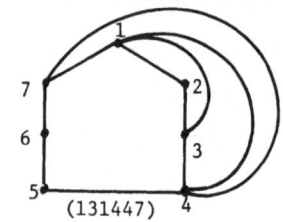
(131447)

Figure 3

There are 162 constrained colorings for a 7-circuit. One of these has all seven vertices colored differently (and thus corresponds to an unconstrained coloring of a graph containing a complete graph on 7 vertices), and the other 161 divide themselves into 23 groups of 7 colorings each. The 7 occurs since it makes a difference which vertex of our graph we use as the initial vertex of our coloring scheme. All of these coloring schemes are shown in Table 1, which is taken from Birkhoff and Lewis.

Coloring Schemes for the Seven Circuit

	v_i	v_{i+1}	v_{i+2}	v_{i+3}	v_{i+4}	v_{i+5}	v_{i+6}	No. of Colors
A_i	a	b	c	b	c	b	c	3
B_i	a	c	a	b	c	a	b	3
C_i	a	b	a	b	c	a	c	3
D_i	a	b	a	c	d	a	b	4
E_i	a	b	a	b	c	a	d	4
F_i	a	d	a	c	b	a	b	4
G_i	a	b	a	d	b	a	c	4
H_i	a	c	a	b	d	a	b	4
I_i	a	b	c	b	d	c	d	4
J_i	a	b	c	d	b	c	d	4
K_i	a	b	c	d	c	b	d	4
L_i	a	b	c	d	c	d	b	4
M_i	a	b	c	d	b	d	c	4
N_i	a	b	a	c	d	a	e	5
O_i	c	a	b	d	e	b	a	5
P_i	d	e	a	b	a	b	c	5
Q_i	c	a	d	a	b	e	b	5
R_i	b	a	c	b	d	e	a	5
S_i	b	a	c	d	b	e	a	5
T_i	c	a	b	d	e	a	b	5
U_i	c	a	d	b	a	e	b	5
V_i	b	a	c	d	e	f	a	6
W_i	c	b	a	e	f	a	d	6
X_i	a	b	c	d	e	f	g	7

Table 1

Let us examine more closely the constrained colorings N_i , $1 \leq i \leq 7$. We make
the following table where the vertex is indicated by its appropriate label. This is
the cyclic coloring table determined by the constrained coloring abacdae.

	1	2	3	4	5	6	7	(00)	(1625)
N_1	a	b	a	c	d	a	e	1	0
N_2	e	a	b	a	c	d	a	1	1
N_3	a	e	a	b	a	c	d	1	1
N_4	d	a	e	a	b	a	c	1	1
N_5	c	d	a	e	a	b	a	1	1
N_6	a	c	d	a	e	a	b	1	0
N_7	b	a	c	d	a	e	a	1	0
							t	7	4

Table 2

For the colorings indicated in Table 2, where we are using 5 colors, the
constrained chromial of the elementary graph with which we are working must be
either $n(n-1)(n-2)(n-3)(n-4)$ or identically zero. It will be identically
zero if and only if two vertices receiving the same color are joined by an outside
edge. For example, for the map (1625) , N_1 , N_6 and N_7 are identically zero,
while $N_2 = N_3 = N_4 = N_5 = n(n-1)(n-2)(n-3)(n-4)$. It is convenient to make a
column for each graph we are considering (as we have done for the maps (00) and
(1625) in Table 2) and to place a 0 or 1 in this column according as the corresponding constrained chromial is or is not identically zero. The sum of the entries
is then called the t value for the given graph. We now define the constrained
rochromial NR for our given graph by

$$NR = \sum_{i=1}^{7} N_i$$

$$= tn(n-1)(n-2)(n-3)(n-4) \quad ,$$

where t is determined by the given graph. In precisely the same manner we define
the constrained rochromials AR, BR, ..., XR, corresponding to the color schemes
in Table 1. We note that

$$XR = 7n(n-1) \ (n-2) \ (n-3) \ (n-4) \ (n-5) \ (n-6)$$

for every map, since no two vertices of the given open 7-circuit are given the same
color for the color scheme X_i .

We next define our free rochromials. To begin this process we start with an
elementary graph G containing our labeled 7-circuit, and with an integer i satisfying
$1 \leq i \leq 7$. We define graphs Z_i , Y_i , X_i , W_i , U_i , V_i in the following manner.

The graph Z_i is obtained from G by joining vertex i to each vertex of the
open 7-circuit with which it is not adjacent by a new edge inside the 7-circuit.

The graph Y_i is obtained from G by first joining vertices i-1 and i+1 (mod 7)
by an edge inside the 7-circuit and then contracting this edge to a single point p.
We then join p by a new edge to each vertex of our new graph with which it is not
adjacent.

The graph X_i is obtained by joining vertices i and i+3 (mod 7) by an edge, contracting this edge to a point p, and then joining p by a new edge to each vertex of
our new graph with which it is not adjacent.

The graph W_i is obtained by joining the vertex i+2 to each of the vertices i
and i+4 (mod 7) and then contracting each of these edges to a single point.

The graph V_i is obtained by joining vertices i-1 and i+1 by an edge, and joining vertices i-2 and i+2 by an edge, and then contracting each of these edges to a
single point.

The graph U_i is obtained by joining vertices i+1 and i+3 by an edge, and joining vertices i-1 and i-3 by an edge, and then contracting each of these edges to a
single point. Examples of these graphs are shown in Figure 4.

Because of the context there will be no difficulty in distinguishing the graphs
(or the chromials) U_i , V_i , W_i , X_i from the color schemes (or constrained chromials)
denoted by the same letters.

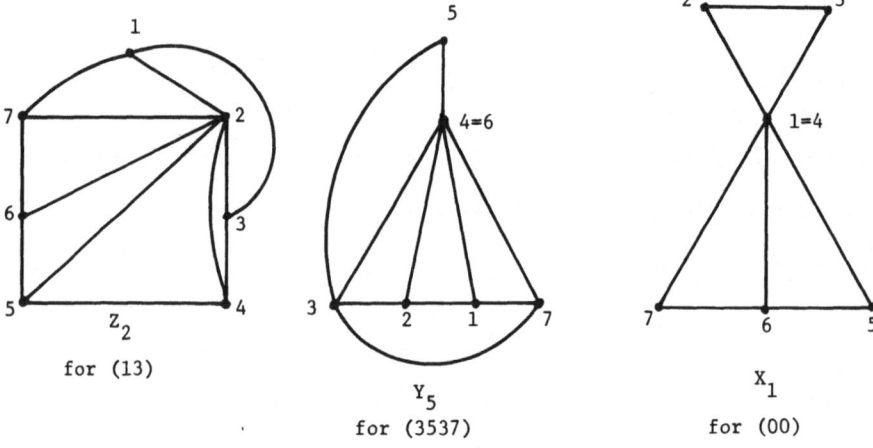

Z₂

for (13)

Y₅

for (3537)

X₁

for (00)

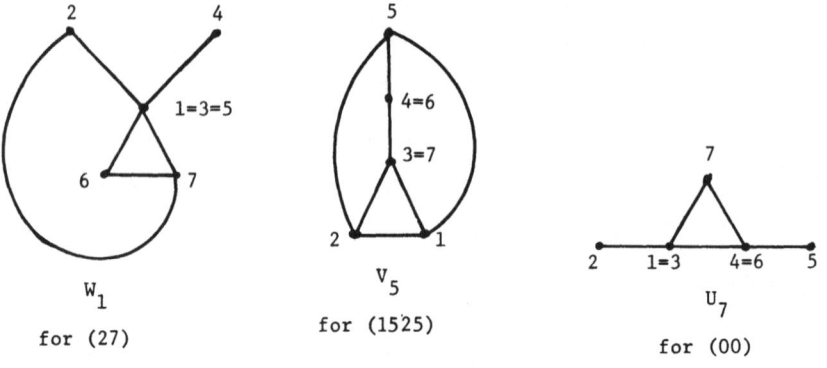

W₁

for (27)

V₅

for (1525)

U₇

for (00)

Figure 4

We now define the rochromials we shall use (for each graph) by

$$ZR = \sum_{i=1}^{7} Z_i, \qquad YR = \sum_{i=1}^{7} Y_i, \qquad XR = \sum_{i=1}^{7} X_i$$

$$WR = \sum_{i=1}^{7} W_i, \qquad VR = \sum_{i=1}^{7} V_i, \qquad UR = \sum_{i=1}^{7} U_i.$$

As in [2] we find it convenient to express our results in terms of $x = n - 2$, n being the number of colors from which our selection is to be made. Our general problem is the following. We wish to find a polynomial

$$K(x) = k_0 + k_1 x + k_2 x^2 + \dots$$

such that given any constrained coloring of our 7-circuit (such as abacdae) there are polynomials

$$A(x) = A_0 + A_1 x + A_2 x^2 + \dots$$
$$B(x) = B_0 + B_1 x + B_2 x^2 + \dots$$
$$C(x) = C_0 + C_1 x + C_2 x^2 + \dots$$
$$D(x) = D_0 + D_1 x + D_2 x^2 + \dots$$
$$E(x) = E_0 + E_1 x + E_2 x^2 + \dots$$
$$G(x) = G_0 + G_1 x + G_2 x^2 + \dots$$

such that given any elementary graph δ, then

(12) $K(x)\delta(x) = A(x)ZR(x) + B(x)YR(x) + C(x)XR(x) + D(x)WR(x) + E(x)VR(x)$

$$ + G(x)UR(x)$$

where $\delta(x)$ is the constrained rochromial determined by the cyclic coloring table and the elementary graph under consideration, and $ZR(x), YR(x), \dots, UR(x)$ are the free rochromials determined by the elementary graph we have chosen.

It is convenient at this point to make several simplifications. Inasmuch as $\delta(x)$ is a constrained rochromial for a specific graph, the t for this graph must be a factor of $\delta(x)$. We write

(13) $$K(x)\delta(x) = t(h_0 + h_1 x + h_2 x^2 + \dots),$$

and assume throughout our discussion that $h_0 \neq 0$. For $n \geq 0$, let

$$A_n = a_0 h_n + a_1 h_{n-1} + \dots + a_n h_0$$
$$B_n = b_0 h_n + b_1 h_{n-1} + \dots + b_n h_0$$
$$\vdots$$

(14) $$G_n = g_0 h_n + g_1 h_{n-1} + \dots + g_n h_0.$$

Also, for the particular elementary graph under consideration, let

$$ZR = z_0 + z_1x + z_2x^2 + \ldots$$

$$YR = y_0 + y_1x + y_2x^2 + \ldots$$

$$\vdots$$

$$UR = u_0 + u_1x + u_2x^2 + \ldots \, ,$$

$$AZ_n = a_0z_n + a_1z_{n-1} + \ldots + a_nz_0$$

$$BY_n = b_0y_n + b_1y_{n-1} + \ldots + b_ny_0$$

$$\vdots$$

$$GU_n = g_0u_n + g_1u_{n-1} + \ldots + g_nu_0$$

$$R_n = AZ_n + BY_n + \ldots + GU_n$$

Substituting (13) in (12), making use of our assumption that $h_0 \neq 0$, and comparing coefficients, gives us the following essential lemma.

<u>Lemma:</u>   (a) $R_0 = t$

(b)   For every positive integer n, $R_n = 0$.

In her Master's thesis [4] my student Frances Hills has made a catalog of the chromials, constrained chromials, rochromials, and constrained rochromials for all elementary graphs for the 4, 5, 6, and 7 circuit. Using this catalog we can find, for any elementary map for the 7-circuit, and any constrained coloring of this circuit, the value of t and the sequences $\langle z_n \rangle$, $\langle y_n \rangle$, $\langle x_n \rangle$, $\langle w_n \rangle$, $\langle v_n \rangle$, and $\langle u_n \rangle$. Using the equations given by the Lemma, we then can set up recursion formulas for our unknown sequences $\langle a_n \rangle$, $\langle b_n \rangle$, $\langle c_n \rangle$, $\langle d_n \rangle$, $\langle e_n \rangle$, and $\langle g_n \rangle$. Six different elementary graphs provide us with sufficient information to solve our problem. The graphs which were actually used were (00), (1314), (1316), (131636), (13163646), (1347). The recursion equations were used in the form

$$a_{n+4} = 5a_{n+3} - 9a_{n+2} + 10a_{n+1} - 5a_n - 2b_{n+3} - b_{n+2} - 5b_{n+1}$$
$$+ 5c_{n+2} - 3c_{n+1} \qquad - d_{n+3} - 2d_{n+2}$$
$$+ 2e_{n+3} - e_{n+2} + e_{n+1} \qquad - 4g_{n+3} - 7g_{n+2} - g_{n+1}$$

$$b_{n+4} = -2a_{n+3} + 5a_{n+2} + 0a_{n+1} - 3a_n + 2b_{n+3} - 7b_{n+2} - 4b_{n+1}$$
$$+ 3c_{n+3} - c_{n+2} - 2c_{n+1} \qquad - 4d_{n+3} - d_{n+2}$$
$$- e_{n+2} + e_{n+1} \qquad + 7g_{n+3} - 7g_{n+2} - g_{n+1}$$

$$c_{n+4} = 36a_{n+3} - 83a_{n+2} + 60a_{n+1} - 24a_n - 21b_{n+3} + 12b_{n+2} - 29b_{n+1}$$
$$- 7c_{n+3} + 50c_{n+2} - 7c_{n+1} \qquad - 4d_{n+3} - 23d_{n+2}$$
$$- e_{n+3} + 2e_{n+2} + 5e_{n+1} \qquad - 27g_{n+3} - 36g_{n+2} - 5g_{n+1}$$

$$d_{n+4} = - 44a_{n+3} + 103a_{n+2} - 60a_{n+1} - 8a_n + 25b_{n+3} - 36b_{n+2} - 7b_{n+1}$$
$$+ 27c_{n+3} - 66c_{n+2} - 21c_{n+1} \qquad - 36d_{n+3} + 13d_{n+2}$$
$$- 59e_{n+3} - 18e_{n+2} - e_{n+1} \qquad - 33g_{n+3} - 4g_{n+2} + g_{n+1}$$

$$e_{n+4} = 12a_{n+3} - 31a_{n+2} + 20a_{n+1} + 0a_n - b_{n+3} + 4b_{n+2} - b_{n+1}$$
$$- 11c_{n+3} + 18c_{n+2} + 5c_{n+1} \qquad + 4d_{n+3} + 5d_{n+2}$$
$$+ 19e_{n+3} + 2e_{n+2} + e_{n+1} \qquad + 9g_{n+3} - 4g_{n+2} - g_{n+1}$$

$$g_{n+4} = 4a_{n+3} - 9a_{n+2} + 5a_{n+1} + 0a_n \qquad - 3b_{n+3} + 3b_{n+2} + 0b_{n+1}$$
$$- 2c_{n+3} + 5c_{n+2} + c_{n+1} \qquad + 2d_{n+3}$$
$$+ 3e_{n+3} + e_{n+2} \qquad + g_{n+3}$$

With the sequences $\langle a_n \rangle, \ldots, \langle g_n \rangle$ at our disposal, (14) enables us to find the sequences $\langle A_n \rangle, \ldots, \langle G_n \rangle$ if we know the sequence $\langle h_n \rangle$. Recalling that

$$f_1 = n^2 - 3n + 1$$
$$= x^2 + x - 1$$

$$f_2 = n^3 - 5n^2 + 6n - 1$$
$$= x^3 + x^2 - 2x - 1$$

I had hoped that the coefficient=of the left member of each of the equations I was seeking would be a constant multiple of $f_1 f_2$. This turned out not to be the case. We need to introduce a new factor

$$f_3 = n^2 - 4n + 2$$
$$= x^2 - 2.$$

The largest zero of f_3 is

$$n = 2 + \sqrt{2}$$

$$= 3.41421.$$

This is the Beraha number B_8 which is quite close to the zero n_3 mentioned at the beginning of this paper.

It is now tedious but not difficult to use a computer to find the constrained rochromials of all elementary maps containing a colored open 7-circuit in terms of the free rochromials of these maps. If we define rochromials and constrained rochromials for an arbitrary planar graph containing a 7-circuit in the obvious way, it follows by the type of induction mentioned near the beginning of this paper that the equations we give here are valid for all planar graphs containing open 7-circuits. My student Brent Arnold has verified on the computer that each of the formulas is valid for all elementary planar graphs. Thus the assumption $h_0 \neq 0$ was not necessary. Hills, in her thesis [4], has used the same method to verify that the equations previously published for the 4, 5, and 6 circuits are correct.

In the equations which follow, the factor f_3 is not present as part of the coefficient of either AR or DR, since each of these equations can be obtained easily from known formulas for coloring 6-circuits.

Three Colors

$$f_1f_2AR = (2x+1)ZR - (4x^2 - 2x - 1)YR - (2x^2 - x - 2)XR$$
$$+ (2x^3 - 8x^2 + x + 1)WR + x(2x^2 - 1)VR + x(x^2 + 2x - 1)UR$$

$$f_1f_2f_3BR = -(4x^2 + 7x + 2)ZR - (2x^4 - 4x^3 - 10x^2 + x + 4)YR$$
$$+ (3x^3 + 0x^2 - 6x+1)XR + (x^5 - 6x^4 + 4x^3 + 25x^2 + 2x - 7)WR$$
$$+ (2x^5 - x^4 - 8x^3 - 2x^2 + x + 1)VR + (2x^4 - 3x^3 - 13x^2 - x+4)UR$$

$$f_1f_2f_3CR = (x^3 - x^2 - 5x - 2)ZR - (4x^4 - 3x^3 - 13x^2 + 5x + 4)YR$$
$$+ (x^3 + 0x^2 - x - 1)XR + (x^5 - 6x^4 + 18x^3 - x - 3)WR$$
$$+ (x^4 + 0x^3 - 6x^2 + 0x + 1)VR + x(3x^4 + 0x^3 - 11x^2 + 0x + 3)UR$$

Four Colors

$$f_1f_2DR = -2(2x + 1)ZR + 2(4x^2 - 2x - 1)YR - (x^3 - 3x^2 + 0x + 3)XR$$
$$+(x^4 - 4x^3 + 13x^2 - x - 1)WR + x(x^3 - 3x^2 - 2x + 1)VR$$
$$-2x(x^2 + 2x - 1)UR$$

$$f_1f_2f_3ER = f_1f_2f_3FR$$
$$= -(3x^3 + 0x^2 - 9x - 4)ZR - (x^5 - 7x^4 + x^3 + 19x^2 - 5x - 4)YR$$
$$+ (2x^4 - 2x^3 - 6x^2 + 3x + 5)XR + (x^6 - 3x^5 + 9x^4 + 3x^3$$
$$- 28x^2 + x + 3)WR$$
$$- (2x^5 + x^4 - 5x^3 - 6x^2 + 2x + 1)VR + x(x^5 - 3x^4 - 6x^3$$
$$+ 11x^2 + 8x - 3)UR$$

$$f_1f_2f_3GR = f_1f_2f_3HR$$
$$= -(2x^3 - 3x^2 - 11x - 4)ZR - (x^5 - 5x^4 + 2x^3 + 16x^2 - x - 4)YR$$
$$+ (2x^4 - 4x^3 - 6x^2 + 8x + 3)XR + (x^6 - 4x^5 + 8x^4 + 4x^3$$
$$- 32x^2 - 6x + 5)WR$$
$$+ (x^6 - 3x^5 - 3x^4 + 10x^3 + 6x^2 - x - 1)VR - (3x^4 - 2x^3 - 14x^2$$
$$- 3x + 2)UR$$

Four Colors (Cont'd)

$$f_1 f_2 f_3 IR = (x^4 + 4x^3 - 2x^2 - 11x - 4)ZR + (7x^4 - 2x^3 - 20x^2 + 3x + 6)YR$$

$$+ (x^5 + x^4 - 6x^3 - 2x^2 + 7x + 1)XR - (2x^5 - 12x^4 + 31x^2$$

$$+ 7x - 5)WR$$

$$- (x^5 + 2x^4 - x^3 - 8x^2 - 2x + 1)VR + (x^6 - 3x^5 - 7x^4 + 9x^3$$

$$+ 11x^2 + x - 2)UR$$

$$f_1 f_2 f_3 JR = - (2x^4 + 4x^3 - 7x^2 - 15x - 4)ZR + (4x^4 + 0x^3 - 15x^2 - 7x + 6)YR$$

$$+ (3x^5 + x^4 - 16x^3 - x^2 + 20x - 3)XR + (11x^4 + x^3 - 37x^2$$

$$- 16x + 11)WR$$

$$- (3x^5 + 2x^4 - 9x^3 - 9x^2 - x + 3)VR - (5x^4 + 2x^3 - 20x^2$$

$$- 11x + 6)UR$$

$$f_1 f_2 f_3 KR = f_1 f_2 f_3 MR$$

$$= - (x^4 + 3x^3 - 5x^2 - 13x - 4)ZR + (5x^4 - 3x^3 - 17x^2 - x + 6)YR$$

$$+ (x^5 + x^4 - 8x^3 - 2x^2 + 12x - 1)XR - (3x^5 - 11x^4 - 6x^3$$

$$+ 35x^2 + 14x - 7)WR$$

$$+ (x^6 - 2x^5 - 4x^4 + 6x^3 + 8x^2 + 3x - 1)VR - (4x^4 + 0x^3 - 17x^2$$

$$- 7x + 4)UR$$

$$f_1 f_2 f_3 LR = (-2x^3 + 2x^2 + 10x + 4)ZR + (-x^5 + 7x^4 - 2x^3 - 23x^2 + 6x + 6)YR$$

$$+ (-2x^3 + 0x^2 + 2x + 2)XR + (-4x^5 + 10x^4 + 6x^3 - 30x^2 - 6x + 2)WR$$

$$+ (x^5 - x^4 - 4x^3 + 9x^2 + 4x)VR + (2x^6 - 4x^5 - 8x^4 + 16x^3 + 8x^2 - 2x)UR$$

Five Colors

$$f_1 f_2 f_3 NR = (x-2)\{(9x^2 + 17x + 6)ZR + (4x^4 - 10x^3 - 21x^2 + 5x + 4)YR$$

$$+ (x^4 - 5x^3 - 6x^2 + 9x + 7)XR + (x^6 - x^5 + 6x^4 - 13x^4$$

$$- 41x^2 - 3x + 3)WR$$

$$- (3x^5 - x^4 - 14x^3 - 9x^2 + 2x + 1)VR - x(2x^4 + x^3 - 14x^2$$

$$- 16x - 1)UR\}$$

$$f_1 f_2 f_3 OR = (x-2)\{2(x^3 + 7x^2 + 10x + 3)ZR + 2(x^4 - 6x^3 - 11x^2 + x + 3)YR$$

$$- (x^4 + 7x^3 + 0x^2 - 13x - 2)XR - (3x^5 - 6x^4 + 9x^3 + 47x^2$$

$$+ 17x - 4)WR$$

$$+ x(x^5 - x^4 - x^3 + 10x^2 + 12x + 5)VR + 2(5x^3 + 11x^2 + 4x - 1)UR\}$$

Five Colors (Cont'd)

$$f \; f_2 f_3 PR = (x-2)\{ \; - \; (x^4 + 3x^3 - 6x^2 - 16x - 6)ZR + (2x^5 + 7x^4 - 13x^3 - 24x^2$$
$$+ \; 10x + 6)YR$$
$$+ \; (x^4 - 3x^3 - 5x^2 + 5x + 6)XR + (12x^4 - 8x^3 - 39x^2 - 3x + 2)WR$$
$$- \; x(2x^4 + x^3 - 7x^2 - 9x - 1)VR - x(4x^4 + 3x^3 - 16x^2 - 13x + 2)UR\}$$

$$f_1 f_2 f_3 QR = (x-2)\{ (x^3 + 11x^2 + 18x + 6)ZR + (4x^4 - 11x^3 - 25x^2 + 6x + 6)YR$$
$$- \; (x^4 + 5x^3 + 0x^2 - 8x - 4)XR - (2x^5 - 7x^4 + 10x^3 + 43x^2$$
$$+ \; 10x - 2)WR$$
$$+ \; x(x^3 + 5x^2 + 12x + 4)VR + x(x^5 - 3x^4 - 3x^3 + 19x^2 + 16x + 2)UR\}$$

$$f_1 f_2 f_3 RR = f_1 f_2 f_3 SR$$
$$= (x-2)\{ \; - \; (x^4 + x^3 - 11x^2 - 19x - 6)ZR + (x^5 + 4x^4 - 11x^3 - 22x^2$$
$$+ \; 3x + 6) \; YR$$
$$+ \; (x^5 + 0x^4 - 9x^3 - 2x^2 + 13x + 3)XR + (9x^4 - 11x^3 - 44x^2$$
$$- \; 11x + 5)WR$$
$$- \; (2x^5 + 0x^4 - 10x^3 - 11x^2 - 2x + 1)VR - (2x^5 + 2x^4 - 12x^3$$
$$- \; 19x^2 - 5x + 2)UR\}$$

$$f_1 f_2 f_3 TR = f_1 f_2 f_3 UR$$
$$= (x-2)\{ \; - \; (x^4 - x^3 - 16x^2 - 22x - 6)ZR + (x^4 - 9x^3 - 20x^2$$
$$- \; 4x + 6)YR$$
$$+ \; x(2x^4 - x^3 - 15x^2 + x + 21)XR + (6x^4 - 14x^3 - 49x^2 - 19x + 8)WR$$
$$- \; (2x^5 - x^4 - 13x^3 - 13x^2 - 3x + 2)VR - (x^4 - 8x^3 - 25x^2$$
$$- \; 12x + 4)UR\}$$

Six Colors

$$f_1 f_2 f_3 VR = (x-2)(x-3)\{ 2(2x^3 + 10x^2 + 13x + 4)ZR + (x^5 + x^4 - 20x^3 - 27x^2$$
$$+ \; 6x + 6)YR$$
$$- \; 2(x^4 + 4x^3 + 0x^2 - 7x - 3)XR + 2(2x^4 - 13x^3 - 28x^2 - 7x + 1)WR$$
$$- \; x(x^4 - 3x^3 - 14x^2 - 15x - 4)VR - 2x(x^4 - x^3 - 11x^2 - 12x - 3)UR\}$$

Six Colors (Cont'd)

$$f_1 f_2 f_3 WR = (x-2)(x-3)\{(6x^3 + 25x^2 + 29x + 8)ZR - (2x^4 + 18x^3 + 25x^2 + x - 6)YR$$

$$+ (x^5 - 3x^4 - 14x^3 + 3x^2 + 22x + 3)XR + x^4 - 29x^3 - 61x^2 - 22x + 5)WR$$

$$- (x^5 - 4x^4 - 17x^3 - 17x^2 - 5x + 1)VR + (x+1)(3x^3 + 15x^2 + 15x - 2)UR\}$$

Seven Colors

$$f_1 f_2 f_3 XR = (x-2)(x-3)(x-4)\{(x^4 + 11x^3 + 34x^2 + 36x + 10)ZR$$

$$- (5x^4 + 27x^3 + 30x^2 - 2x - 6)YR - (5x^4 + 13x^3 - 5x^2 - 23x - 6)XR$$

$$- (4x^4 + 44x^3 + 73x^2 + 25x - 2)WR + 7x(x+1)^3 VR$$

$$+ 7x(x+1)^2 (x+2)UR\}$$

Acknowledgment. I am deeply grateful to W. T. Tutte for the opportunity to learn from him during the year I spent at the University of Waterloo and for the constant encouragement he has given me every since to solve the problem which is the subject of this paper.

REFERENCES

1. Birkhoff, George D. and Lewis, D. C., "Chromatic polynomials", Trans. A.M.S. 60 (1946), 355-451.

2. Hall, D. W., and Lewis, D. C., "Coloring six-rings", Trans. A.M.S. 64 (1948), 184-191.

3. Hall, D. W., Siry, J. W., and Vanderslice, B. R., "The chromatic polynomial of the truncated icosahedron", Proc. A.M.S. 16 (1965), 620-628.

4. Hills, F. D., "Chromials of n-circuits, $4 \leq n \leq 7$," Master's thesis, SUNY at Binghamton, 1973.

5. Rector, R. W., "Fundamental linear relationships for the seven-ring", Doctoral Dissertation, University of Maryland, 1956.

6. Tutte, W. T., "On chromatic polynomials and the golden ratio", J. Combinatorial Theory 9 (1970), 289-296.

7. Tutte, W. T., "More about chromatic polynomials and the golden ratio", Combinatorial Structures and their Applications, (R. K. Guy, ed.) Gordon and Breach, New York, 1970, 439-453.

8. Tutte, W. T., "Chromatic sums for planar triangulations, V: Special equations", Department of Combinatorics and Optimization, University of Waterloo, Research Report CORR 73-2, January 23, 1973.

ABSOLUTE RETRACTS IN GRAPHS

Pavol Hell
University of British Columbia

ABSTRACT

If Y is a subgraph of X, any homomorphism r: X → Y satisfying r(y) = y for all y ε V(Y) is termed a retraction of X onto Y, and Y a retract of X. Let C denote a class of graphs, together with a class of homomorphisms; we investigate the class AR(C) of all absolute retracts of C, i.e., the class of all graphs Y ε C that are retracts of any supergraph X ε C for which the inclusion Y ⊆ X is a homomorphism in C.

We consider various classes C, e.g., the class P of all finite planar graphs and all homomorphisms, the class B of all bipartite graphs and metric homomorphisms, etc. In particular, we show that the Four Color Conjecture is equivalent with the statement AR(P) ≠ ∅, and if the Four Color Conjecture is true we have a complete description of AR(P). We give a "good characterization" of AB(B), and obtain partial results for other interesting classes of graphs.

RETRACTS IN GRAPHS

1. We study retractions of graphs in analogy to retractions of topological spaces. This notion was introduced in topology by Borsuk, [1,2]; he defined a retraction of the topological space X onto a subspace Y as a continuous mapping r: X → Y satisfying r(y) = y for all y ∈ Y. It seems that, in the case of graphs, homomorphisms should assume the role played by continuous mappings in the definition of retraction.

Recall that a homomorphism f (cf. [5,6,14]) of the graph X to the graph Y is a mapping of the vertex-sets, f: V(X) → V(Y), such that f(x) is adjacent to f(x') whenever x is adjacent to x'. We note, that a homomorphism X → K_n is a mapping of the vertices of X to symbols 1, 2, ..., n, such that adjacent vertices of X are mapped upon adjacent, i.e., different symbols. Hence a homomorphism X → K_n is an n-coloring of X and vice versa. In particular, a graph is n-colorable iff it admits a homomorphism to K_n. If f: X → Y is a homomorphism and Y admits a homomorphism c: Y → K_n, then cf is a homomorphism of X to K_n. Consequently, if there is a homomorphism of X to Y, then $X(Y) \geq X(X)$.

There is a natural relationship between homomorphisms and the notion of a subgraph. Let the vertex-set of the graph Y be included in the vertex-set of the graph X; then Y is a subgraph of X if and only if the inclusion mapping i: Y → X (defined by i(y) = y for all y) is a homomorphism. Indeed, i is a homomorphism if and only if (y, y') ∈ E(Y) implies (y, y') ∈ E(X), i.e., E(Y) ⊂ E(X). More generally Y is a subgraph of X, if and only if there is a one-to-one homomorphism of Y into X.

Invertible homomorphisms (homomorphisms f such that f^{-1} exists and is also a homomorphism) coincide with the isomorphisms in the usual sense. We also note that for V(Y) ⊂ V(X), Y is a full (induced) subgraph of X if and only if the inclusion mapping is an _isomorphism_ of Y into X.

Let Y be a subgraph of X. We define a retraction of X onto Y, as a homomorphism r: X → Y satisfying r(y) = y for all vertices y of Y. If a retraction of X onto Y exists, Y is called a retract of X.

An example of a retraction is shown below.

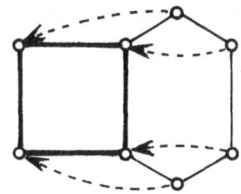

Figure 1

2. The notion of retract is not an artificial one, as we shall illustrate with
the following example: A country which had experiences a spontaneous and chaotic
industrial development at the end of the last century was the subject recently of a
study by industrial management specialists. They found that the chaotic placement
of the factories constituted a major obstacle in increasing productivity, and suggested decreasing the number of factories, augmenting their productions. They proposed to keep all the more recent factories and move the remaining productions into
these. The rearrangement is subject to preserving transportation lines, i.e., two
factories with a direct transportation line shall be moved into factories possessing such a line again. If X is the graph with factories as vertices and direct
transportation lines as edges (it would, of course, have all loops), and Y the
subgraph induced by the remaining factories, what we are looking for is a retraction of X to Y. Similar examples can be encountered, where the graphs are without
loops, or bipartite, planar, etc. However, these are not to serve as a justification for the study of retracts. In a parallel to the situation in topology, the
justification is mostly based on the following necessary and sufficient condition
for extendability of homomorphisms (in particular, of colorings):

Proposition 1. Let X be a subgraph of X', f: X → Y a homomorphism. Let Z be
the graph obtained from the disjoint union of Y and X' by identifying x with f(x)
for all vertices x of X. Then a necessary and sufficient condition for the existence of a homomorphism f': X' → Y such that f is a restriction of f' is that Y is
a retract of Z.

Proof. Let j be the mapping of V(X') into V(Z) defined by j(x) = x for all

$x \epsilon V(X')$. Note that j is not necessarily an inclusion mapping, as two distict vertices x,x' of X' may be identified in Z (if and only if $f(x) = f(x')$). However j
is always a homomorphism $X' \to Z$.

When r: $Z \to Y$ is a retraction, then $f' = rj$ is a homomorphism $X' \to Y$. Moreover f' restricted to the vertices of X is equal to f. Indeed, for all vertices
x of X, $f'(x) = r(j(x)) = r(x)$, but x and $f(x)$ are identified in Z, thus
$r(x) = r(f(x))$; finally $f(x) \epsilon V(Y)$, i.e., $r(f(x)) = f(x)$.

Let f': $X' \to Y$ be an extension of f: $X \to Y$. We define r: $Z \to Y$ by
$r(z) = f'(z)$ if $z \epsilon V(X')$ and $r(z) = z$ if $z \epsilon V(Y)$. If z belongs to both X' and Y,
then z is the result of an identification of $x \epsilon V(X)$ and $f(x) \epsilon V(Y)$. Since $r(x)$
$= f'(x) = f(x) = r(f(x))$, there is no contradiction in the definition of r. Also,
if x,x' are different vertices of X' corresponding to the same z in Z, then
x, x' $\epsilon V(X)$ and $f(x) = f(x')$; hence $f'(x) = f(x) = f(x') = f'(x')$. Thus r is a
well-defined mapping. Since f' and j are homomorphisms, and since each edge of Z
is either an edge of Y or an edge of $j(X')$, r is a homomorphism and consequently a
retraction of Z onto Y.

3. If Y is a retract of X, then we have two homomorphisms occurring: the
inclusion i: $Y \to X$, and the retraction r: $X \to Y$. According to what was said in
section 1, $\chi(Y) = \chi(X)$. Another necessary condition for the existence of a retraction
is related to the metric of a graph. A subgraph Y of X is called _isometric_ if the
distance it induces coincides with the distance in X, i.e., if $d_Y(y,y') = d_X(y,y')$
for all vertices y, y' of Y. The minimum number of edges that have to be crossed on
any path from y to y' cannot be increased by an edge-preserving mapping (homomorphism); hence if Y is a retract of X, Y is an isometric subgraph of X.

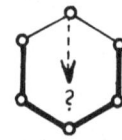

A subgraph which is not isometric cannot be
a retract.

Figure 2

Note that y, y' are adjacent in Y if and only if $d_Y(y,y') = 1$. Thus an isometric
subgraph is a full (induced) subgraph.

The problem of finding a retraction is very difficult. It belongs, in fact (for finite graphs), to the Karp-Cook class of "hard" problems, [12]. To see this, we first remark, that finding a retraction of the disjoint union of X and Y onto Y is equivalent to finding a homomorphism X → Y (and similarly, finding a retraction of the product*) X x Y onto Y is equivalent with finding a homomorphism Y → X). Hence if the problem of finding a retraction is solvable in polynomial time, then so is the problem of finding a homomorphism, and consequently also the determination of the chromatic number of a graph, which belongs to the Karp-Cook class, [12].

Below, we illustrate the difficulty of finding a homomorphism; the two graphs have very similar graph-theoretic properties, yet they admit no homomorphism in either direction (and thus neither one is a retract of their disjoint union or product).

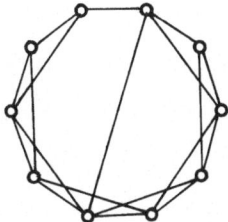 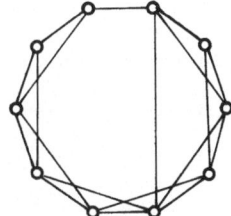

Figure 3

This is the basic difficulty in studying retractions. There is a class of graphs in which homomorphisms between any two graphs exist (except when the second graph is totally disconnected, i.e., consists of isolated vertices), namely $\mathcal{B}$, the class of all bipartite graphs. This is so, since any bipartite graph X admits a 2-coloring c: X → K_2, and if Y has at least one edge then there is a homomorphism h: K_2 → Y, and a homomorphism hc: X → Y. Let us exclude the totally disconnected graphs from $\mathcal{B}$. It seems then that $\mathcal{B}$ is an easier class to handle. We note that K_2 is a retract of any X ε $\mathcal{B}$; indeed any 2-coloring X → K_2 can serve as a retraction, since K_2 is a subgraph of X. Which other graphs X have this property, of being a retract of any graph in $\mathcal{B}$ that contains X? As an example, P_5, the path with 5 vertices does not have the property, as illustrated in Figure 2. However,

---

*The product (conjunction in [4]) of X and Y has vertex set V(X) x V(Y) and (x,y) adjacent to (x',y') if and only if x is adjacent to x' in X and y to y' in Y, cf. [5,13].

the hexagon should have not been admitted as a candidate, since it does not contain P_5 as an isometric subgraph!

Let $\mathcal{C}$ denote a class of graphs, $\mathcal{P}$ a property of inclusion. We denote by AR($\mathcal{C}, \mathcal{P}$) the class of all graphs Y ε $\mathcal{C}$ such that Y is a retract of X whenever X ε $\mathcal{C}$ and Y is a subgraph of X such that the inclusion i: Y $\rightarrow$ X has the property $\mathcal{P}$. The elements of AR($\mathcal{C}, \mathcal{P}$) are called <u>absolute retracts</u> in $\mathcal{C}$ with respect to $\mathcal{P}$.

4. A paper on retracts of bipartite graphs, in which AR($\mathcal{B}, \mathcal{M}$) is determined, is under preparation ($\mathcal{M}$ is the property of being an isometric subgraph). Here we shall summarize the main results and make a few observations.

<u>Proposition 2</u>. AR($\mathcal{B}, \mathcal{M}$) consists of all graphs that are obtained from finite paths by first taking their products and then a retractions, i.e., Y ε AR($\dot{\mathcal{B}}, \mathcal{M}$) if and only if Y is a retract of a product of finite paths.

For infinite graphs, the products may be of infinitely many paths, but each of them being finite. In the proof of Proposition 2, the following observation is implicit:

<u>Corollary</u>. Each bipartite graph is an isometric subgraph of a suitable product of finite paths.

<u>Proposition 3</u>. Y $\notin$ AR($\mathcal{B}, \mathcal{M}$) if and only if there exists an independent set S of diameter 2 in Y, such that either the set U(S) of all vertices adjacent to each vertex of S is empty, or there exists a function f of U(S) into Y such that for each u ε U(S) we have:

$\qquad$ f(u) $\notin$ S, f(u) not adjacent to any vertex of S,

$\qquad$ d(u, f(u)) $\geq$ d(s, f(u)) for all s ε S,

and $\qquad$ d(f(u), f(u')) $\leq$ d(f(u), u) + d(f(u'), u') - 4 for all u, u' ε U(S).

The Propositions 2 and 3, in conjunction, provide us with a "good characterization" (in the sense of Edmonds, [3]) of AR($\mathcal{B}, \mathcal{M}$): either Y ε AR($\mathcal{B}, \mathcal{M}$) and we can exhibit a retraction of some product of paths onto Y, or Y $\notin$ AR($\mathcal{B}, \mathcal{M}$) and we can exhibit S with U(S) = $\emptyset$ or with a function f: U(S) $\rightarrow$ V(Y) satisfying the above three properties. Two characteristic instances of this situation are illustrated below.

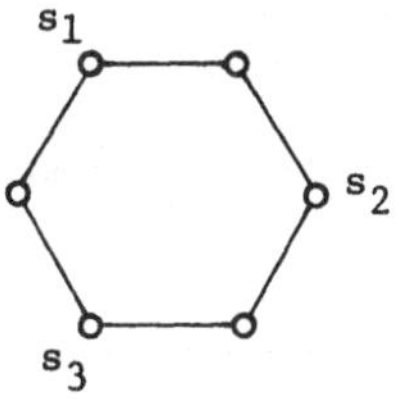

S with U(S) = ∅

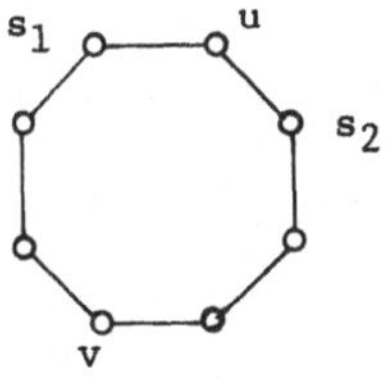

S with U(S) = {u}, f(u) = v

Figure 4

Neither the hexagon, nor the octagon belong to AR($\mathcal{B},\mathcal{M}$).

Corollary. Any tree, or a graph whose all cycles have length 4, is a member of AR($\mathcal{B},\mathcal{M}$).

As an application we prove the following fact, first observed by Sabidussi:

Proposition 4. A cycle of minimum length in a bipartite graph X is a retract of X.

Proof. If C is a cycle of minimum length in X ε $\mathcal{B}$, then C − e, where e is any edge of C, is a path, which is an isometric subgraph of X − e. By the above corollary, C − e is a retract of X − e, and any retraction r: X − e → C − e is also a retraction of X onto C.

5. Another interesting class of graphs to investigate is $\mathcal{P}$, the class of all finite planar graphs. It turns out, that AR($\mathcal{P}$, ∅) = AR($\mathcal{P},\mathcal{M}$) and we shall write AR($\mathcal{P}$) instead. Clearly, if the Four-Color Conjecture (4CC) is true, then K_4 ε AR($\mathcal{P}$), just as we showed K_2 ε AR($\mathcal{B}$, ∅). In [10] we prove that if the 4CC is true, then AR($\mathcal{P}$) consists of all those maximal planar graphs for which each triangle bounding a face belongs to a complete graph with 4 vertices. Of special interest is the following proposition (also proved in [10]), that gives an equivalent form of 4CC in terms of "algebraic" properties of the class $\mathcal{P}$:

Proposition 5. 4CC is true if and only if AR($\mathcal{P}$) ≠ ∅.

6. Applications. In the first place we must mention again extensions of homomorphisms (colorings in particular). Besides Proposition 1, giving a criterion for
extendability of a particular homomorphism in terms of retracts, we have, as in topology, the following properties: Let Y be a subgraph of X. Then

(1) Y is a retract of X if and only if each homomorphism Y → Z can be extended
to a homomorphism X → Z.

(2) Y is a retract of X if and only if each homomorphism Z' → Y (for Z' a subgraph of Z) which can be extended to a homomorphism Z → X, can also be extended to a homomorphism Z → Y.

Let us prove (1). If r: X → Y is a retraction, then any homomorphism f: Y → Z
can be extended to fr: X → Z, as (fr)(y) = f(r(y)) = f(y) for all y ε V(Y). If
every homomorphism Y → Z can be extended to a homomorphism X → Z, then consider the
identity homomorphism 1_Y: Y → Y; its extension r: X → Y is a retraction. The proof
of (2) is analogous.

The properties (1) and (2) are useful, when we know that Y is a retract of X,
e.g., when Y ε AR($\mathcal{B},\mathcal{M}$), X ε $\mathcal{O}$ and Y is an isometric subgraph of X. In this vein,
Proposition 2 yields another application of retracts to extendability of homomor-
phisms:

(3) Y ε AR($\mathcal{B},\mathcal{M}$) if and only if any homomorphism Z' → Y can be extended to a
homomorphism Z → Y, where Z' is any isometric subgraph of Z ε $\mathcal{B}$.

As a further application we shall study rigid graphs. A graph is rigid, cf.
[6, 7, 11] if it has no homomorphisms into itself, except the identity. These
graphs play an important role in full embeddings of certain categories, and in
finding graphs with given group (monoid), [7, 11, 15]. To show that a graph is
rigid usually requires a lot of brute force, cf. [6, 7]. For a finite graph X, one
often starts by showing that X is asymmetric (has no automorphisms, except 1_X);
this is generally easier, as automorphisms preserve degrees, cycles, distances,
etc. But after that there is no general method for concluding the proof. The
following proposition is useful:

Proposition 6. A finite graph X is rigid if and only if

(1) X is asymmetric.

(2) X does not contain a proper retract of itself.

<u>Proof</u>. A rigid graph is asymmetric and does not admit a proper retraction.

If (1) is satisfied, then any homomorphism (except 1_X) of X into itself has image

Y which is a <u>proper</u> subgraph of X. The set of all homomorphisms X → Y forms a

finite non-empty semigroup under composition. Such a semigroup necessarily con-

tains an idempotent, i.e., there is a homomorphism r: X → Y with r r = r. Let

r(X) denote the full subgraph of X induced by r(V(X)). Then r(X) is a subgraph of

Y, thus a proper subgraph of X, and r: X → r(X) is a retraction.

<u>Remark</u>. If the deletion of the set V' $\subset$ V(X) results in decreasing the chro-

matic number of X, then V' intersects the vertex-sets of all retracts of X.

Indeed, if a retract Y of X is disjoint from V' then V(Y) $\subset$ V(X) - V' and

$\chi(Y) < \chi(X)$, which is impossible. Thus the critical vertices of X (vertices x such

that $\chi(X - x) < \chi(X)$) belong to all retracts of X.

    <u>Example</u>. Let n be even, n ≥ 8.

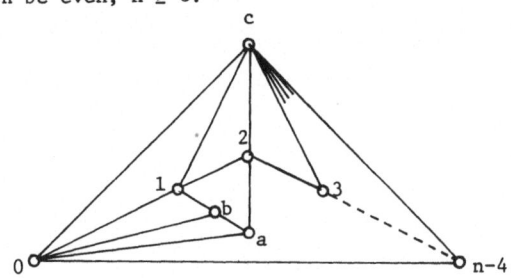

<u>Figure 5</u>

This graph was shown rigid in [6]. Our remark and Proposition 6 facilitate the

proof (similarly, most other proofs of rigidity of finite graphs can be shortened

using this method): To see that X is asymmetric we note with [6], that c is the

only vertex such that its neighborhood is a cycle. The cycle, i.e., 0, 1, ...,

n-4, 0 is odd, and so $\chi(X) = 4$; but $\chi(X - x) = 3$ for x = 0, 1, ..., n-4, c. Hence

each retract of X would contain 0, 1, ..., n-4, c; we see that X contains no

proper retract of itself and therefore is rigid.

    Next we shall make the following observation. Each bipartite graph can be

mapped (by a homomorphism) onto a path. Let $\ell(X)$, for X ϵ $\mathcal{B}$ denote the maximum

length (i.e., number of edges) of a path that is a homomorphic image of X.

<u>Proposition 7</u>. $\ell(X)$ = diam X, for each connected X ε $\mathcal{B}$.

<u>Proof</u>. Since, as we noted before, the distance between two vertices cannot be increased by a homomorphism, $\ell(X) \leq$ diam X. Let d(x, x') = diam X. Then any path joining x and x', with diam X edges, is an isometric subgraph of X, and by the Corollary of Proposition 3, it is a retract of X. Hence $\ell(X)$ = diam X.

Finally we shall discuss a possible argument for the decidability of the 4CC. The following statement was suggested by Z. Hedrlin: (ZH) There exists an integer k such that each planar graph X has a homomorphic image Y ε $\mathcal{P}$ satisfying $|V(Y)| \leq k$

If (ZH) could be proved, and such a k found, then a systematic examination of all planar graphs with at most k vertices would yield (in theory) a final answer to the 4CC. As an application of Proposition 5, one can prove (see [10]), that the criterion (ZH) of decidability of the 4CC is equivalent to the 4CC itself!

In [9], most of the above, and also other related results can be found. Recently, new partial results on AR$(\mathcal{H}, \mathcal{m})$, where $\mathcal{H}$ is the class of all k-chromatic graphs, were found jointly by E. Mendelsohn and the author.

REFERENCES

1. Borsuk, K., Doctoral Thesis, Warsaw, 1929.

2. Borsuk, K., Sur les rétractes, _Fund. Math_. 17 (1931),152-170.

3. Edmonds, J., "Minimum Partition of a Matroid into Independent Subsets",
    J. Res. Nat. Bur. Standards, 69B (1965), 67-72.

4. Harary, F., _Graph Theory_, Addison-Wesley, Reading, Mass., 1969.

5. Hedetniemi, S. T., "Homomorphisms of Graphs", University of Michigan Technical
    Report 03105-42-T, December 1965.

6. Hedrlín, Z. and Pultr, A., "On Rigid Undirected Graphs", _Canad. J. Math_. 18
    (1966) 1237-1242.

7. Hell, P., "Rigid Undirected Graphs with Given Number of Vertices", _Comment._
    Math. Univ. Carolinae, 9 (1968), 51-69.

8. Hell, P. "Full Embeddings into Some Categories of Graphs", _Algebra Universalis_
    2 (1972), 129-141.

9. Hell, P., "Retractions de Graphes", Ph.D. Thesis, Université de Montréal, 1972.

10. Hell, P., "Absolute Planar Retracts and the Four Color Conjecture", _J. Combi-_
    natorial Theory, submitted.

11. Hell, P., and Nešitril, J.,"Groups and Monoids of Regular Graphs (and of
    graphs with Bounded Degrees)", _Canad. J. Math_. 25 (1973), 239-251.

12. Karp, R. M., "Reducibility Among Combinatorial Problems", _Complexity of_
    Computer Computations (R. E. Miller et al; eds.), Plenum Press, New York,
    1972, 85-103.

13. Miller, D. J., "The Categorical Product of Graphs, _Canad. J. Math_. 20 (1968),
    1511-1521.

14. Sabidussi, G., "Graph Derivatives", _Math. Z_. 76 (1961), 385-401.

15. Sichler, J., "Nonconstant Endomorphisms of Lattices", _Proc. Amer. Math. Soc_.
    34 (1972), 67-70.

CANCELLING EULERIAN GRAPHS

Joan P. Hutchinson
Dartmouth College

ABSTRACT

If G is an (undirected) Eulerian graph, we label the edges of G and define
the sign of an Eulerian path on G to be the sign of the associated permutation
of the edges of the graph which is given by the Eulerian path. A path is positive
if its sign is +1, negative if -1. One asks when a graph contains the same number
of positive and negative Eulerian paths.

A vertex of a graph G is said to cancel if there are an equal number of
positive and negative Eulerian paths which begin at the vertex. A graph is said
to cancel if every vertex cancels. In fact, whether a graph cancels or not is
independent of the choice of labels for the edges of the graph.

Properties of cancelling graphs are explored. Using results obtained by
Swan for directed graphs, it can be shown that a graph with at least twice as
many edges as vertices always cancels. The relevance of cancelling graphs to
the study of polynomial identities for skew-symmetric and symmetric matrices will
also be presented.

CANCELLING EULERIAN GRAPHS

1. INTRODUCTION

This paper presents some properties of Eulerian graphs which arise from the study of polynomial identities for sets of matrices.

We shall consider graphs which may have loops and multiple edges, and we shall call a graph Eulerian if the graph contains either 0 or 2 vertices of odd degree. Euler proved that it is precisely these graphs on which a path* can be traced which crosses every edge once and only once. We call such a path an Eulerian path, and if the terminal vertex of the path is the same as the initial vertex, we may call the path an Eulerian circuit. In fact, Euler's theorem tells us that if V is a vertex of a graph G with all vertices of even degree, then there are Eulerian circuits which begin and end at V; however, if G has two vertices of odd degree, there is an Eulerian path which begins at V if and only if V is one of the vertices of odd degree.

Suppose G is an Eulerian graph with m edges. If we label the edges of G, then every Eulerian path on G gives us a permutation on m elements.

Definition 1. The sign of an Eulerian path is the sign of the associated permutation of the edges of the graph.

An Eulerian path is positive (or negative) if its sign is +1 (or -1). The question which we wish to explore is that of determining when a graph contains an equal number of positive and negative Eulerian paths which begin at a fixed vertex.

Definition 2. A vertex of a graph cancels (or is said to be a cancelling vertex) if there is an equal number of positive and negative Eulerian paths which begin at that vertex.

A graph cancels (or is called a cancelling graph) if every vertex cancels. For example, if G is the path of m edges, G does not cancel, for each vertex of

---

*Editors Note: This is sometimes called a trail.

degree 1 is the initial vertex of precisely one Eulerian path. If G is the cycle of length m, every vertex is the initial vertex of precisely 2 Eulerian circuits. If $m \equiv 0$ or $1 \pmod 4$, the 2 circuits have the same sign; if $m \equiv 2$ or $3 \pmod 4$, they have opposite signs and thus each vertex cancels as does the entire graph.

The interest in these questions arises from the study of polynomial identities for sets of matrices. If we define the standard identity of degree m by

$$[x_1, \ldots, x_m] \equiv \sum_{\sigma \varepsilon s_m} \text{sgn}\sigma \; x_{\sigma 1} \; \cdots \; x_{\sigma m} = 0 \; ,$$

then Amitsur and Levitzki were the first to prove that any set of m $n \times n$ matrices over a field of characteristic 0 satifies the standard identity of degree m if $m \geq 2n$, and that $2n$ is the best possible bound [1]. Swan [2] recognized the essentially combinatorial nature of their proof and gave another proof by establishing an equivalent result about directed Eulerian graphs. His important result is stated as Theorem 1.

The question was also posed of determining the minimum standard identity for the set of all $n \times n$ skew-symmetric matrices. Kostant [3] proved that if n is even, any m $n \times n$ skew-symmetric matrices satisfy the standard identity of degree $m \geq 2n-2$. Smith and Kumin conjectured that Kostant's result was true for all n and gave the best possible bound [4]. The conjecture is true, and the proof given by this author also is a graph-theoretic argument about cancelling Eulerian graphs [5]. The result is stated in Theorems 2 and 3 without proof. The proof builds upon a variety of properties of cancelling Eulerian graphs, which in themselves are not difficult or deep, but which are presented here as an introduction to the subject and hopefully as a help in understanding their further implications. Results for skew-symmetric matrices have been obtained by Rowen [6] and Owens [7].

2. PROPERTIES OF CANCELLING EULERIAN GRAPHS

Notice that whether a graph cancels or not is independent of the labelling; that is, if a graph cancels with one labelling, it cancels with any other since the interchanging of two edge labels of a graph causes the associated permutation of an Eulerian path to be multiplied by a transposition. Thus each positive Eulerian path becomes negative and vice versa. If we began with an equal number of positive and negative Eulerian paths, we will still have an equal number after interchanging labels.

Also if a graph has multiple edges, it cancels, for we may pair Eulerian paths by pairing two paths which are identical except that the two multiple edges are interchanged. Two such paired paths will have opposite signs since the corresponding permutations differ by a transposition.

There are a number of sufficient conditions for a graph to cancel which we can determine.

Proposition 1. If G is a graph with all vertices of even degree and if $|E(G)| \equiv 2,3$ (mod 4), then G cancels.

Proof: If V is a vertex of G, we pair the Eulerian circuits at V by associating with one circuit that one obtained by reversing the first. They have opposite signs as Eulerian paths since $|E(G)| \equiv 2,3$ (mod 4).

Proposition 2. If G has 2 vertices of odd degree, G cancels if and only if one of the vertices of odd degree cancels.

Proof: By definition if the graph cancels, all vertices cancel. Conversely, suppose V and W are the vertices of odd degree and suppose V cancels. Of course, all vertices of even degree are automatically cancelling vertices since no Eulerian path on G can begin at such a vertex. Thus we need only check that W cancels. There is a one-to-one correspondence between the Eulerian paths which begin at V and those which begin at W obtained by reversing each Eulerian path. Depending upon what the number of edges of G is congruent to modulo 4, the signs of such corresponding paths are either always the same or always opposite, but in either case W will also have an equal number of positive and negative Eulerian paths if V does, and thus G cancels.

Sometimes we can determine when a graph cancels by discovering a cancelling subgraph. We mention one such example, omitting the straightforward proof.

Proposition 3. Suppose G has an isthmus e such that $G-\{e\} = F \cup F'$. Then G cancels if either F or F' is a cancelling graph. Suppose G is 2-line connected and that

$G-\{e,e'\} = F \cup F'$. Then a vertex $V \epsilon F$ cancels if F' is a cancelling graph.

It is often convenient to divide the Eulerian paths into subsets and show that each subset cancels. Suppose G has all vertices of even degree and let V be a vertex of G. We define the rotation equivalence classes of Eulerian circuits at V as follows. If C is a circuit at V with associated permutation $(x_1,\ldots,x_m)$ and C' another circuit at V with permutation $(y_1,\ldots,y_m)$, then C and C' are in the same equivalence class if and only if

$$(y_1,\ldots,y_m) = (x_i,\ldots,x_m,x_1,\ldots,x_{i-1})$$

for some $i \epsilon \{2,3,\ldots,m\}$. If V has valence $2p$, there are p circuits in each equivalence class.

<u>Proposition 4</u>. If G has all vertices of even degree and an odd number of edges, then G cancels if and only if at least one vertex cancels.

<u>Proof</u>: We use the fact that when the number of edges of G is odd, each rotation class at a vertex V of valence $2p$ not only has p elements, but also all members of one class have the same sign as Eulerian paths. This is so since the permutations

$$(x_1,\ldots,x_m) \quad \text{and} \quad (x_i,\ldots,x_m,x_1,\ldots,x_{i-1})$$

have the same sign when m is odd. Thus a vertex cancels if and only if there is a set of representatives of the equivalence classes which contains an equal number of positive and negative Eulerian paths.

Suppose V is a cancelling vertex and suppose W is another vertex of G. Let e be an edge incident with V. We pick as our set R of representatives of the Eulerian circuits at V those circuits which either begin on or end on edge e. We know that R contains an equal number of positive and negative Eulerian circuits. We pick the set S of representatives of the Eulerian circuits at W to be those which upon leaving W, traverse edge e before returning to W for the first time. There is a one-to-one, sign-preserving correspondence between the elements of R and S since an element of R can be cyclically rotated to a unique element of S and conversely. Thus S cancels and so does the vertex W.

Swan proved a very strong result (Theorem 1) for directed Eulerian graphs. On a directed graph, we now require an Eulerian path to follow the given edge orientations. We then can define a cancelling vertex and a cancelling directed graph as in Definitions 1 and 2.

Theorem 1. If G is a directed graph with n vertices and m edges, and if $m \geq 2n$, then G cancels.

We can immediately obtain the same result for undirected graphs.

Corollary 1. If G is an undirected graph with n vertices and m edges and if $m \geq 2n$, then G cancels.

Proof: With every Eulerian path E on G we obtain an orientation of the edges of G, and thus define G_E to be the resulting directed graph.

We divide the set of all Eulerian paths which begin at a vertex V of G into equivalence classes by assigning two paths E and E' to the same class if the graphs G_E and G_E are the same directed graph. By Swan's theorem each G_E is a cancelling graph so that each equivalence class cancels. Thus V cancels as does G.

Several people have noted that this is the best possible bound when we allow graphs to have loops, for the graph in Figure 1 has n vertices and $2n-1$ edges, but neither of the two vertices of odd degree cancels since each vertex of odd degree is the initial vertex of precisely one Eulerian path.

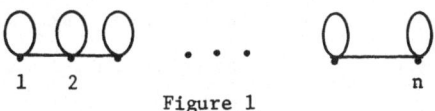

1   2   n

<u>Figure 1</u>

Possibly this is also the best possible bound for loopless graphs since for $n = 6,7$ there are non-cancelling, loopless graphs with n vertices and $2n-1$ edges, but we ask

Question 1. Is $2n$ the best possible bound for loopless, undirected graphs? That is, for every n is there a loopless graph with n vertices and $2n-1$ edges which does not cancel?

For instance, it is known that for every n and $k \leq 2n-3$, there is a loopless graph with n vertices and k edges which does not cancel, but no sharper results are known in general.

We have seen some sufficient conditions for a graph to cancel. It is not hard to characterize the graphs on n vertices and with at most n edges which cancel. In fact, all graphs with n vertices and k edges, $k \le n$, cancel except for the path with k edges and the graphs shown in Figure 2 where the number of edges in each cycle is congruent to 0 or 1 modulo 4. In general, no more is known about graphs with fewer than 2n edges.

Question 2. Can one find necessary and sufficient conditions such that a graph with n vertices and k edges, n < k < 2n , cancels?

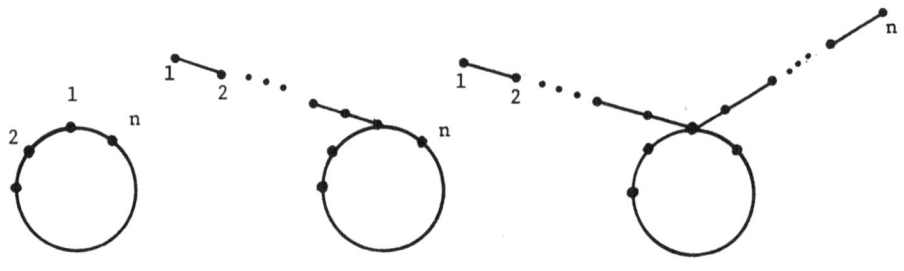

Figure 2

3. TOTALLY CANCELLING GRAPHS

There is another question which is reasonable to ask. Suppose we have an Eulerian graph G and suppose we make a list of all Eulerian paths on G without re-gard to their initial point. That is, we write down all permutations which give us an Eulerian path and we ask when the list contains an equal number of positive and negative paths. Of course, the list has this property if the graph cancels.

Definition 3. A graph G cancels totally if there is an equal number of positive and negative Eulerian paths on G.

This concept also arises from the study of matrices and the standard identity. Given matrices $A_1, \ldots, A_m$ such that $[A_1, \ldots, A_m] \ne 0$, we can determine some information about when the matrix $[A_1, \ldots, A_m]$ has 0 trace and when it is skew-symmetric or symmetric.

We can characterize graphs which cancel totally in terms of what we know about graphs which cancel.

Proposition 5. If G has two vertices of odd degree and if G does not cancel, then G cancels totally if and only if $|E(G)| \equiv 2,3 \pmod 4$.

Proof: If V and W are the two vertices of odd degree, then we pair an Eulerian path beginning at V with the reversed path which begins at W. Two such paired paths have the same sign if and only if $|E(G)| \equiv 2,3 \pmod 4$.

Proposition 6. If G has all vertices of even degree and an even number of edges, then G cancels totally.

Proof: Note that the permutations

$$(x_1,\ldots,x_m) \quad \text{and} \quad (x_2,\ldots,x_m,x_1)$$

have opposite signs.

We define the total equivalence classes of Eulerian circuits on G as follows. We pick a vertex V of G and a set R of representatives of the rotation equivalence classes of Eulerian circuits at V. Each element E of R determines a total equivalence class [E], and an arbitrary Eulerian circuit F is in [E] if and only if E and F have respectively the following associated permuations:

$$(x_1,\ldots,x_m) \quad \text{and} \quad (x_i,\ldots,x_m,x_1,\ldots,x_{i-1})$$

for some $i \varepsilon \{2,3,\ldots,m\}$. Then every Eulerian path on G belongs to precisely one total equivalence class.

In the equivalence class [E], we pair paths with permutations

$$(x_{2k-1},\ldots,x_m,x_1,\ldots,x_{2k-2}) \quad \text{and} \quad (x_{2k},\ldots,x_m,x_1,\ldots,x_{2k-1})$$

for $k \varepsilon \{1,2,\ldots,m/2\}$. Each pair of paths have opposite signs, each equivalence class cancels, and G cancels totally.

Proposition 7. If G has all vertices of even degree and an odd number of edges, then G cancels totally if and only if G cancels.

Proof: When m is odd, the set of permutations

$$\{(x_i, \ldots, x_m, x_1, x_{i-1})\}, \ 1 \le i \le m$$

all have the same sign. Thus each total equivalence class, as defined in the proof
of Proposition 6, contains $|E(G)|$ paths, all of the same sign, and G cancels totally if
and only if a set of representatives of the total equivalence classes cancels.

Suppose G does not cancel. By proposition 4, no vertex V cancels. We pick as
a set R of representatives of the rotation equivalence classes of Eulerian circuits
at V those circuits which either begin on or end on a fixed edge e, incident with
V. We know that R does not contain an equal number of positive and negative Eulerian
paths. But R is also a set of representatives of the total equivalence classes, and
by the previous discussion we see that G does not cancel totally.

Thus the concept of total cancellation leads us back to our earlier questions
about when an arbitrary graph cancels.

4. EULERIAN GRAPHS AND SKEW-SYMMETRIC MATRICES

We conclude with a statement of the graph-theoretic result which is equivalent
to matrix results for skew-symmetric matrices.

Definition 4. An E path on a directed graph is a path which traverses every edge
of the graph once and only once and which may travel in either direction on an edge,
regardless of the orientation.

The sign of an E path is the product sgn $\sigma(-1)^z$, where σ is the associated
permutation of the edges of the graph and where z is the number of edges traversed
opposite to their orientation. We call $(-1)^z$ the orientation coefficient of an E
path P, and denote it by $OC(P)$.

In this context we then define a cancelling vertex and a cancelling graph as in
Definition 2, and we can prove the following.

Theorem 2. If G is a directed (loopless) graph with n vertices and m edges and if
$m \ge 2n-2$, then G cancels.

Theorem 3. Given $k < 2n-2$, there is a directed (loopless) graph with n vertices
and k edges which does not cancel.

The proofs [6] are rather intricate, but the main ingredients are the results discussed in this paper.

Results about partially directed graphs can also be obtained as can some results for undirected graphs. For instance, a partial answer to Question 1 is the following corollary of theorem 2.

<u>Corollary 2</u>. If G is a bipartite, undirected graph with n vertices and m edges and if $m \geq 2n-2$, then G cancels.

<u>Proof</u>: Given such a graph G we shall orient the edges of G in a special way so that not only does the oriented graph G' cancel by Theorem 2, but also all the E paths at a fixed vertex V will have orientation coefficient +1 . Since V will be a cancelling vertex of G' , we shall see that there are an equal number of E paths at V with associated positive permutation as there are with negative permutation. Thus V will also be a cancelling vertex of the original graph G.

Pick a vertex V of G which we wish to show cancels. We create an oriented graph G' by orienting the edges of G in the following way. Let W be an Eulerian walk on G which begins at V. (If there are none, $V \epsilon G$ automatically cancels.) Then if e joins vertices X and Y in G, then edge e of G' is oriented from X to Y if and only if X is the initial vertex of e with respect to W. Let the flux of a vertex Z in a directed graph be given by the difference between the number of in-directed edges at Z and the number of out-directed edges. Then with this orientation of G either all vertices of G' will have flux 0 or all but two will have flux 0, and of these two, one has flux +1 and the other −1 .

The E path W on G' has orientation coefficient +1 and we shall see that any other path P, beginning at V, also has $OC(P) = +1$. Since G is bipartite, we know that we can partition the vertices of G into two disjoint sets V_1 and V_2 such that every edge joins a vertex of V_1 with a vertex of V_2 . Thus to calculate $OC(P)$ we may calculate the orientation coefficient contribution to P at every vertex of V_1 ; that is, if $A \epsilon V_1$, we determine the contribution to the orientation coefficient of all the edges incident at A and we do this for every vertex in V_1 . In this way we will check the orientation coefficient contribution of every edge of G' once and only once since G' is bipartite. We shall see that at each vertex A of

V_1, the contribution to the orientation coefficient is +1 .

Suppose the valence of A is 2p , $A \epsilon V_1$. The flux of A is then 0. Consider the following numbers concerning P:

$\quad$ x_1 = the number of times P comes in to A on an edge directed

$\quad\quad$ to A,

$\quad$ y_1 = the number of times P goes out of A on an edge directed

$\quad\quad$ in to A,

$\quad$ x_2 = the number of times P comes in to A on an edge directed

$\quad\quad$ out of A,

$\quad$ y_2 = the number of times P goes out of A on an edge directed

$\quad\quad$ out of A.

Then $x_1 + y_1 = p = x_2 + y_2$ and $x_1 + x_2 = p = y_1 + y_2$.

Thus $y_1 - x_2 = x_2 - y_1$, i.e., $y_1 = x_2$. At A we see that $x_1 + y_2$ edges make no contribution to the orientation coefficient of P. But on the remaining $y_1 + x_2$ edges, P does travel opposite to the edge orientation so that the orientation coefficient of P receives a contribution of $(-1)^{y_1 + x_2} = (-1)^{2y_1} = +1$ at the vertex A.

Suppose B is a vertex of V_1 with flux +1 and the valence of B is $2q + 1$. Let x_1, x_2, y_1 and y_2 be defined as before, and we see $x_1 + y_1 = q = x_2 + y_2 - 1$, and $x_1 + x_2 = q = y_1 + y_2 - 1$, whence $y_1 = x_2$. Again the contribution to OC(P) at B is

$$(-1)^{y_1 + x_2} = (-1)^{2y} = +1 .$$

The same argument holds for a vertex with flux -1 , and therefore $OC(P) = +1$.

Since G' is a directed, loopless, graph with n vertices and m edges, $m \geq 2n-2$, G' cancels by Theorem 2, and, in particular, the vertex V cancels. Since all E paths at V have the same orientation coefficient, +1 , there must be the same number of paths with associated positive permutation as those with negative permutation. This means that V is a cancelling vertex of G, the original, undirected graph, and since V was arbitrary, G is a cancelling graph.

REFERENCES

1. Amitsur, S. A., and Levitzki, J., "Minimal Identities for Algebras", _Proc. Amer. Math. Soc._, 1 (1950) 449-463.

2. Swan, R. G., "An Applicationof Graph Theory to Algebra", _Proc. Amer. Math. Soc._, 14 (1963) 367-373; Correction 21 (1969) 379-380.

3. Kostant, B., "A Theorem of Frobenius, A Theorem of Amitsur-Levitzki, and Cohomology Theory", _J. Math. Mech._, 7 (1958) 237-264.

4. Smith, K. C., and Kumin, H. G., "Identities on Matrices", _Amer. Math. Monthly_, 79 (1972) 157-158.

5. Hutchinson, J. P., "Eulerian Graphs and Polynomial Indentities for Sets of Matrices", _Proc. Nat. Acad. Sci._ U. S. A.

6. Rowen, L. H., "Standard Polynomials in Matrix Algebras", to appear.

7. Owens, F., "Applications of Graph Theory to Matrix Theory", to appear.

A GRAPHICAL REALIZATION PROBLEM

Sukhamay Kundu
University of Texas

ABSTRACT

Let G be a finite r-graph where two (distinct) vertices are joined by at most r lines and let [G] denote the degree sequence of G. If s is an integer, $s \geq r+1$, then we prove that a degree sequence $[K_i]$ has a realizing s-graph containing graph G if and only if the sequence $[k_i] - [H]$ is s-realizable for all subgraphs H of G.

A GRAPHICAL REALIZATION PROBLEM

1. INTRODUCTION

In this note, we consider the following graphical realization problem. Given n natural numbers $[k_i: 1 \le i \le n]$ and an r-graph G, when does there exist an s-graph $(s \ge r + 1)$ containing G whose degree sequence is given by $[k_i]$? The realization problem reduces to a factorization problem of an s-graph when stated in terms of complement graphs: Does there exist a subgraph (factor) of $K_n(s) - G$ with degree sequence $[s(n-1) - k_i]$? Here $K_n(s)$ denotes the complete s-graph on n vertices, each vertex having degree $s(n-1)$. A necessary and sufficient condition for the existence of a $[k_i]$ factor in an arbitrary s-graph G has been given by Tutte [3]. It is the purpose of this note to provide a simple alternate characterization of sequences $[k_i]$ that are realizable by an s-graph containing a given r-graph G. The price of such simplifications in those conditions is that our theorem does not hold, for example, if we let $s = r$ and G arbitrary.

The principal result (Theorem 1) of this note can be derived from a more general realizability theorem (see Appendix). However, the proof of that theorem requires application of a few important theorems from theory of linear inequalities. Therefore, we feel that a direct combinatorial proof as given here should be of interest, particularly because of its simplicity.

2. NOTATIONS AND DEFINITIONS

By an r-graph $G = (V(G), E(G))$, we mean a set of vertices $V(G)$ and a set of lines $E(G)$ where each line joins a pair of distinct vertices and no r+1 lines join the same vertices. The vertices are denoted by $v_1, v_2, \ldots, v_n$ and (v_i, v_j), denotes a line joining v_i to v_j. The number of lines joining v_i to v_j is called the _multiplicity_ of (v_i, v_j). An r-graph H is said to be a _subgraph_ of G if $V(H) = V(G)$ and $E(H) \subseteq E(G)$. The _degree_ of a vertex is the number of lines incident with that vertex. If we let d_i denote the degree of vertex v_i in G, then the sequence $[d_i]$ is said to be the _degree sequence_ of G and is denoted by $[G]$. We say that $[d_i]$ is _r-realizable_ and that G is an _r-realization_ of that sequence. (A

necessary and sufficient condition for r-realizability is given in the Appendix.)
The complete r-graph $K_n(r)$ is defined by the degree sequence $d_i = r(n-1)$, $1 \le i \le n$.
In other words, each pair of vertices in $K_n(r)$ are joined by r parallel lines.

Throughout the paper we regard an s-graph as a subgraph of the complete graph $K_n(s)$. The lines of G joining vertices v_i, v_j are labeled, unless otherwise stated, as $(v_i,v_j)_1$, $(v_i,v_j)_2$, ..., $(v_i,v_j)_t$ where t is the multiplicity of (v_i,v_j). It will be convenient to write a sequence $[d_i - k_i]$ as $[d_i] - [k_i]$. Also, the symmetric difference of two sets A and B will be denoted by A $\triangle$ B.

3. RESULTS

We prove the following realizability theorem.

Theorem 1. Let G be an r-graph, $[k_i]$ a sequence of n natural numbers, and $s \ge r + 1$. Then $[k_i]$ has an s-realization containing graph G if and only if the following condition holds:

(1.1) The sequence $[k_i] - [H]$ is s-realizable for all
        subgraphs H of G.

The "only if" part of the theorem is obvious. To show that condition (1.1) is sufficient, we proceed by induction on the number of lines in G.

Proof ("if" part). The theorem is clearly true if $|E(G)| = 0$. Assume the theorem for all r-graphs with q-1 lines, and let $|E(G)| = q$. Let $e = (v_i,v_j)_r$ be a line of G and let H denote the subgraph obtained from B by removing e. It follows from the induction hypothesis that there exists an s-graph F_1 containing H, where $[F_1] = [k_i]$. There also exists an s-graph F_2 containing H such that $[F_2]$ is the same as $[k_i]$ except for the degrees of v_i and v_j which are one less than k_i and k_j, respectively. Without loss of generality, we can assume that $m = |E(F_1) \cap E(F_2)|$ is maximum (among all possible choices of F_1,F_2) and that $e \notin E(F_1)$. Now one can easily show that there exists a sequence of distinct lines $P = [(x_0,x_1), (x_1,x_2), (x_2,x_3), ..., (x_{2p}, x_{2p+1})]$ which has the following properties: (a) $x_0 = v_i$ and $x_{2p+1} = v_j$; (b) the lines (x_i,x_{i+1}) belong alternately to $E(F_1) - E(F_2)$ and $E(F_2) - E(F_1)$ while the first and last lines of P belong to F_1. The vertices x_i

need not all be distinct; by definition, the lines of P are disjoint from the lines of H. The existence of the "alternating path" P follows from the fact that the sequence $[F_1] - [F_2]$ consists of zeros only except for the ith and jth terms which are equal to 1.

First suppose that $e \in P$, and hence e belongs to F_2. If $e = (x_t, x_{t+1})$, then it follows that $x_t = v_i$ and $x_{t+1} = v_j$. For, otherwise, the first t+1 lines of P would form a closed alternating path Q of even length, and we could redefine graph F_1 by $E(F_1') = E(F_1) \Delta Q$, i.e., replace those lines of F_1 appearing in Q by the other half of the lines of Q which belong to graph F_2. The result is a larger value for m. Similar remarks now show that the line $f = (v_i, v_j)_s$ is not in the sequence P. We conclude therefore that not both e and f belong to P.* For brevity, let us assume that $f \notin P$. However, this too leads to a contradiction because of the following: If we let Q denote the augmented sequence $Q = [P, (v_j, v_i)_s]$, then the s-graph F_1' defined by $E(F_1') = E(F_1) \Delta Q$ gives a larger value for m. Thus e belongs to F_1 and the theorem is proved.

- APPENDIX -

1. Let $d_1 \geq d_2 \geq d_3 \geq \ldots \geq d_n$ be n natural numbers. The following conditions are both necessary and sufficient for $[d_i]$ to be s-realizable.

(a) $d_1 + \ldots + d_p \leq sp(t-1) + (d_{t+1} + \ldots + d_n)$ for all $1 \leq p \leq t \leq n$,

(b) $d_1 + d_2 + \ldots + d_n$ is even.

Condition (b) may be rewritten as

(b') $d_1 + d_2 + \ldots + d_p \leq sp(p-1) + \sum_{j \geq p+1} \min(sp, d_j)$.

This result is contained in a theorem of Fulkerson, Hoffman, and McAndrew (Theorem 2.1, [1]). The special case $s = 1$ was proved by Erdös and Gallai.

2. The following theorem is proved in [2] for the case $s = 1$. The same proof holds almost word by word for general s.

---

*Note that $f \in P$ would imply that $f \in F_2$, since $f \in F_1$ implies that $e \in F_1$, by labeling convention.

An s-graph G is said to have <u>odd cycle property</u> if any two odd cycles of G either have a vertex in common, or there exist a pair of vertices, one in each cycle, which are joined by a line. In other words, the distance between two odd cycles of G is at most one. The concept was first introduced by Fulkerson, et al in [1],

<u>Theorem</u>. Let G be an s-graph whose complement in $K_n(s)$ has odd cycle property. Then there exists an s-realization of $[k_i]$ containing G if and only if the sequence $[k_i]$ - [H] is s-realizable for all subgraphs H of G.

Clearly, the complement of an r-graph in $K_n(s)$, where $s \geq r + 1$, has odd cycle property. Thus the theorem proved in this note follows from the general theorem.

<h2 style="text-align:center">REFERENCES</h2>

1. Fulkerson, D. R., Hoffman, A. J., and McAndrew, M. H., "Some Properties of Graphs with Multiple Edges," <u>Canad. J. Math.</u>, 17 (1965), 166-177.

2. Kundu, S., "A Factorization Theorem for Certain Class of Graphs," <u>Discrete Math.</u>, to appear.

3. Tutte, W. T., "A Short Proof of the Factor Theorem for Finite Graphs," <u>Canad. J. Math.</u>, 6 (1954), 347-352.

THE EXISTENCE OF SMALL TACTICAL CONFIGURATIONS

Marc J. Lipman
Dartmouth College

ABSTRACT

A symmetric rank two tactical configuration is a regular bi-partite graph of girth 6. If it is $(n+1)$-regular on $2(n^2+n+1+k)$ points it is denoted $T(n,k)$. There is a polynomial, $P(n) \cong n^4$, such that $T(n,k)$ exist for all $k > P(n)$. This paper discusses the cases $k < P(n)$, and some methods for finding such "small" $T(n,k)$, including computational methods and the use of difference sets.

THE EXISTENCE OF SMALL TACTICAL CONFIGURATIONS

A <u>tactical configuration of rank</u> r is an r-partite graph with $\binom{r}{2}$ lines between pairs of points in different parts, which has constant degree in each part. It is <u>symmetric</u> if the relations have equal constant degree on the parts. In this paper all tactical configurations will have girth six.

A symmetric tactical configuration of rank two is therefore a regular bipartite graph of girth six, and if it is (n+1)-regular on $2(n^2+n+1+k)$ points it is denoted T(n,k), a generic term for that class of graphs. The smallest possible (n+1)-regular bipartite graph of girth six, if it exists, contains $n^2 + n + 1$ points in each part, corresponds to an nth order projective plane, and is a T(n,0), hence the notation is reasonable.

It is difficult to determine whether or not T(n,k) exists for arbitrary n and k. As a special case consider the existence problem for T(n,0), where the results of Singer, Bruck-Ryser-Chowla, and others have produced infinite classes of existence and infinite classes of non-existence, and there is still an infinite class for which existence is an open question.

For general T(n,k) very little is known. Payne [3] has shown that there is an infinite class of T(n,1) which do not exist. The only known T(n,1) are for n = 2 and n = 3. Longyear [2] has given a polynomial in n, $P(n) \cong n^4$, such that T(n,k) exist for all n and all k > P(n).

Call a T(n,k) <u>small</u> if k < P(n), that is if it is smaller than the configuration guaranteed by Longyear's bound. Aside from the nonexistence of many T(n,1) the existence of small T(n,k) is an open question.

A T(n,k) is <u>cyclic</u> if for some listing of the two sets, P, L, the sets of points of the bi-partite T(n,k), the map:

$$\begin{cases} P_i \rightarrow P_{i+1} \\ \ell_i \rightarrow \ell_{i+1} \end{cases} \quad (\text{mod } N = n^2 + n + 1 + k)$$

is a collineation, that is, considering the two sets as points and lines, the relation of adjacency is preserved by the map. This is equivalent to saying that the adjacency matrix has a circulant (see Figure 1).

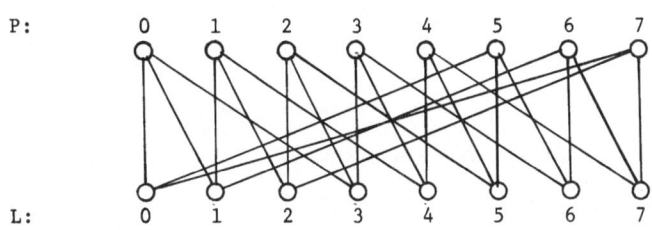

Figure 1. A Cyclic T(2,1) With $0 \leftrightarrow \{0,1,3\}$

Take a cyclic listing of a cyclic T(n,k) and look at the set of lines touching P_0, say $\{\ell_0, \ell_1, \ldots, \ell_n\}$. This completely determines the T(n,k) by the cyclic map. A complete enumeration of the relation is:

$$P_i \leftrightarrow \{\ell_{0+i}, \ldots, \ell_{n+i}\} \quad (\text{mod } N - n^2 + n + 1 + k).$$

This is called the cyclic expansion of the T(n,k).

Writing the set L as integers 1 through $n^2 + n + 1 + k$ in the good listing gives the following:

The set of integers $\{\ell_0, \ldots, \ell_n\}$ is a difference set mod $N = n^2 + n + 1 + k$. Further, any set of n+1 integers forming a difference set mod N, N as above, determines a cyclic T(n,k). This follows immediately from girth 6. Note that these are not perfect difference sets if $k \neq 0$, since there are only $n^2 + n$ differences taken from $n^2 + n + k$ choices.

Therefore the existence of cyclic T(n,k) is equivalent to the existence of difference sets of size $n + 1$ and modulus $n^2 + n + 1 + k$. Thus in the time-honored tradition of mathematics we translate the problem.

The remainder of this paper is devoted to straightforward methods of generating non-perfect difference sets and some preliminary results.

A brute force computer search is viable for low n and k. Note that merely checking all strictly increasing (n+1)-tuples is not feasible as there are: $[N]^{n+1}/(n+1)! > n^n$ possible candidates.

As a slight improvement note that the first element may be chosen to be 0. If k is small enough, at least one of the $n^2 + n$ differences will be a unit mod N, so that the second element may be chosen to be 1, also without loss of proof of existence, although not all difference sets will be generated.

Ewing and I have created several computer programs for the building of non-perfect difference sets, using various heuristics to limit the search procedure in cases where we were reasonably sure of existence beforehand. Table 1 gives known results from these programs, and other work.

A second approach generalizes the work of Hall on Cyclic Projective Planes [1]. He worked with perfect difference sets. Note that $a \equiv b \pmod{N}$ if and only if $x^a \equiv x^b \pmod{x^N - 1}$.

n	k:	0	1	2	3	4	5	6	7	8	9		14
3		Y	Y→										
4		Y	N	Y→									
5		Y	N	N	N	Y→							
6		N	N	N	N	N	Y	Y	Y	Y→			
7		Y	N	N	N	N	N	Y	Y	Y	0	...	Y→
8		Y	0	0	0	0	0	0	Y	0	...		

Y: T(n,k) exist

N: T(n,k) do not exist

0: Open question

Arrows imply T(n,m) exist for all m>k.

Table 1: Existence of cyclic T(n,k)

So, if $\{d_0, \ldots, d_n\}$ is a difference set for a $T(n,k)$, and $N = n^2 + n + 1 + k$, then define

$$t(x) = x^{d_0} + x^{d_1} + \ldots + x^{d_n}$$

$$\overline{t}(x) = x^{N-d_0} + x^{N-d_1} + \ldots + x^{N-d_n}$$

Then: $t(x)\overline{t}(x) = n + 1 + x + x^2 + \ldots + x^{N-1} - \Sigma x^{m_i} \pmod{x^N - 1}$ where m_i is not a difference. For perfect difference sets the term Σx^{m_i} is 0.

Given this equation, results paralleling Hall's are obtainable. For example, if N is even let $x = -1$, and it follows that $t(x) = \overline{t}(x) = m$ for some integer

$-n-1 \leq m \leq n+1$ and:

$$n + 1 + x + x^2 + \ldots + x^{N-1} - \Sigma x^{m_i} = n - \Sigma(-1)^{m_i}.$$

Since -1 satisfies $x^N - 1$ we have:

$$m^2 = n - \Sigma(-1)^{m_i}$$

an equation of numbers. For $T(n,1)$, having only the one difference $N/2$ missing we have: $T(n,1)$ exist only if $n + 1$ or $n - 1$ is a square.

In general, let $x = \xi$, an Nth root of 1. Then $(d_0 + \ldots + d_n)(-d_0 + \ldots + -d_n)$ $= n - \Sigma\xi^{m_i}$ is an equation of numbers, since ξ satisfies both $x^N - 1$ and $1 + x + x^2 + \ldots x^{N-1}$. If $d_0 = 0$ and $d_1 = 1$ the equation is further reduced. Note that the m_i occur in pairs symmetric about $N/2$, since $d_i - d_j = N - (d_j - d_i)$. For example, if k is odd, $N/2$ is a missing difference.

Finally consider the following definition.

A difference set $\{d_0, \ldots, d_n\}$ (mod $N=n^2+n+1+k$) is <u>eternal for</u> (n,k) if it is also a difference set modulo $N + j$ for all $j > 0$.

Then $\{d_0,\ldots,d_n\}$ is eternal for (n,k) if and only if $d_i < N/2$ for all i, since then all differences $d_{i+j} - d_i > N/2$ and $d_i - d_{i+j} > N/2$, hence no coincidences occur at larger N. If $\{d_0,\ldots,d_n\}$ is eternal for (n,k) it follows that $T(n,m)$ exist for all $m \geq k$. Hence in Table 1 the arrows mark eternal difference sets. Note that for every n there are eternal difference sets, but small ones are not trivial to find.

If we denote the differences $p_i = d_i - d_{i-1}$ as the primary differences for an eternal difference set, then all "positive" differences can be written as sums of consecutive primary differences, and if $d_0 = 0$ the sequence of primary differences determines the difference set, which in turn determines the $T(n,k)$.

As examples of results from these considerations, we have:

<u>Theorem.</u>   1) $n > 3$ implies: There are no eternal $(n,1)$ difference sets.

2) $n > 4$ implies: There are no eternal $(n,2)$ or $(n,3)$ difference sets.

3) $N > 5$ implies: There are no eternal $(n,4)$ or $(n,5)$ difference sets.

<u>Proof of 1).</u>   Let $\{d_0,\ldots,d_n\}$ be an eternal difference set for $(n,1)$, $n > 3$, $d_0 = 0$. Let $\{p_1,\ldots,p_n\}$ be the set of primary differences. The p_i are distinct positive integers. Then $d_n = \Sigma p_i < N/2 = (n^2 + n + 1 + 1)/2$. So $\Sigma p_i \leq (n^2 + n)/2$

and since p_i are distinct $\{p_1,\ldots,p_n\} = \{1,2,\ldots,n\}$ in <u>some</u> order. We show that <u>no</u> order will do. Suppose 1 is next to $r < n$. Then $r + 1$ is a difference $\leq n$, so $r + 1 = p_j$ for some j, which is impossible. Therefore 1 can only be next to n, giving a difference of $n + 1$. Similarly 2 can only be next to n and/or $n - 1$. But if 2 is next to $n - 1$, and $n - 1 \neq 2$ by assumption, then $(n - 1) + 2 = n + 1$ is a coincidence, which is impossible. Therefore 2 can only be next to n.

But that means the order of the primary differences must be .1, n, 2. or .2, n, 1. There is no place for the remaining differences, a contradiction to $n > 3$.

The proofs of 2) and 3) are similar but correspondingly more tedious.

<u>Conjecture</u>: For $n > 4$ the smallest cyclic $T(n,k)$ exist at $k = n - 1$, except possibly at $k = 0$.

REFERENCES

1. Hall, M., Jr., Cyclic Projective Planes, <u>Duke Mathematical Journal</u>, Vol. 14 (1947), 1079-1090.

2. Longyear, J. Q., <u>Tactical Configurations</u>. Doctoral Dissertation, Pennsylvania State University (1972).

3. Payne, S. E., On the Non-Existence of a Class of Configurations Which are Nearly Generalized n-Gons, <u>J. Combinatorial Theory</u>, 12 (1972), 268-282.

TACTICAL CONFIGURATIONS: AN INTRODUCTION

Judith Q. Longyear
Dartmouth College

ABSTRACT

The original definition of tactical configuration was given by E. H. Moore in 1896, but the definition now in use is in terms of graph theory. A tactical configuration of rank r is a collection of r disjoint vertex sets $A_1,\ldots,A_r$ called bands and a relation of incidence among these vertices, so that each vertex in band A_i is incident with the same number of vertices in A_j. This constant number, say $d_{i,j}$, is called the i-j degree, and the collection of all the i-j degrees is called the set of degrees for the configuration. Note that $d_{i,j}$ need not be equal to $d_{j,i}$. The numbers $d_{i,i}$ are not defined, since each band is composed of independent vertices. Thus a tactical configuration may be regarded as a multiregular, r-partite graph. The girth of a graph, or of a tactical configuration regarded as a graph, is the number of vertices in any smallest polygon in the graph. This paper describes the important questions concerning the construction and existence of tactical configurations.

TACTICAL CONFIGURATIONS: AN INTRODUCTION

The original definition of tactical configuration was given by E. H. Moore in
1896, but the definition now in use is in terms of graph theory. A tactical configuration of rank r is a collection of r disjoint vertex sets $A_1, \ldots, A_r$ called
bands and a relation of incidence among these vertices, so that each vertex in band
A_i is incident with the same number of vertices in A_j. This constant number, say
d_{ij}, is called the i-j degree, and the collection of all the i-j degrees is called
the set of degrees for the configuration. Note that $d_{i,j}$ need not be equal to $d_{i,j}$.
The numbers $d_{i,j}$ are not defined, since each band is composed of independent vertices.
Thus a tactical configuration may be regarded as a multiregular, r-partite graph.
The girth of a graph, or of a tactical configuration regarded as a graph, is the
number of vertices in any smallest polygon in the graph.

Examples of rank 2 configurations are abundant. A projective plane of order n
is a configuration with degree set $\{n,n\}$, girth 6, and both bands of cardinality
$n^2 + n + 1$. Again identifying points or treatments as one band and lines or blocks
as the other, BIBD which parameters (b,v,r,k,λ) are configurations with degree set
$\{r,k\}$ and girth 6 or 4 as $\lambda = 1$ or $\lambda > 1$. There are many others, such as Steiner
systems, generalized quadrangles and generalized polygons. In more practical areas
tactical configurations occur in traffic flow patterns, herd geneologies, the study
of protein decompositions and coding theory ; see the bibliography.

The question of which configurations exist is still largely unresolved. A set
of degrees is called <u>constructive</u> if there are positive integers $a_1, \ldots, a_r$ which
satisfy the obvious requirements $d_{i,j} a_i = d_{j,i} a_j$. It is known that if D is a constructive set of degrees, and if the girth is chosen either arbitrarily, or arbitrarily even in the case of rank 2, then there are very large configurations having
the chosen girth and the degree set D. Some simple arguments show that there are
lower bounds for the possible cardinalities of the bands, thus the major unresolved
question is:

(1) For which band cardinalities do configurations with given girth and degree

set exist?

This question is not easy, since it includes the question of the projective

plane of order 10, but it has some relatively amenable parts. For example, all the

numerically possible rank 2 configurations of girth 6 and degrees {3,4} exist. The

smallest of these is a BIBD with parameters (12,9,3,4,1). It is of some interest to

determine the least cardinalities for which configurations of rank 2 exist.

The numerical lower bound for $|A_1|$ is $1 + t + st$, and that for $|A_2|$ is

$1 + s + st$. Clearly, unless $s = t$, $|A_1|$ must be slightly larger than $1 + t + st$

(if one makes the usual assumption that $s \le t$). If we let r be the excess in $|A_1|$

over $1 + t + st$ and let $|A_2| = 1 + s + st + w$, then r and w satisfy:

(A) $$(1+s)r = st(t-s) + (1+t)w .$$

Projective planes and BIBD's occur when (A) can be satisfied with $w = 0$ and a

configuration constructed on $|A_1| + |A_2|$ vertices. The Bruck-Ryser-Chowla theorem

gives conditions on s and t under which this numerical lower bound cannot be

achieved. No research seems to have been done on those remaining extremal cases

where $1 + s \not\equiv 0 \pmod{st(t-s)}$, and so the most interesting theoretical challenge is:

(2) Find and prove the analogues to the Bruck-Ryser-Chowla theorem, that is,

give conditions on the degrees and girth which guarantee that the smallest configu-

ration with these parameters has larger band sizes than is numerically necessary.

As an example of the need for such a theorem, the numeric lower bound on a

girth 6 configuration with degrees $d_{1,2} = d_{2,1} = d_{2,3} = d_{3,2} = 3$ and $d_{1,3} = d_{3,1} = 0$

is 7 vertices in each band, but the smallest so far known has 21 vertices in each

band.

Almost nothing is known for $r \ge 3$ or for girth ≥ 8, and if $r \ge 3$ then nothing

is known for girth ≥ 5. Indeed, the major problem in construction is to prove that

the constructed configuration has the desired girth. In order to show that a con-

figuration has girth in excess of g, one must demonstrate that it is impossible to

have any polygon of smaller size with vertices in the various bands. For girth 6, this means checking one case for rank 2, 13 cases for rank 3 and 67 cases for rank 4. In fact, it is a combinatorial problem of some interest to determine how many cases there are.

Since all the practical applications of configurations, and most of the classic forms, have the property that only bands with consecutive subscripts have positive degrees, it is certainly practical to confine the investigation of configurations to these "pathlike" ones, at least until someone has developed a more powerful technique for examining girth. Again for girth 6, the number of cases to determine in the pathlike case is one for rank 2, 3 for rank 3, and 5 for rank 4. As an example of the power of the pathlike restriction, we construct girth 4 configurations of every size and every pathlike constructive degree set D. Given that D is constructive, it is an easy exercise in number theory to determine which band cardinalities are possible. Since the degree set is pathlike, the construction will automatically produce a configuration of girth of at least 4, since no triangle is possible in a pathlike configuration. Thus, we need only exhibit a construction for rank 2, and the finished configuration consists of r-1 of these incidence relations used in conjunction. Suppose that the selected pair of degrees are m and k, that $(m,k) = d$ and that $m = dM$, and $k = dK$ where $(M,K) = 1$. The possible band cardinalities are $M(d+i)$ and $K(d+i)$, where i is any nonnegative integer. Label the vertices of one band to the numbers less than $M(d+i)$, and have a vertex j from the first band adjacent a vertex n from the second band iff j is congruent with one of $n+1, n+2, \ldots, n+m$ (modulo $K(d+k)$), where $m \leq k$.

Another technique which avoids proving girth cases is a doubling technique. Let T be configuration of girth g and degrees D, then the double of T has as bands two disjoint copies of each band of T. The copies are called first and second, and a vertex from a first copy is adjacent with a vertex from a second copy iff they were adjacent vertices in the original configuration. It is clear that this cannot lower girth, and in some cases raises girth, since it takes odd polygons into polygons of twice the length. Doubling may also be used to produce a rank 2 configuration from a regular graph. Used this way, the double of the 3-regular Moore graph of girth 5 is a configuration of girth 6, since the hexagon not in the skeleton doubles into a pair of hexagons. Again, the 3-regular McGee graph of girth 7

yields a configuration of girth 8, due to an octagon in the original.

There are many other constructions for small configurations, particularly for

BIBD, but very few are known for larger configurations.

REFERENCES

1. Longyear, J. Q., "Large Tactical Configurations", _Discrete Math._ 4 (1973)
    379-382.

2. Longyear, J. Q., "Non-Existence Criteria for Small Configurations", _Canad._
    J. Math. 25 (1973) 213-215.

3. Moore, B. H., "Tactical Memoranda I, II, III", _Amer. J. Math_. 18 (1896)
    254-303.

4. Payne, S. E., and Tinsley, M. F., "On v_1 x v_2(n,s,t) Configurations",
    J. Combinatorial Theory 7 (1969), 1-14.

APPLICATIONS

5. Busacker, R. G. and Saaty, T. L., _Finite Graphs and Networks_,
    McGraw Hill, New York, (1965)

6. Mycielski, J. and Ulam, S. M., "On the pairing process and the Notion of
    Generalized Distance", _J. Combinatorial Theory_ 6, p. 227-234.

7. Pless, Vera, "On the Uniqueness of the Golay Codes", _J. Combinatorial_
    Theory 5, p. 215-228.

STRUCTURAL CHARACTERIZATIONS OF STABILITY
OF SIGNED DIGRAPHS UNDER PULSE PROCESSES*

Fred S. Roberts
Rutgers University

ABSTRACT

The notion of a pulse process on a signed digraph arose in connection
with studies of energy demand and other societal problems. Two notions of
stability of a signed digraph D under a pulse process, pulse and value
stability, have been characterized in terms of the eigenspace of the transpose
of the signed adjacency matrix of D. In this paper, conditions for stability
stated in terms of the structure of the signed digraph are studied.

---

*This research was supported by NSF Grants GI-34895 and GI-44

STRUCTURAL CHARACTERIZATIONS OF STABILITY
OF SIGNED DIGRAPHS UNDER PULSE PROCESSES

1. PULSE PROCESSES

In applying signed digraphs to the study of energy demand and other societal problems, one encounters the notion of a pulse process (see Roberts [5,6], and Brown and Roberts [1]). In this paper, we study two notions of stability of a signed digraph under a pulse process and discuss possible characterizations of stability which relate to the structure of the signed digraph.

We shall use the digraph terminology of Harary, Norman, and Cartwright [4]. A digraph D with a sign (+ or -) on each arc will be called a __signed digraph__. The __signed adjacency matrix__ $A = A(D) = [a_{kj}]$ is defined as expected. The __sign__ of a path, sequence, cycle, or semicycle in D is the product of the signs of its arcs.

Let $x_1, x_2, \ldots, x_n$ be the nodes of the signed digraph D and suppose that at each discrete time t, each node k_j attains a __value__ $v_j(t)$. In a __simple pulse process__ __starting at node__ x_i, one assumes $v_j(0) = 0$, all j, and defines $v_j(t+1)$ as follows:

$$v_j(t+1) = v_j(t) + \sum_{k=1}^{n} a_{kj}\, p_k(t) ,$$

where $p_k(t)$ is $v_k(t) - v_k(t-1)$ if $t > 0$, 1 if $t = 0$ and $k = i$, and 0 if $t = 0$ and $k \neq i$. The number $p_k(t)$ is called the __pulse__ at node x_k at time t. For a discussion of the definition of simple pulse process, the reader is referred to Roberts [5]. Sometimes it is convenient to refer to the starting node. Thus in a simple pulse process starting at x_i, we denote $v_j(t)$ by $v^t(x_i, x_j)$ and $p_j(t)$ by $p^t(x_i, x_j)$. These numbers can be calculated using A(D).

__Theorem 1.__ (__[5]__) (a) $p^t(x_i, x_j)$ is given by the difference between the number of positive sequences from x_i to x_j of length t and the number of negative sequences from x_i to x_j of length t, which in turn is given by the i,j entry of A^t.

(b) $v^t(x_i, x_j)$ is given by the difference between the number of positive sequences from x_i to x_j of length at most t and the number of negative sequences from x_i to x_j of length at most t, which in turn is given by the i,j entry of

$A + A^2 + \ldots, + A^t$.

A node y in D is said to be <u>pulse stable</u> (under all simple pulse processes) if the sequence $\{|p^t(x,y)|: t = 0,1,\ldots\}$ is bounded for every x, and <u>value stable</u> (under all simple pulse processes) if the sequence $\{|v^t(x,y)|: t = 0,1,\ldots\}$ is bounded for every x. Finally, D is said to be <u>pulse</u> or <u>value stable</u> if each node is.

2. CHARACTERIZATIONS OF STABILITY

Questions of stability can be reduced to the eigenspace of the matrix S, the transpose of the signed adjacency matrix A. We shall refer to the eigenvalues of S as the eigenvalues of D. Suppose J is the Jordan Canonical Form corresponding to S. We say that an eigenvalue λ of S is <u>linked</u> if in J there appears a 2 x 2 submatrix of the form:

$$\begin{bmatrix} \lambda & 1 \\ 0 & \lambda \end{bmatrix}$$

The following characterizations of stability, proved in Brown and Roberts [1], can now be stated:

Theorem 2. A signed digraph D is pulse stable if and only if every nonzero eigenvalue of D has magnitude equal to unity and no nonzero eigenvalue of D is linked.

Theorem 3. A signed digraph D is value stable if and only if it is pulse stable and unity is not an eigenvalue of D.

Unfornately, neither theorem gives conditions in terms of the structure of the digraph or signed digraph. We would like, therefore, to pose the following questions.

<u>Question 1</u>: What structural condition on a signed digraph corresponds to the condition that all nonzero eigenvalues have magnitude unity?

<u>Question 2</u>: What structural condition corresponds to the condition that there are linked eigenvalues?

<u>Question 3</u>: What structural condition corresponds to pulse stability?

Question 4: What structural condition corresponds to value stability?

Similar questions could be asked for <u>weighted digraphs</u>, digraphs where the sign of an arc is replaced by a weight (a positive or negative real number). Although we shall not answer these questions totally, we shall mention some partial results concerning Questions 3 and 4.

3. ROSETTES

A (signed) digraph D is a <u>rosette</u> if it has more than one node and either D consists of a single cycle or D is strongly connected and has exactly one node x, the <u>central node</u>, with more than one arc leading into it. Rosettes are easily seen to be exactly those digraphs with a central node x and nonintersecting cycles leading out of x. Some rosettes are shown in Figure 1.

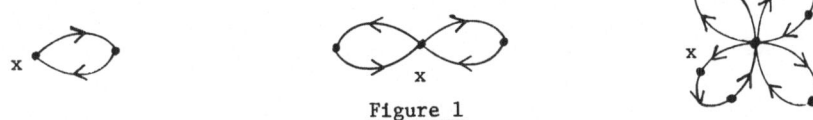

<u>Figure 1</u>

Stability theory for rosettes settles a number of questions about stability in general. It has also turned out to be useful in practice (see Roberts [6]). Stability in rosettes reduces to stability at the central node x. In fact, the situation is even simpler than that. Let us say a node x of a signed digraph is <u>self-stable in pulse</u>(or <u>in value</u>)if the sequence $|p^t(x,x)|$ (or the sequence $|v^t(x,x)|$) is bounded. We have

<u>Theorem 4</u>. A rosette with central node x is pulse or value stable if and only if x is self-stable in pulse or value respectively.

<u>Proof</u>. We use Theorem 1 to calculate $v^t(y,z)$. Suppose first that every path from y to z goes through x. Suppose that the unique path P_{yx} from y to x has length r and the unique path P_{xz} from x to z has length s. Then $v^t(y,z) = \pm v^{t-r-s}(x,x)$, since every sequence from y to z consists of P_{yx}, a sequence from x to x, plus P_{xz}. If there is a path from y to z not going through x, let this path have length r and sign σ. Then $v^t(y,z) = (\sigma)1+(\sigma)v^{t-r}(z,z)$, which reduces to the first case. Similar reasoning applies to $p^t(y,z)$.

This result generalizes to what we shall call advanced rosettes. A (signed) digraph D is an _advanced rosette_ if it has more than one node and either D is a single cycle or D is strongly connected and there is exactly one node x, the _central node_, which is on all cycles in D.

Theorem 5. An advanced rosette with central node x is pulse or value stable if and only if x is self-stable in pulse or value respectively.

Proof. The result follows from Theorem 4 if D is a single cycle. If D is not a single cycle, note that for all y and z in D, there are only finitely many sequences from y to z not going through x. Then, to go from y to z, except for finitely many sequences, one first goes from y to x (in finitely many ways), then x to x, and then x to z (in finitely many ways.)

It should be remarked that the notion of local self-stability is not in general sufficient even for local stability, though it is in advanced rosettes.

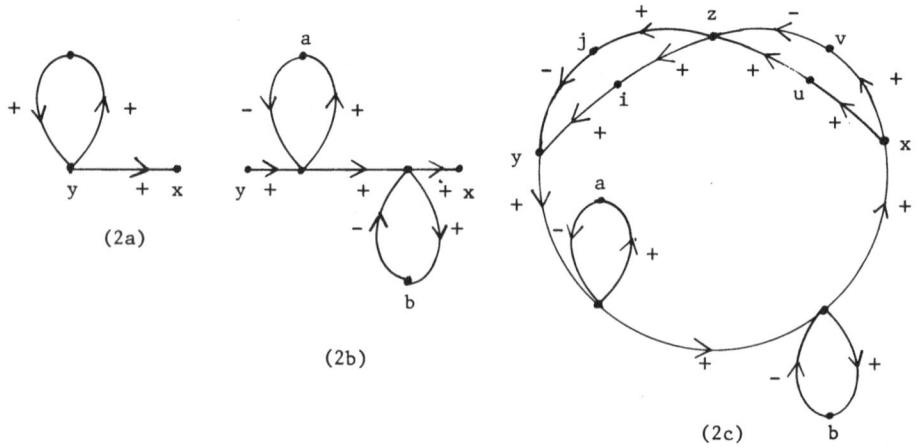

(2a)

(2b)

(2c)

Figure 2

To give a counterexample, note that in the signed digraph D of Figure 2a, node x is value self-stable but $v^t(y,x)$ tends to $+\infty$. Even global self-stability (each node self-stable) is not sufficient for stability. For, consider the signed digraph D of Figure 2b. Each node is value self-stable. But node x is not value stable. It is not even pulse stable, for $p^{3+2k}(y,x) = (-1)^k(k+1)$. This follows since to get from y to x in 3+2k steps, one goes m times through node a and then k-m times through node b, and each choice of m = 0,1,...,k gives a different sequence. It might be conjectured that the additional assumption that D be strongly connected is sufficient for global (local) self-stability to imply global (local) stability. But it follows from a counterexample, due to Joel Spencer, that this conjecture is also false. Let D be the signed digraph of Figure 2c. Then each node is value self-stable. For, to get from any point $\alpha \neq u,v$ to α , one first goes from α to x and then, after hitting x for the last time, one returns to α. But for each such sequence whose last return goes through u, there is a corresponding one of equal length and opposite sign whose last return goes through v. Thus, α is value stable. To get from u to u, one first goes from u to y and then y to u. The first trip to y can be either through i or through j. Thus there is again a one-to-one correspondence between these sequences, and so u is value self-stable. Similarly, v is value self-stable. Finally, the node x is not value stable, or even pulse stable. The sequences from y to x which go through x more than once divide into those using u and those using v, and these cancel out as before. The calculation of $p^t(y,x)$ thus reduces to considering sequences which go from y to x without using x more than once, and this calculation is analogous to that for the signed digraph of Figure 2b.

The theory of pulse and value stability in advanced rosettes has been studied in detail in a Rand Corporation Report by T. A. Brown, F. S. Roberts, and J. Spencer. The results obtained are stated in terms of the <u>rosette sequence</u>, the sequence $(a_1,a_2,...,a_s)$, where a_i is the sum of the signs of the cycles of length i and s is the maximal integer so that $a_s \neq 0$. The following theorems are proved in the that report for rosettes, and it is observed that the same proofs apply to advanced rosettes.

<u>Theorem 6</u>. If an advanced rosette is pulse stable, then $a_s = \pm 1$ and

$$a_i = (-a_s)(a_{s-i}), \quad 1 \leqq i \leqq s-1 .$$

Theorem 7. A pulse stable advanced rosette is value stable if and only if $\sum_i a_i \neq 1$, i.e., if and only if the number of positive cycles is not exactly one more than the number of negative cycles.

4. RELATIONS BETWEEN BALANCE AND STABILITY

It was pointed out in Roberts [5] that positive cycles in a signed digraph D, otherwise known as <u>balanced cycles</u> after the work of Harary [3] and Cartwright and Harary [2], contribute to the instability of D under pulse processes. In this section we summarize some results about the relations between balance and stability.

A signed digraph is <u>(cycle) balanced at the node x</u> if each cycle through x is balanced, and it is <u>balanced</u> if each node is balanced. We shall use the terms <u>local</u> and <u>global balance</u>, respectively, for these two notions.

To begin with, we observe that global balance (under certain circumstances) implies (global) value instability, but local balance does not imply local value instability.

Theorem 8. If a signed digraph D is (globally) balanced and has at least one cycle, then D is value unstable.

Proof. If the node x is on a cycle, then note that all sequences from x to x are positive and there are infinitely many.

Theorem 9. There is a signed digraph D which is balanced at a node x and value stable at x.

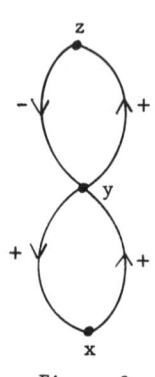

Figure 3

Proof. Let D be the signed digraph of Figure 3. Note that D is balanced at node x. Since D is a rosette, it is by Theorem 4 sufficient to show that D is value self-stable at y. But to get from y to y in t steps, one can go through z a certain number of times and through x a certain number of times, in any order. A simple counting argument shows that there are exactly as many positive sequences as negative ones.

In certain balanced signed digraphs, we can give a more complete description
of the value unstable nodes. Let us say that D is <u>semicycle balanced</u> if each
semicycle of D is positive. Let us say that node x is <u>reachable</u> from node y if
there is a path in D from y to x.

<u>Theorem 10</u>. If a signed digraph D is semicycle balanced, then a node x of D is
value unstable if and only if it is reachable from a node y which is on a cycle.

<u>Proof</u>. Note that for all nodes a and b, all sequences from a to b have the same
sign, by Theorem 13.2 of Harary, Norman, and Cartwright [4]. Now suppose x is
reachable from a node y which is on a cycle. Then there are infinitely many
sequences from y to x. Each such sequence has the same sign. Thus, x is value
unstable. If x is not reachable from any node y which is on a cycle, then for all
y, the number of sequences from y to x is the same as the number of paths, a finite
number.

It should be noted that cycle balance (rather than semicycle balance) is not
a sufficient condition in Theorem 10. To give a counterexample, consider the signed
digraph D of Figure 4. D is cycle balanced but the semicycle y,a,x,b is unbalanced.

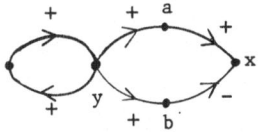

The node y is on a cycle and x is reachable from y,
but x is value stable since the number of positive
sequences of length t from any node to x is the
same as the number of negative sequences.

Figure 4

According to the proof of Theorem 10, it is sufficient to assume in place of
semicycle balance the following sequence condition: for all pairs of nodes a and b,
all sequences from a to b have the same sign. Thus, it is also sufficient to assume
in place of semicycle balance the conditions that D be cycle balanced and that for
all pairs of nodes a and b, all paths from a to b have the same sign. For it is
easy to show that the sequence condition follows from these two assumptions.

In closing, let us define a signed digraph as <u>totally unbalanced at x</u> if all
cycles through x are negative, and (globally) <u>totally unbalanced</u> if every node is.
Although global balance implies value instability, at least for signed digraphs with
cycles, global total unbalance does not imply value stability. To give a counter-

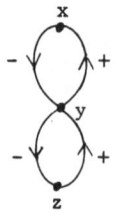

Figure 5

example, consider the signed digraph D of Figure 5. Note that $v^t(y,y) = (-1)^s 2^s$, where $t = 2s$, since to get from y to y, one may go through either x or z, and there are two choices for each two steps.

REFERENCES

1. Brown, T. A., and Roberts, F. S., "Pulse Processes on Weighted Digraphs," to appear.

2. Cartwright, D., and Harary, F., "Structural Balance: A Generalization of Heider's Theory," Psych. Rev. 63 (1956), 277-293.

3. Harary, F., "On the Notion of Balance of a Signed Graph," Michigan Math. J. 2 (1954), 143-146.

4. Harary, F., Norman, R. Z., and Cartwright, D., Structural Models, Wiley, New York, 1965.

5. Roberts, F. S., "Signed Digraphs and the Growing Demand for Energy," Environment and Planning 3 (1971), 395-410.

6. Roberts, F. S., "Building and Analyzing an Energy Demand Signed Digraph," Environment and Planning 5 (1973), 199-221.

THE CHARM BRACELET PROBLEM AND ITS APPLICATIONS*

Paul K. Stockmeyer
College of William and Mary

ABSTRACT

The necklace problem has proved to be both a sound pedagogical device in teaching enumeration theory and a valuable counting tool with several graphical applications. In this paper we solve the more general charm bracelet problem and provide two applications for which the necklace problem in not sufficient.

We set the stage in Section 1 by providing a brief review of the necklace problem. This serves as a basis for comparison in Section 2, where we discuss the charm bracelet problem and derive its solution. Sections 3 and 4 contain nontrivial graphical applications of the results of Section 2.

Definitions for all graphical terms and concepts can be found in [3]. For further background and broader treatment of topics of an enumerative nature, [5] should be consulted.

*This research was supported by the Office of Naval Research under contract N00014-73-A-0374-0001, NR044-459. Reproduction in whole or in part is permitted for any purpose of the United States Government.

THE CHARM BRACELET PROBLEM AND ITS APPLICATIONS

1. NECKLACES

The necklace problem asks for the number N of closed necklaces with n equally spaced beads, each of which is one of m colors or types. Two necklaces are considered the same if one can be rotated or reflected into the other. The 6 distinct necklaces for n = 4, m = 2 are illustrated in Figure 1.

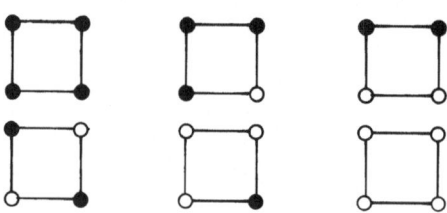

Figure 1. Examples of Necklaces

A full discussion of the solution, using the enumeration methods of Pólya [11], can be found either in [5, page 44], [3, page 183], or [4]. Essentially one uses the cycle index $Z(D_n)$ of the dihedral group of degree n, defined by

$$(1) \qquad Z(D_n) = Z(D_n; z_1, z_2, z_3, \ldots, z_n)$$

$$= \frac{1}{2n} \sum_{i|n} \phi(i) z_i^{n/i} + \begin{cases} 1/2 \; z_1 z_2^{(n-1)/2} & \text{n odd} \\ 1/4 \; (z_1^2 z_2^{(n-2)/2} + z_2^{n/2}), & \text{n even} \end{cases}$$

The answer is obtained by replacing each variable z_i with the integer m:

$$(2) \qquad N = Z(D_n; m, m, \ldots, m) \; .$$

We note that $Z(D_4; 2,2,2,2) = 6$, in agreement with Figure 1.

In many applications, each type of bead is assigned a weight, typically a non-negative integer. The weight of a necklace is defined to be the sum of the weight of its beads. If b_i is the number of types of beads with weight i, then $b(x) = \Sigma b_i x^i$ is the generating function for beads by weight, frequently called the figure counting series. Similarly, if N_i is the number of necklaces with weight i, then the generating function $N(x) = \Sigma N_i x^i$ is often called the configuration counting series. Pólya's Theorem [11] relates these two series to obtain the formula

(3) $$N(x) = Z(D_n;\ b(x), b(x^2), \ldots, b(x^n))\ .$$

Equation (2) is clearly a special case of (3), obtained by setting x equal to 1.

A typical graphical application of the necklace problem is the enumeration of unicyclic graphs, described in [12, page 148]. In this situation a 'bead' is a rooted tree, whose weight is its number of points. If $t(x)$ is the generating function for rooted trees, then it follows that the generating function $U_n(x)$ for unicyclic graphs with cycle length n is

(4) $$U_n(x) = Z(D_n;\ t(x), t(x^2), \ldots, t(x^n))\ .$$

2. CHARM BRACELETS

The charm bracelet problem is similar to the necklace problem except that flat charms are attached to the frame instead of beads. These charms are assumed to be firmly fixed to the frame and cannot be rotated or turned over unless the corresponding operation is performed on the entire necklace. Some charms are asymmetric; that is, a reflection of the necklace (and hence its charms) does not affect their appearance. They can be fastened to the frame in essentially only one way. Other charms are non-symmetric. They are different from their mirror images, and can be attached to the frame in either of two possible orientations. As with necklaces, two charm bracelets are considered the same if one can be either rotated or reflected into the other. We illustrate these concepts in Figure 2, which contains the seven bracelets with three charms drawn from a stock of two types, one symmetric and one non-symmetric.

Suppose we have a types of symmetric charms and b types of non-symmetric charms. Since each non-symmetric charm can be attached in two ways, there is a total of $c = a + 2b$ possibilities at each position on the bracelet. Thus there are c^n different "fixed" or "positioned" bracelets. In order to determine the number of inequivalent bracelets, we use the famous orbit counting formula due to

Burnside [1, page 191].

Figure 2. Examples of Charm Bracelets

<u>Burnside's Theorem</u>. Let G be a permutation group acting on a set S, and for each

$g \epsilon G$, let H(g) be the number of elements of S fixed by g. The number O(G) of orbits

of G is given by

(5)
$$O(G) = \frac{1}{|G|} \sum_{g \epsilon G} H(g).$$

In applying this theorem, we take S to be the set of c^n different fixed

bracelets. For G we take the group induced on S by the dihedral group D_n of de-

gree n acting on the bracelet frame. Thus for each of the n rotations and n

reflections of D_n we must determine the number of bracelets that are left unchanged

by such action.

We consider first the n rotations. As is well known, for each integer i

dividing n, there are $\phi(i)$ rotations each consisting of d/i cycles of length i.

Clearly a bracelet will be unchanged by such rotation if and only if the same

charm is attached, with the same orientation, to each place in any cycle. Thus

there are $c^{n/i}$ bracelets fixed by each of the $\phi(i)$ rotations, for i dividing n.

In considering the n reflections, it is convenient to consider first the case n odd. In this case a reflection consists of $(n-1)/2$ transpositions and one fixed place. In order for a bracelet to be fixed by a reflection, the same charm must occur on both places of any transposition. Moreover, a symmetric charm must be attached at the fixed place. Thus for each of the n reflections there are $a \cdot c^{(n-1)/2}$ bracelets fixed by that action.

In the even case there are two types of reflections. Half of them consist of $n/2$ transpositions, while the other half contain $(n-2)/2$ transpositions and 2 fixed places. Arguing as above, there are $c^{n/2}$ bracelets fixed by the first type of reflection and $a^2 \cdot c^{(n-2)/2}$ bracelets unchanged by the second.

The results of the above discussion can be summarized in a form that will be of use later. We define the modified cycle index $Z(D_n*)$ of the group D_n by

$$(6) \qquad Z(D_n*) = Z(D_n*; z_1, z_2, \ldots, z_n, y)$$

$$= \frac{1}{2n} \sum_{i|n} \phi(i) z_i^{n/i} + \begin{cases} \frac{1}{2} y z_2^{(n-1)/2}, & n \text{ odd} \\[2mm] \frac{1}{4} y^2 z_2^{(n-1)/2} + z_2^{n/2}, & n \text{ even} \end{cases}$$

The number of charm bracelets with n charms, drawn from a store of a symmetric charms and b non-symmetric charms, with $c = a + 2b$, is then

$$(7) \qquad CB = Z(D_n*; c, c, \ldots, c, a) .$$

We observe that $Z(D_3*; 3, 3, 3, 1) = \frac{1}{6}(3^3 + 2 \cdot 3^1 + 3 \cdot 1 \cdot 3) = 7$, in agreement with Figure 2.

If one wishes to use generating functions to count charm bracelets by weight, the following weighted form of Burnside's theorem is needed. A proof can be found in [3, page 180].

Burnside's Theorem, Weighted Form. Let G be a permutation group acting on a set S, and let w be a function that assigns a weight to each orbit of G. Each element $s \epsilon S$ is assigned the weight of the orbit in which it is contained. Then the generating function for orbits of G by weight is

$$(8) \qquad \frac{1}{|G|} \sum_{g \epsilon G} \Pi x^{w(s)}$$

where the product is over all s∈S fixed by the permutation g.

Again we take S to be set of c^n different fixed bracelets. The generating functions for symmetric and non-symmetric charms are denoted $a(x) = \Sigma a_i x^i$, and $b(x) = \Sigma b_i x^i$, respectively, and we define $c(x) = a(x) + 2b(x)$. CB(x) will denote the generating function for charm bracelets by weight. It is an easy exercise to show that in the case of charm bracelets, formula (8) can be computed from the modified cycle index of the group D_n by replacing each variable z_i with the series $c(x^i)$ and the variable y by the series a(x). This yields the following main result:

<u>Charm Bracelet Theorem</u>. The generating function CB(x) for charm bracelets with n charms is given by

(9) $$CB(x) = Z(D_n^*;\ c(x,),\ c(x^2),\ldots,c(x^n),a(x)).$$

If all the charms happen to be symmetric, we have $b(x) = 0$ and thus $c(x) = a(x)$. In this case equation (9) is identical to (3). Hence we see that the necklace problem can be viewed as a special case of the more general charm bracelet problem.

3. TRIANGULATION OF POLYGONS

Our first application provides a new solution to a problem first solved by R. Guy [2] and subsequently solved by Moon and Moser [9] and by Harary and Palmer [5]. The problem is to determine the number of ways of dissecting a regular (n + 2)-gon into n triangles by n-1 non-intersecting diagonals. Two dissections that differ only by a rotation or reflection will be considered the same. It is easy to verify that there is a unique dissection for n = 1, 2, and 3, while for n = 4 there are 3 distinct triangulations.

We first must count various classes of rooted triangulations. For $n \geq 1$, let a_n and b_n denote the number of triangulations of an (n + 2)-gon rooted at a symmetric and non-symmetric exterior edge, respectively. For convenience we set $a_0 = 1$, representing a degenerate 2-gon, or edge. Further, we set $c_n = a_n + 2b_n$, so that c_n is the number of triangulations rooted at an oriented exterior edge. We define a(x), b(x), and c(x) to be the generating functions corresponding to the sequences a_n, b_n, and c_n respectively.

It is easily seen (see, for example, [5]), that the function $c(x)$ satisfies the equation

(10) $$c(x) = 1 + x \cdot c^2(x)$$

which, when solved for $c(x)$, yields

(11) $$c(x) = \frac{1 - (1 - 4x)^{1/2}}{2x}$$

Expanding, one obtains

(12) $$c(x) = \frac{(2n)!}{n!(n+1)!} x^n$$

$$= 1 + x + 2x^2 + 5x^3 + 14x^4 + 42x^5 + 132x^6 + 429x^7 + 1430x^8$$
$$+ 4862x^9 + 16{,}796x^{10} + \ldots,$$

a result apparently known to Euler.

In order to determine $a(x)$, we note that a triangulated polygon rooted at a symmetric exterior edge can be constructed by first placing a triangle on the edge and then attaching mirror image exterior-edge-rooted triangulated polygons to the two new edges. Translated into generating functions, this implies

(13) $$a(x) = 1 + x \cdot c(x^2)$$
$$= 1 + x + x^3 + 2x^5 + 5x^7 + 14x^9 + \ldots .$$

Then we have

(14) $$b(x) = \frac{c(x) - a(x)}{2}$$
$$= x^2 + 2x^3 + 7x^4 + 20x^5 + 66x^6 + 212x^7 + 715x^8 + 2{,}424x^9$$
$$+ 8{,}398x^{10} + \ldots .$$

We now use the charm bracelet theorem to determine the generating function $F(x)$ for the number F_n of triangulated $(n+2)$-gons rooted at one of the triangles. Clearly such a polygon can be considered a bracelet of three charms, each of which is an exterior-edge-rooted triangulated polygon. Thus we have

(15) $$F(x) = x \, Z(D_3^*; c(x), c(x^2), \ldots, c(x^n), a(x))$$
$$= \frac{x}{6} (c^3(x) + 2c(x^3) + 3a(x)c(x^2))$$
$$= x + x^2 + 2x^3 + 6x^4 + 16x^5 + 52x^6 + 170x^7 + 715x^8$$
$$+ 2{,}424x^9 + 8{,}398x^{10} + \ldots .$$

Another intermediate result we need is the number H_n of triangulated polygons rooted at an interior edge. The corresponding generating function is again found by using the charm bracelet theorem, this time with two charms. We have

(16) $\qquad H(x) = Z(D_2^*; c(x)-1, a(x)-1)$

$$= x^2 + x^3 + 5x^4 + 12x^5 + 45x^6 + 143x^7 + 511x^8 + 1768x^9$$

$$+ \ldots .$$

We will call an interior edge of a triangulated polygon **symmetric** if the polygon possesses an automorphism that interchanges the two triangles incident with the edge. The remaining intermediate result we need is the number J_n of polygons rooted at a symmetric interior edge. The generating function for this sequence can be shown to be

(17) $\qquad J(x) = c(x^2) - 1$

$$= x^2 + x^4 + 5x^6 + 14x^8 + 42x^{10} + \ldots$$

In order to determine the number K_n of unrooted triangulated polygons we follow the standard method for unrooted trees, developed by Otter [10] and applied repeatedly in [6]. In this case, the method yields

(18) $\qquad K(x) = F(x) - H(x) + J(x)$

$$= x + x^2 + x^3 + 3x^4 + 4x^5 + 12x^6 + 27x^7 + 82x^8 + 228x^9 + \ldots$$

We can find a closed form for the coefficients of $K(x)$ by using equations (10), (13,), (15), (16), and (17) to express $K(x)$ in terms of $c(x)$ only, and then using the closed form for the coefficients of $c(x)$ given by (12). If one interprets as zero any term containing a nonintegral factorial, then the number K_n can be expressed as

(19) $\qquad K_n = \dfrac{(2n-1)!}{(n-1)!\,(n+2)!} + \dfrac{3(n-1)!}{2((n-2)/2)!\,((n+2)/2)!} + \dfrac{(n-2)!}{((n-3)/2)!\,((n+1)/2!}$

$$+ \dfrac{((2n-2)/3)}{3((n-1)/3)!\,((n+2)/3)!}$$

Incidentally, the number K_{16} is given incorrectly as 1,046,609 in both [2] and [5]. The correct number is 983,244.

Asymptotically, the first term of (19) is clearly the dominant one. Using Sterling's formula, we find that

(20) $\qquad\qquad\qquad\qquad K_n \sim 2^{2n-1}\pi^{-1/2}n^{-5/2}.$

4. PROJECTIVE PLANE TREES

A plane tree is a tree that has been embedded in the (Euclidean) plane. Two plane trees are ismorphic if there exists an orientation-preserving homeomorphism of the plane onto itself that maps one onto the other. Plane trees have been counted by Harary, Prins, and Tutte [7].

We define a projective plane tree, or PPT, as a tree that has been embedded in the real projective plane. Two PPT's are isomorphic if any homeomorphism of the projective plane onto itself maps one onto the other. Thus while a PPT is always isomorphic to its mirror image, a plane tree might not be. Consequently there are fewer PPT's than plane trees on n points, for $n \geq 7$. In this section we count the number of isomorphism classes of PPT's on n points for each positive integer n.

As usual, we must first obtain a few preliminary results. For $n \geq 2$ we let a_n and b_n denote now the number of planted PPT's on $n + 1$ points that are symmetric and non-symmetric, respectively. Again we set $c_n = a_n + 2b_n$, so that now c_n is the number of planted plane trees on $n + 1$ points. Also, $a(x)$, $b(x)$, and $c(x)$ will again denote the corresponding generating functions.

The somewhat surprising fact that the numbers c_n are again the Catalan numbers of Section 3 was noted in [7]:

$$(21) \quad c(x) = \frac{(2n-2)!}{n!(n-1)!} x^n$$
$$= x + x^2 + 2x^3 + 5x^4 + 14x^5 + 42x^6 + 132x^7 + 429x^8 + 1430x^9 + \ldots$$

A planted PPT can be constructed by identifying the roots of any number of smaller planted PPT's. In particular, a symmetric planted PPT can be constructed from either none or one smaller symmetric planted PPT together with any number of pairs of arbitrary planted plane trees. Thus

$$(22) \quad a(x) = x(1 + a(x))(1 + c(x^2) + c^2(x^2) + \ldots)$$
$$= \frac{x(1 + a(x))}{1 - c(x^2)} \quad .$$

Solving for $a(x)$, we have

$$23) \quad a(x) = \frac{x}{1 - x - x(x^2)}$$
$$= x + x^2 + 2x^3 + 3x^4 + 6x^5 + 10x^6 + 20x^7 + 35x^8 + 70x^9 + \ldots$$

from which we obtain

$$(24) \qquad b(x) = \frac{c(x) - a(x)}{2}$$

$$= x^4 + 4x^5 + 16x^6 + 56x^7 + 197x^8 + 680x^9 + \ldots$$

A rooted PPT can clearly be considered as a charm bracelet in which the charms are planted plane trees. Letting R_n denote the number of rooted PPT's with n points and $R(X)$ the corresponding generating function, the charm bracelet theorem yields

$$(25) \qquad R(x) = x \, Z(D_n^*; \, c(x), \, c(x^2), \, \ldots, \, c(x^n), a(x)),$$

where the sum is taken from n = 0 to ∞. Explicitly, we have

$$(26) \qquad R(x) = x + x^2 + 2x^3 + 4x^4 + 9x^5 + 21x^6 + 56x^7 + 155x^8 + 469x^9$$
$$+ 1,480x^{10} + \ldots$$

A line-rooted PPT can be considered a bracelet with two charms. Thus the generating function $L(x)$ for line-rooted PPT's satisfies

$$(27) \qquad L(x) = Z(D_2^*; \, c(x), \, c(x^2), \, a(x))$$

$$= x^2 + x^3 + 3x^4 + 6x^5 + 17x^6 + 44x^7 + 133x^8 + 404x^9 + 1319x^{10}$$
$$+ 1319x^{10} + \ldots$$

Further, it is easily seen that the generating function $S(x)$ for PPT's rooted at a symmetry edge is given by

$$(28) \qquad S(x) = c(x^2)$$

$$= x^2 + x^4 + 2x^6 + 5x^8 + 14x^{10} + \ldots$$

To obtain the generating function $T(x)$ for unrooted PPT's, we again utilize the Otter formula,

$$(29) \qquad T(x) = R(x) - L(x) + S(x)$$

$$= x + x^2 + x^3 + 2x^4 + 3x^5 + 6x^6 + 12x^7 + 27x^8 + 65x^9$$
$$+ 175x^{10} + \ldots .$$

CONCLUSIONS

Neither of the two preceeding problems is new. At least three distinct solutions to the triangulation problem have been published, and the number of projective plane trees can be obtained from results in [8], although the answer is not given explicitly. However, the method is new, and offers a unified approach to problems previously solved on an ad hoc basis. The formulas can be used, for example, to solve almost any problem involving counting of configurations embedded in the plane, where rotations and reflections of configurations are not considered distinct.

REFERENCES

1. Burnside, W., _Theory of Groups of Finite Order_. Second Edition, Cambridge University Press, 1911. Reprinted Dover, 1955, New York.

2. Guy, R. K., "Dissecting a Polygon into Triangles", _Research Report_, University of Calgary, 1960.

3. Harary, F., _Graph Theory_. Addison-Wesley, 1969, Reading.

4. Harary, F., "Enumeration Under Group Action: Unsolved Graphical Enumeration Problems, IV." _J. Comb. Theory_, 8 (1970) 1-11.

5. Harary, F., and Palmer, E. M., _Graphical Enumeration_, Academic Press, 1973, New York.

6. Harary, F., and Prins, G., "The Number of Homomorphically Irreducible Trees and Other Species", _Acta Math_. 101 (1959) 141 - 162.

7. Harary, F., Prins, G., Tutte, W. T., "The Number of Plane Trees", _Indag. Math_ 26 (1964) 319-329.

8. Harary, F., and Robinson, R. W., "The Number of Achiral Trees", _J. Reine Angew. Math._, to appear.

9. Moon, J. W., and Moser, L., "Triangular Dissections of n-gons" _Canad. Math._ Bull. 6 (1963) 175-177.

10. Otter, R., "The Number of Trees", _Ann. of Math_. 49 (1948) 583-599.

11. Pólya, G., "Kombinatorische Anzehlbestimmungen für Gruppen, Graphen und chemische Verbindungen", _Acta Math_. 68 (1937) 145-254.

<u>ON TUTTE'S FACTORIZATION THEOREM</u>

David P. Sumner
University of South Carolina

ABSTRACT

Tutte characterized graphs without 1-factors by showing that such graphs never contain a set S whose removal results in more odd components than the cardinality of S. In this paper, such a set S is called an <u>antifactor set</u>. The minimal (with respect to containment) antifactor sets are studied, and we answer the questions: (1) What type of points belong to a minimal antifactor set? (2) Where are the minimal antifactor sets of a graph G located? and (3) What bounds can be placed on the size of a minimal antifactor set? As a consequence of the theorems that answer these questions, we obtain several sufficient conditions for the existence of a 1-factor in a graph G.

ON TUTTE'S FACTORIZATION THEORM

1. INTRODUCTION

In this paper, all undefined terminology will conform with that in Harary [4].
In particular, a graph G will consist of a finite set $V(G)$ of points and a set $E(G)$
of unordered pairs of distinct points. The elements of $E(G)$ are called lines.

If G is a graph, then a 1-factor of G is a spanning subgraph that is regular
of degree one. It is worthwhile to observe that a 1-factor of a graph G having
order p may alternately be viewed as a partition of the points into two-element subsets in such a way that the elements of each subset are adjacent.

In [6], Tutte characterized graphs that possess a 1-factor. An odd (even)
component of a graph is a component having an odd (even) number of points.

Theorem 1. (Tutte): A graph G has a 1-factor iff there does not exist a set
$S \subseteq V(G)$ such that $G - S$ has more than $|S|$ odd components.

When given a graph G and asked if it has a 1-factor, we need only exhibit any
particular 1-factor to verify that the answer is yes. On the other hand, if G does
not have a 1-factor, how can we prove it without simply asserting that we tried
everything and failed? As a consequence of Tutte's theorem, we need only exhibit
an appropriate set S and any skeptic may easily check for himself that S satisfies
the conditions of Tutte's characterization and thereby see that the graph has no
1-factor. It is because Tutte's result yields such a convenient way of verifying
that a graph does not have a 1-factor that the characterization must be considered
a good one in the sense of Edmonds. A particularly nice discussion of this may be
found in Chvátal [3].

As an example, consider the graph

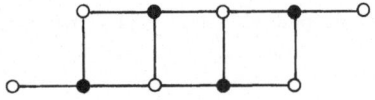

This graph does not have a 1-factor as may be seen by choosing the darkened points
for the set S. Moreover, in this case, no other choice for S would serve.

It is our purpose in this paper to attempt to come to a fuller understanding of the properties of the set S in Tutte's theorem. In this way it is hoped that the theorem will be enhanced somewhat. First we give such sets a name.

Definition: If G is a graph and $S \subseteq V(G)$ such that $G - S$ has more than $|S|$ odd components, then S will be called an antifactor set. If in addition no proper subset of S is also an antifactor set, then S will be called a minimal antifactor set.

For a minimal antifactor set $S \subseteq V(G)$, we shall consider the following three questions:

(1) What types of points can S contain?

(2) What is the location of S within G?

(3) What bounds can be placed on the cardinality of S?

The proofs of many of the results in this paper are omitted and will appear elsewhere.

2. THE TYPE OF POINTS IN A MINIMAL ANTIFACTOR SET

Definition: A point v of a graph G will be called a star center iff v is the center of an induced $K_{1,3}$. Thus v is a star center iff v is adjacent to three independent points.

Theorem 2. If G is a connected graph of even order that does not have a 1-factor and if S is a minimal antifactor set for G, then every element of S is a star center.

As a consequence of this result, we may restate Tutte's theorem as follows.

Theorem 3. A graph G has a 1-factor iff there does not exist a set S of star centers of G such that $G - S$ has more than $|S|$ odd components.

Corollary 1. If G is a connected graph with no induced $K_{1,3}$, then G has a 1-factor iff G has even order.

This corollary is also a special case of the following theorem that is proved in Sumner [5].

Theorem 4. If G is a connected graph of even order $p \geq 4$ that does not contain a 1-factor, then for every k where $4 \leq k \leq p$, G contains a connected, induced subgraph of order k that also fails to possess a 1-factor.

Since the only connected graph of order four that does not have a 1-factor is $K_{1,3}$, we have Corollary 1 as the special case $k = 4$ of this theorem. In general, the contrapositive version of this result gives a kind of forbidden subgraph condition for the existence of a 1-factor in a graph.

It is well known that a line graph cannot contain an induced $K_{1,3}$ and hence

Corollary 2. A connected line graph has a 1-factor iff it has even order.

Corollary 2 has also been established by Chartrand, Stewart, and Polimeni in [2] and by Sumner in [5].

If G is a connected graph of even order that does not have a 1-factor, then for any antifactor set S, $|S| \geq \kappa(G)$, the connectivity of G.

Corollary 3. If a connected graph G of even order has fewer than $\kappa(G)$ star centers, then G has a 1-factor.

We can say a little more about the minimal antifactor sets of cubic graphs.

Theorem 5. Let G be a connected cubic graph that does not possess a 1-factor. Then for any minimal antifactor set S of G,

(i)   S is an independent set of star centers;

(ii)   G − S has exactly $|S| + 2$ odd components and no even components;

(iii)   If $|S| \neq 1$, then S contains at least three cutpoints of G; and

(iv) $|S| \neq 2$.

It was already observed by Tutte [6] that Petersen's theorm on the existence of a 1-factor in a cubic block follows as a direct consequence of this theorem. However, we remark that Petersen's theorem is a special case of Theorem 5 (iii).

Also from Theorem 5 and Theorem 2 it follows that any cubic graph without a 1-factor must contain a cutpoint that is also a star center. Thus we have

Corollary 4. If G is a cubic graph in which every cutpoint lies in a triangle, then G has a 1-factor.

3.   THE LOCATION OF A MINIMAL ANTIFACTOR SET

Our major result in this section is

Theorem 6. Let G be a connected graph of even order that does not have a 1-factor.

Then if S is any minimal antifactor set for G, then S is properly contained in a single block of G.

Since every block of a nontrivial tree is K_2, we obtain

Corollary 5. Let T be a tree of even order. Then T has a 1-factor iff for every point v of T, T - {v} has at most one odd component.

For a direct proof of Corollary 5 see Bernhart [1].

For cubic graphs we can say a little more.

Theorem 7. Let G be a connected, cubic graph with no 1-factor, and let S be a minimal antifactor set for G. Then S is not contained in a single cycle of G.

4. BOUNDS ON THE SIZE OF A MINIMAL ANTIFACTOR SET

We first observe the following simple bounds.

Theorem 8. If S is a minimal antifactor set for a connected graph G having even order but no 1-factor, then

$$(i) \qquad \kappa(G) \leq |S| \leq (p - 2)/2 \qquad and$$

$$(ii) \qquad |S| \leq \beta_0(G) - 2.$$

The next lemma is an easy consequence of Hall's theorem on the existence of an SDR (set of distinct representatives).

Lemma 1. If S is a minimal antifactor set for a connected graph G with $|S| = k \geq 1$ and A_1, A_2,..., A_k are any k odd components of G - S, then there exists an ordering v_1, v_2,..., v_k of the elements of S such that for each i = 1,2,...,k, v_i is adjacent to some point in A_i.

Hence we obtain

Corollary 6. If S is a minimal antifactor set for a connected graph G, then $|S| \leq \beta_1(G)$.

A more interesting bound is given by the next statement.

Theorem 9. If S is a minimal antifactor set for a connected graph G, then $|S| \leq \max_B\{q(B) - p(B) + 2\}$ where q(B) and p(B) are respectively the number of lines and points of the block B and the maximum is taken over all the blocks of G.

Corollary 7. (i) If G is a connected (p,q)-graph, then for any minimal antifactor set S, $|S| \leq q - p + 2$; and (ii) If G is a connected, planar graph having r regions, then $|S| \leq r$.

<u>Remark</u>. All of the bounds in this section are attainable. In fact, the graph in the example at the beginning of this paper achieves all of the upper bounds at once.

REFERENCES

1. Bernhart, F. R., "On Trees with Perfect Matchings", unpublished.

2. Chartrand, G., Polimeni, A. D., and Stewart, M. J., "The Existence of 1-factors in Line Graphs, Squares, and Total Graphs", <u>Indag. Math</u>., to appear.

3. Chvátal, V., "New Directions in Hamiltonian Graph Theory" in <u>New Directions in the Theory of Graphs</u>, (Frank Harary, ed.) Academic Press, 1973, 65-66.

4. Harary, F., <u>Graph Theory</u>, Addison-Wesley, 1969.

5. Sumner, D. P., "Graphs with 1-factors", <u>Proc. Amer. Math. Soc</u>., to appear.

6. Tutte, W. T., "The Factorization of Linear Graphs", <u>J. London Math. Soc</u>. 22 (1947) 107-111.